# *Nerve and Muscle*
## MEMBRANES, CELLS, AND SYSTEMS

# Nerve and Muscle
## MEMBRANES, CELLS, AND SYSTEMS

## Richard B. Stein

*University of Alberta*
*Edmonton, Alberta, Canada*

*PLENUM PRESS · NEW YORK AND LONDON*

Library of Congress Cataloging in Publication Data

Stein, Richard B. 1940-
  Nerve and muscle.

  Includes index.
  1. Electrophysiology. 2. Nerves. 3. Muscle. I. Title. [DNLM: 1. Electrophysiology.
2. Nervous system—Physiology. 3. Muscles—Physiology. WL102 N454]
QP341.S78                        599'.01'852                        80-15028
ISBN-13: 978-1-4684-3799-7      e-ISBN-13: 978-1-4684-3797-3
DOI: 10.1007/ 978-1-4684-3797-3

First Printing—November 1980
Second Printing—September 1981
Third Printing—July 1983

© 1980 Plenum Press, New York
Softcover reprint of the hardcover 1st edition 1980
A Division of Plenum Publishing Corporation
233 Spring Street, New York, N.Y. 10013

# Preface

There has been a convergence in recent years of people from the physical and biological sciences and from various engineering disciplines who are interested in analyzing the electrical activity of nerve and muscle quantitatively. Various courses have been established at the graduate level or final-year undergraduate level in many universities to teach this subject matter, yet no satisfactory short text has existed.

The present book is an attempt to fill this gap, and arises from my experience in teaching this material over the past fifteen years to students on both sides of the Atlantic. Although covering a wide range of biophysical topics from the level of single molecules to that of complex systems, I have attempted to keep the text relatively short by considering only examples of the most general interest. Problems are included whenever possible at the end of each chapter so the reader may test his understanding of the material presented and consider other examples which have not been included in the text.

No attempt has been made to write a comprehensive textbook. That would have required a longer book, probably a multiauthor work by specialists in each area who can weigh the evidence for or against current concepts. Instead, I have tried to start where these books leave off, with the current conceptual picture, and to elaborate the mathematical and physical bases of these concepts. Care has been taken to define both biological and mathematical terms whenever they are introduced, to aid those who are unfamiliar with either field. Standard units and abbreviations according to the Système Internationale (SI) have been used throughout, although with their common American spellings. Many of these will be familiar, such as the meter (m), gram (g), and second (s). Others may need some introduction: the siemen (S) for conductance, rather than the mho, and the newton (N) for force, rather than the kilogram (kg), which is a unit of mass. The two units are related, according to Newton's law ($F = ma$), by the acceleration of gravity, 9.8 m/s$^2$. The standard prefixes are also used: kilo (k) $= 10^3$, centi (c) $= 10^{-2}$, milli (m) $= 10^{-3}$, micro ($\mu$) $= 10^{-6}$, nano (n) $= 10^{-9}$, and pico (p) $= 10^{-12}$. Thus, micrometer ($\mu$m) is used rather than micron and nanometer (nm $= 10^{-9}$ m) rather than Ångstrom (Å $= 10^{-10}$ m).

Readers with a modest knowledge of biology or of mathematics should be able to appreciate the major themes of the book. Wherever mathematical notions or proofs are introduced, the essence and implications are carefully explained for the benefit of those with a limited mathematical background. The mathematics required for a full understanding involves calculus, including the use of Laplace and Fourier transforms, and some knowledge of stochastic processes.

Interestingly, the same mathematical approaches are useful in describing such diverse pairs of phenomena as (1) the movements of ions through membrane channels and the attachment of bonds between actin and myosin molecules in muscle, (2) the enzyme reactions involved in transport of ions and the conductance changes involved in synaptic transmission, (3) the spread of charge along a nerve and muscle fiber and the transfer function of a sensory receptor, and (4) the quantal release of transmitter and the stochastic pattern of nerve impulses.

These few examples suggest that there is a biophysics of nerve and muscle which is not merely a collection of unrelated topics that happened to attract the attention of people from the physical sciences and engineering. These underlying similarities, arising from physical principles and mathematical approaches, have been emphasized, rather than the diversity and complexity of biological systems. This emphasis also requires a book which is short enough to read in its entirety, rather than a more comprehensive reference source.

To the extent that I have succeeded in these aims, I am indebted to Professor Walter Rosenblith of the Massachusetts Institute of Technology, who first introduced me to the field of biophysics, and to Doctor Denis Noble and Doctor Peter Matthews of Oxford University, who aroused my interests in the problems of biological membranes and in motor control. These fine teachers, by their enthusiasm and example, produced a lasting effect. This book would not have been possible without their efforts on my behalf in years past. The book also owes much to colleagues at the University of Alberta and to former students such as Parveen Bawa, Robert Wong, and Ted Milner, whose comments and suggestions have improved the text greatly. Fred Loeffler and Ken Burt prepared many of the figures, while Tella White typed numerous drafts of the book. Finally, my wife, Suzanne, and other members of my family have provided the support to bring this work to a successful conclusion.

R. B. Stein

# Contents

# Membrane Organization

Membranes are ubiquitous components of cells. Not only are they found surrounding cells and separating one cell from another, but they are also present within cells. They surround organelles such as mitochondria, which provide energy for the many functions that cells undertake, and they surround the cell nucleus, which gives overall direction to the activities of cells. Membranes are also found as endoplasmic reticulum, a system of tubules which forms a meshwork upon which protein synthesis can take place throughout the cell. Understanding the relationship between the structures and the functions of biological membranes has been a long and difficult task. This continuing struggle to analyze the basic properties of membranes illustrates the problems which have been encountered in many areas of biophysics. In briefly reviewing this area, material will be emphasized which is important for understanding the developments in later chapters.

Already in the last century, Overton (1899) suggested that fats or lipids were an important constituent of cell membranes but he did not do any detailed chemical analysis. He simply noted that substances which preferred the oil side of an oil–water interface (*hydrophobic* substances) penetrated cells readily. Substances which preferred the water side (*hydrophilic* substances) did not penetrate readily. The ease of penetration or permeability depended on the size of a molecule and its partition coefficient between oil and water (the fraction of the substance which ended up in the oil phase compared to that in the water phase).

The other important constituent of membrane, protein, was also identified from indirect evidence. Danielli and Harvey (1935) showed that the low surface tension of natural membranes, compared to artificial membranes composed of lipids alone, could be accounted for by the adsorption of a thin layer of protein onto the surface of the lipid structure. Later analysis of many diverse types of membrane indicated that there was a basic unifying feature in that all membranes contained these two main constituents, although in varying proportions (Korn, 1966). Other substances such as carbohydrates are present, but in smaller amounts. The carbohydrates combine with the other two constituents to form *glycoproteins* and *glycolipids*, which are generally found on the outside of membranes (Malhotra and van Harreveld, 1968).

## Phospholipids

The roles of the lipid constituents of cell membranes were identified first. The lipids in biological membranes tend to align themselves in a monolayer at an oil–water interface and tend to form a bilayer if the lipids are placed between two aqueous compartments and thinned out as much as possible. The reasons for this behavior are easily understood from the structure of lipids. Lipids are derived from an alcohol such as glycerol by interaction with a fatty acid as indicated in Fig. 1.1. Most of the lipids in membranes are not simple lipids, but contain a phosphate group, and are therefore known as *phospholipids*. The phosphorous and associated atoms form what is known as a *phosphatide*, which replaces one of the OH groups of the alcohol. The other two OH groups of glycerol are replaced by fatty acid chains. The phospholipids are *amphipathic*, which simply means that one end of the molecule (the one containing the phosphorous and other polar or charged groups) prefers water to oil (hydrophilic), while the other end (containing the hydrocarbon chains of the fatty acids) prefers oil to water (hydrophobic). Thus, at an oil–water interface the molecules tend to line up so that the hydrophilic ends face the water and the hydrophobic ends face the oil. Indeed, Singer and Nicholson (1972) emphasized that to be thermodynamically stable a membrane structure must maximize the interactions between hydrophobic parts of molecules and also maximize the interactions between hydrophilic parts of molecules.

If two aqueous solutions are separated by oil, the phospholipids will line up at the two interfaces in opposite directions (the hydrophobic tails will extend into the oil at each boundary and point toward each other). If the oil is removed, a double or bimolecular layer of phospholipids can be produced. A bilayer of phospholipid separating two aqueous solutions does maximize hydrophobic and hydrophilic interactions for phospholipids under many conditions, although exceptions have been noted (Lucy, 1964). By thinning out a phospholipid to a bilayer, many of its properties can be studied in an isolated system. These bilayers are so thin that they do not show diffraction colors and are often referred to as *black lipid membranes*. If high-energy sound (sonication) is applied to such membranes, they tend to form into spherical particles known as *liposomes*. Black lipid membranes

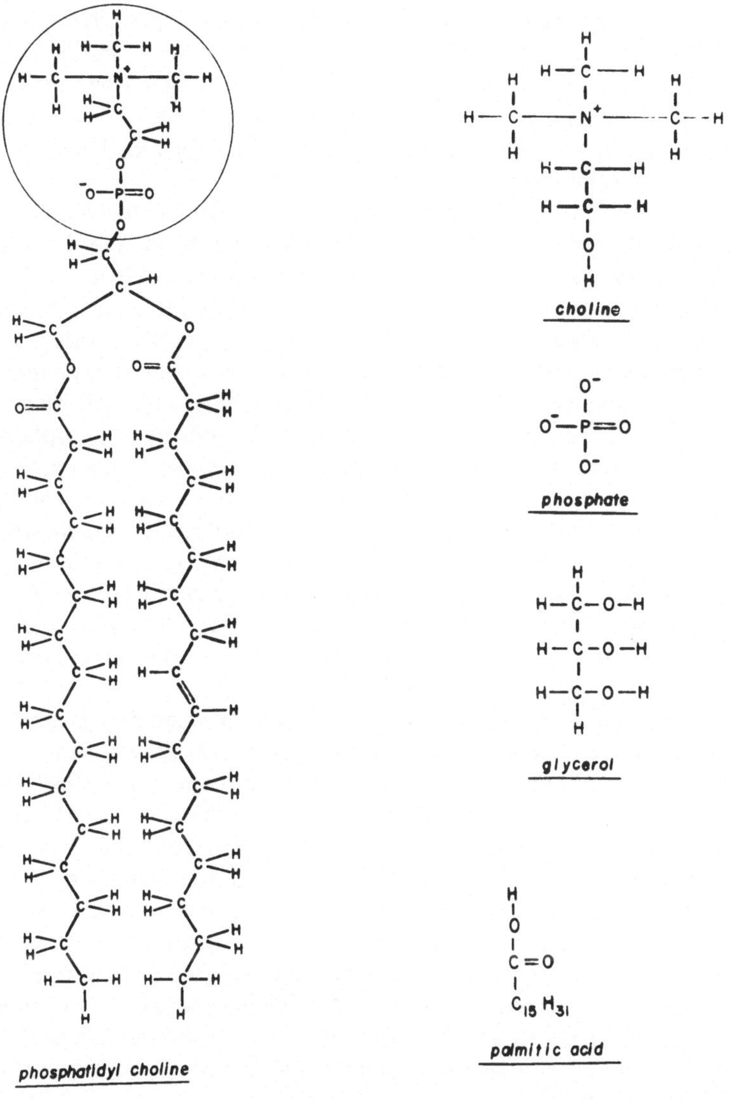

*phosphatidyl choline*

*choline*

*phosphate*

*glycerol*

*palmitic acid*

Fig. 1.1. The structure of a typical phospholipid (phosphatidyl choline) has been likened to a clothespin (Ganong, 1975). The head end (enclosed by a circle) contains phosphorous (known as a phosphatide) and usually a nitrogen group (in this case choline). The head end is ionic and attracted toward an aqueous medium (hydrophilic). The fatty acid chains (the tails or prongs of the clothespin) may be saturated (left) with a hydrogen atom linked to every possible site on carbon atoms or unsaturated (right) with carbons double-bonded to one another. These chains are repelled by an aqueous medium (hydrophobic). The structure is derived from an alcohol (in this case glycerol), the three carbon atoms of which form the central links of the molecule. (Adapted from Capaldi, 1974.)

and liposomes have been very useful models for studying membranes (Bangham, 1972; Finkelstein, 1972). Together with studies on natural membranes, evidence has accumulated for the suggestions by Gorter and Grendel (1925) and Danielli and Davson (1935) that many of the phospholipids in membranes are in the form of a bimolecular leaflet (see for example the review by Hendler, 1971).

Phospholipids can occur in a solid state similar to fats, or in a liquid state similar to oils. By careful measurements of the heat generated (calorimetry) as membranes are heated and cooled (Steim *et al.*, 1969), it has been established that much of the phospholipid in membranes is in a liquid state. Thus, phospholipids can diffuse laterally within the plane of the membrane, although the hydrophobic ends will always remain away from the aqueous solution and the hydrophilic ends will always remain pointing towards the aqueous solution. The different phospholipid constituents have different melting points, and experiments have indicated that the lipids composing membranes can be altered depending on the ambient temperatures. For example, a reindeer has a temperature gradient along its leg, the hoof being normally at the lowest temperature. The composition of the cell membranes varies systematically to maintain the lipid portion of the membranes in a fluid state. Thus, phospholipids with lower melting points are incorporated at higher concentrations nearer the hoof (Fox, 1972).

Cholesterol, a sterol which is present in membranes and therefore often grouped with the phospholipids, may also play a beneficial role in regulating the liquidity of membranes (Ladbrooke *et al.*, 1968). The liquid state is in general much more permeable than the solid state to the penetration of substances. Cholesterol is a hydrophobic molecule which packs well between phospholipid molecules in the membrane, but in doing so hinders their tendencies to form a solid, crystalline array. With substantial concentrations of cholesterol a sharp transition between liquid and solid is not observed and the membrane may remain in an intermediate or gel state. An interesting example of a membrane where the phospholipid molecules appear to be in the gel state is *myelin* (see last section of this chapter), which is a series of membranes wrapped around parts of nerve cells, thus forming an insulating barrier.

Although in a liquid state, the phospholipids in membranes are essentially confined to a two-dimensional surface. Even movement from one surface of the bilayer to the other (known as *flip–flop*) proceeds at very slow rates (Rothman and Lenard, 1977). The slow rate of flip–flop presents a problem, since the phospholipids are synthesized on the cytoplasmic side of bilayers forming the surface membrane or the endoplasmic reticulum (a network of membranes within the cytoplasm). However, the newly synthesized phospholipids in growing membranes equilibrate rapidly between the two sides of the bilayer (Lodish and Rothman, 1979), which suggests that growing membrane contains an enzyme to facilitate flip–flop (Bretscher, 1973).

## Membrane Proteins

The role of proteins and their location in the membrane has been more controversial. Robertson (1960), in advancing his *unit membrane* theory based on the uniform appearance of many membranes under the electron microscope, felt that the proteins were mainly in an extended or $\beta$-configuration and provided a surface coating which appeared more dense on the inner or cytoplasmic edge, compared to the outer edge of the cell membrane. Danielli and Davson (1935) had also suggested that the proteins were on the periphery of membranes, although perhaps in a globular form. More recently, evidence has grown that many membrane proteins, like a number of other proteins, are in the form of alpha helices (Glaser and Singer, 1971). Other proteins are *peripheral* to the membrane, in that they are weakly attached to one side or the other of the membrane, peripheral proteins and can be separated by fairly mild biochemical treatments without grossly disrupting the membrane. This is in contrast to proteins which are an *integral* or *intrinsic* part of the membrane. According to experiments using radioactive labeling techniques, integral proteins often extend completely through the membrane from one side to the other (reviewed by Capaldi, 1974). Integral proteins often have large numbers of hydrophobic or nonpolar amino acids which would tend to lie deep within the membrane rather than on its surface.

Protein molecules are found in the same orientation in the membrane and do not flip from one side to the other (Rothman and Lenard, 1977). Thus, the problem of inserting integral proteins into a membrane is more severe than for phospholipids. Lodish and Rothman (1979) discuss a model in which proteins are inserted, as they are synthesized, through the membrane of the endoplasmic reticulum with the help of a membrane transport protein. Only after synthesis do the molecules coil into their three-dimensional structure, which locks them into the membrane. Vesicles containing newly synthesized proteins then fuse with the surface membrane, as shown in Fig. 1.2. Note that the portion of the protein molecule on the cytoplasmic side of the vesicle remains on this side throughout. The portion of the protein in the lumen of the vesicle emerges on the exterior surface of the membrane without requiring flip–flop. A similar process can be imagined for release of transmitter chemicals from vesicles stored in the synaptic regions of nerve cells (see Chapter 7). The only difference would be that the transmitter would have to be detached from the membrane so that it could be released into the extracellular space.

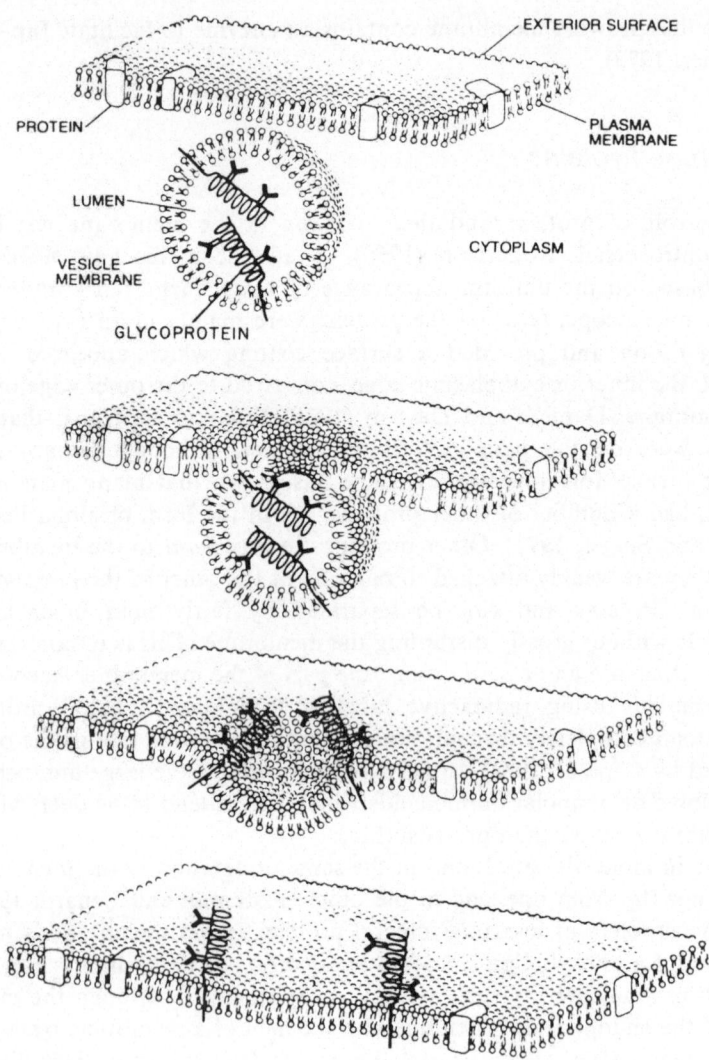

Fig. 1.2. Fusion of a vesicle with the plasma membrane of a cell preserves the orientation of the proteins embedded in the phospholipid bilayer of the vesicle. The protein shown in this example has a coiled or helical portion to which two carbohydrate segments are attached. As described in the text, a similar mechanism could account for the release of chemical transmitters by a nerve cell. (From Lodish and Rothman, 1979.)

Singer and Nicholson (1972) assembled evidence and pieces of previous models to develop a *fluid mosaic model* of the cell membrane. In a fluid the hydrophobic portions of the lipid are able to move freely. Also, as mentioned above, the phospholipid molecules are able to diffuse laterally in the two-dimensional plane of the membrane, although retaining their orientation perpendicular to this surface. The protein molecules can also diffuse freely along the surface of the membrane according to several studies, including the elegant work of Frye and Edidin (1970). The term *mosaic* refers to the fact that at any one time, single protein molecules or groups of protein molecules will be scattered along the membrane in a fairly random way, as shown in Fig. 1.3. The grouping of protein molecules is thought to be particularly significant for complex enzymatic activity such as the oxidative phosphorylation (production of high-energy phosphate molecules) which takes place, for example, on mitochondrial membranes. The high-energy phosphate molecule *adenosine triphosphate* (ATP) is then used for many purposes thoughout the cell.

Protein molecules may serve a variety of functions depending on the role of the different types of membranes (Guidotti, 1972), for example:

(1) Proteins may be assembled in groups to perform complex enzymatic functions as suggested above. The membrane structure simply provides a suitable framework for setting up this sort of "assembly line" operation.

(2) Proteins on the external surface can react with antibodies and therefore have an immunological function (Singer and Nicholson, 1972). These proteins may also be able to recognize other cell types and so play

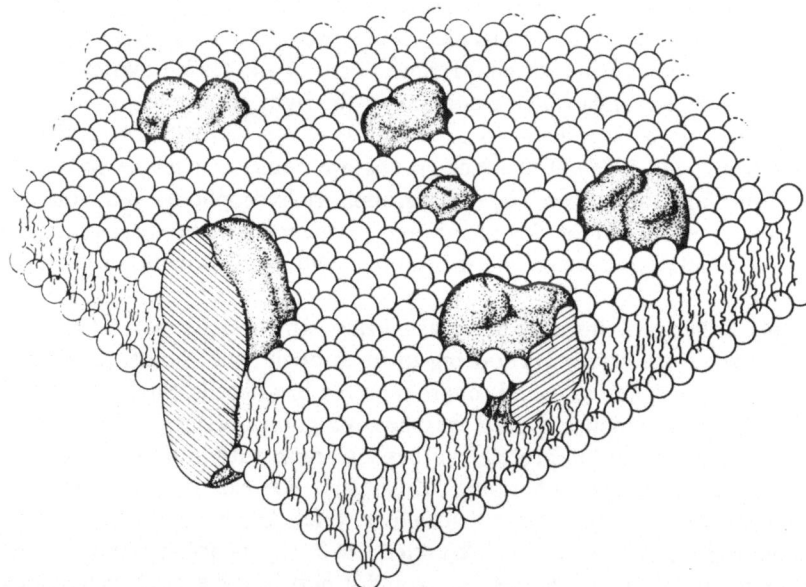

Fig. 1.3. The lipid-globular protein mosaic model of Singer and Nicholson (1972). Single protein molecules (with stippled surfaces) or small groups of molecules are free to move laterally in a fluid lipid matrix. Different protein molecules may be quite peripheral to the membrane or be an intrinsic part of it.

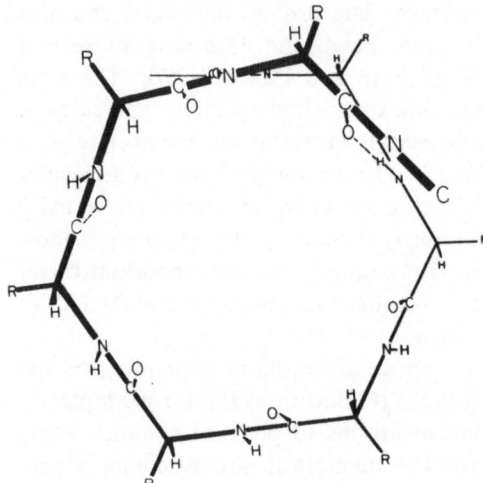

Fig. 1.4. End view of one turn of a helical molecule (gramicidin A) which could function as a transmembrane channel or water-filled pore. The size of the pore is such that $K^+$ ions would pass through more easily than $Na^+$ ions. (From Urry *et al.*, 1971.)

an important role in limiting growth and proliferation and in determining the normal structural relationships of different cells in a given tissue. This function of cell recognition requires further elucidation.

(3) The presence of large fractions of the protein in an $\alpha$-helical form, and the fact that some of these proteins extend right through the membrane, suggest that these molecules could provide a *pore* through which ions and other substances could flow. Urry *et al.* (1971) synthesized a molecule (see Fig. 1.4) which has many of the right properties to serve as a pore, although it is unlikely to be identical to any actual biological pore (Smythies *et al.*, 1974). The inner groups are polar, and hence consistent with water and ionic flow. The outer groups are hydrophobic and would fit well into the interior of the membrane. Evidence will be discussed later that some small ions cross cell membranes without dissolving in the lipid portion.

(4) Other proteins may serve as *carriers* which assist in the transport of substances which (a) are too big to fit through small pores, (b) do not dissolve in phospholipids, and (c) are required inside cells. Such substances include sugars, which are used in oxidative phosphorylation, and amino acids, which are used in protein synthesis. Certain antibiotics seem to have the properties required of biological carriers. For example, the molecule *nonactin* shown in Fig. 1.5 has a structure which has been referred to as a "tennis ball seam" (Eigen and Winkler, 1970). By bending, the molecule can open and close as a tennis ball would if part of it were cut away along the seam. A potassium ion is also shown with four waters of hydration in this figure. A hydrated potassium ion (but not a hydrated sodium ion) can fit snugly, and with good hydrogen bonding, into the interior of the

nonactin molecule. The outside of the molecule is nonpolar and should remain dissolved in the phospholipid. Although not a natural carrier, this molecule represents a model of how potassium might be "carried" across a cell membrane. If nonactin is introduced into artificial lipid membranes, the permeability to potassium is increased selectively. By adjusting the

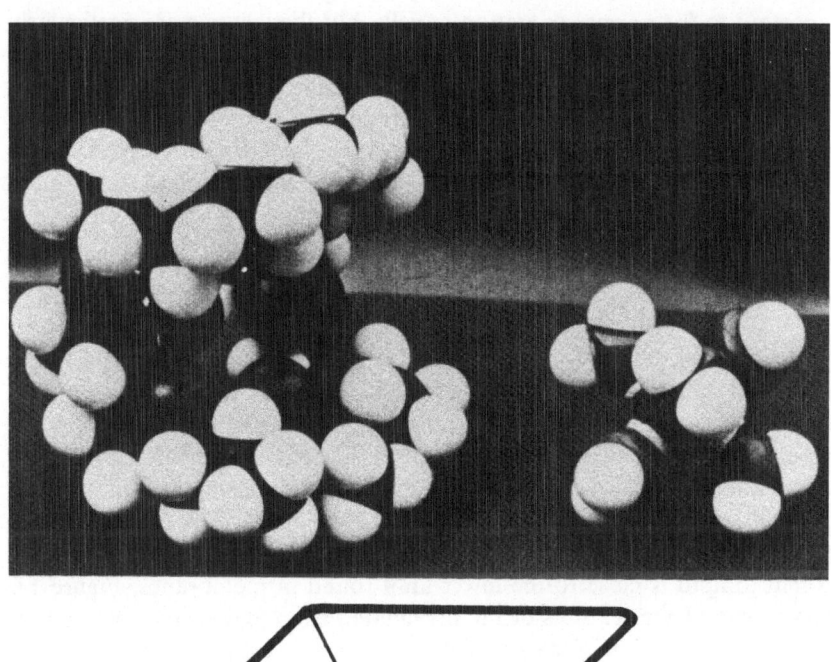

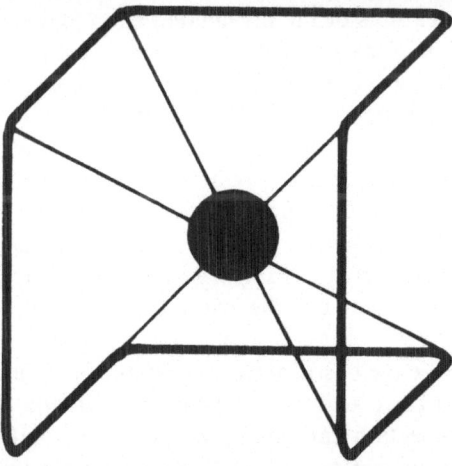

Fig. 1.5. Molecular model of nonactin (upper left) which could serve as a carrier for $K^+$ ions (upper right). The $K^+$ ion is shown with four waters of hydration. The structure of nonactin has been likened to a tennis ball seam (below) and can open up to accommodate the $K^+$ ion in the center. The outside of the nonactin molecule is hydrophobic, so that the complex could diffuse readily across a lipid membrane. (Modified from Eigen and Winkler, 1970.)

concentration of nonactin, the permeability can be set to that normally found in cell membranes.

Other functions of proteins will no doubt be found, but already the reason why the percentage of proteins varies widely in different membranes (from 20% to 80% of their weight) is obvious. Different membranes require each of these functions to varying degrees. All these functions are important to the normal operation of cells, but the last two are particularly important for the subject matter of this book. They indicate that there are at least three different ways that substances can cross cell membranes: (1) by dissolving in the phospholipid and diffusing across; (2) by passing through small, water-filled pores; or (3) by means of carriers. Later chapters will deal in more detail with each of these methods.

## Myelin

Various specializations of membranes also occur. One of the most interesting and best studied of these specializations occurs in the membranes of Schwann cells and glial cells which form the myelin around the nerve axons in the peripheral and central nervous systems of vertebrates. As has already been mentioned, the membrane composing myelin is not in the fluid state (Singer and Nicholson, 1972) because of its function as an insulator. There are few pores through these membranes, and the ratio of protein to lipid is close to the lower limit found in membranes. Figure 1.6 shows some of the features of the myelinated axon viewed in cross section (Berthold, 1978). The Schwann cell wraps many times around the axon and except for the outer layer and sometimes the inner layer, the wrapping is so tight that the cell fluid or cytoplasm of the Schwann cell is extruded. The two cytoplasmic edges of the membrane fuse to form what appears as the *major dense line* in the repetitive structure, while the fusion of the outer edges produces the *minor dense line*. There is also fusion at the inner and outer boundaries of the Schwann cell (the *inner* and *outer mesaxon* in Fig. 1.6) to form what are referred to as *tight junctions* (Schnapp and Mugnaini, 1978). These junctions seal off the axon further from the extracellular space.

A Schwann cell envelops a particular length of an axon (from 0.1 to 10 mm), and toward the end of the Schwann cell the classical circular shape seen in cross section can change to quite convoluted shapes as the axon thins down at the gap or *node of Ranvier* between two Schwann cells (Fig. 1.6B). Further specializations are observed in these regions. The individual wrappings or lamellae of Schwann cell membranes separate off and form structures which can have up to seven layers. At high power, occlusions or *septa* can be seen (Fig. 1.7a) in the space between the myelin and the axon. A tentative model of these *septate junctions* is shown in Fig.

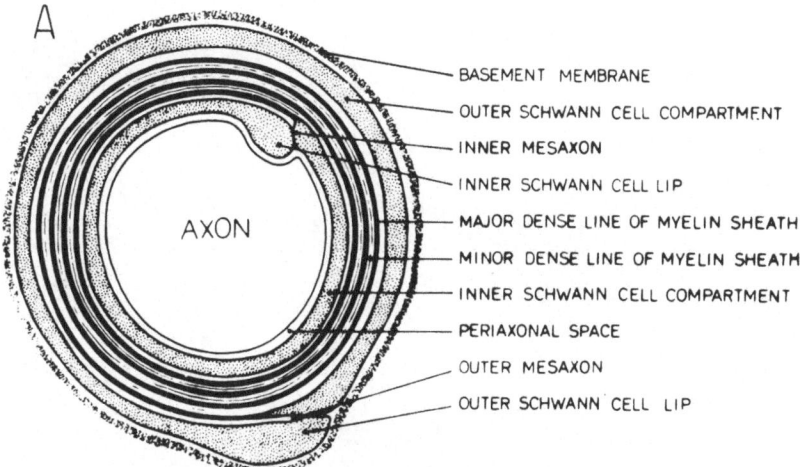

BASEMENT MEMBRANE
OUTER SCHWANN CELL COMPARTMENT
INNER MESAXON
INNER SCHWANN CELL LIP
MAJOR DENSE LINE OF MYELIN SHEATH
MINOR DENSE LINE OF MYELIN SHEATH
INNER SCHWANN CELL COMPARTMENT
PERIAXONAL SPACE
OUTER MESAXON
OUTER SCHWANN CELL LIP

AXON

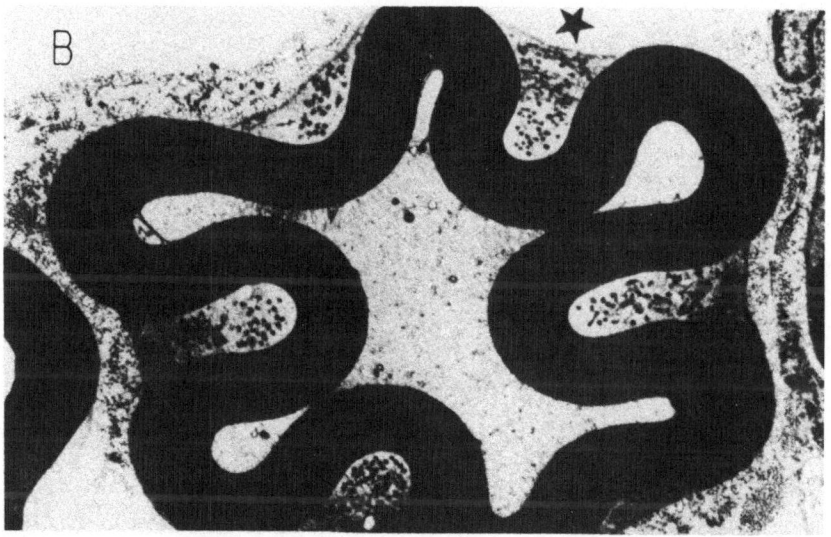

Fig. 1.6. (A) Cross section of a small, myelinated axon, showing some of the specializations that occur and the terms used to describe them. Further details are in the text. (B) Cross section of a large myelinated fiber. The staining of this electron micrograph is intense enough that the many turns of myelin appear as a single black band. Close to the nodes of Ranvier, the axon thins down and the myelin becomes convoluted, as shown here. Schwann cell cytoplasm occupies the external space (star). (From Berthold, 1978.)

1.7b. Some evidence for this model was obtained with the technique of *freeze fracturing*, in which the tissue is frozen and cleaved in such a way as to show some of the internal structure of the membrane (see, for example, Schnapp and Mugnaini, 1978). These junctions also seal off the axon membrane from the general extracellular space.

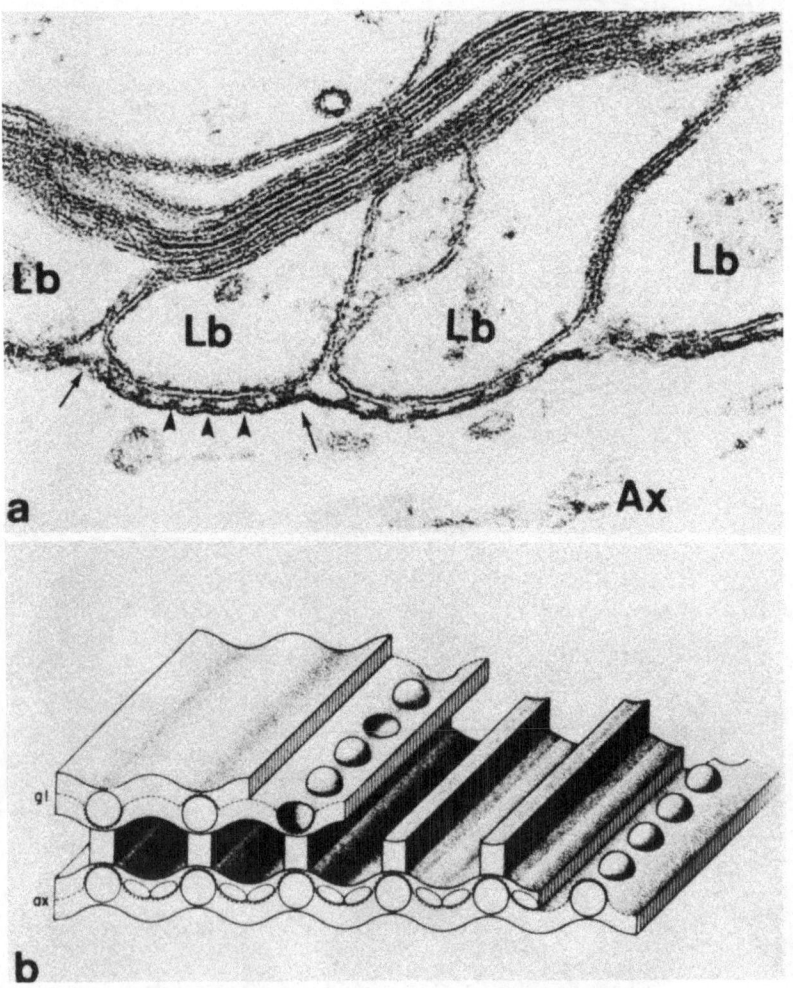

Fig. 1.7. (a) High-power electron micrograph showing what is referred to as the lateral belt (Lb) where individual turns of myelin end near the axonal (Ax) membrane. Note the region of membrane between the thin arrows over which there are regularly spaced occlusions, or *septa*, between the membranes. (b) Tentative model of these connections based on electron micrographs from thin sections and from freeze-fractured material. (From Schnapp and Mugnaini, 1978.)

Septate and tight junctions, in contrast to gap junctions which will be considered later (Chapter 7), are designed to prevent current flow. Other membrane specializations will be discussed in later chapters, but first some general thermodynamic constraints imposed on the functions of cell membranes will be considered. Their importance in arriving at a suitable structure for cell membranes has already been mentioned briefly.

# 2

## Membrane Thermodynamics

Classically, thermodynamics dealt mainly with equilibrium states in which all the state variables such as temperature, volume, concentrations of substances, free energy, etc., were constant (Holman, 1969). This last variable is particularly important since it distinguishes equilibrium states from some nonequilibrium steady states where free energy is continually being used. For a nonequilibrium steady state to be maintained, a source of energy is continuously required to replace the energy expended. A "resting" cell can be considered to be in just such a *nonequilibrium* steady state because it requires a continual source of energy for its metabolism and to maintain the gradients of substances that normally occur across cell membranes.

### Equilibrium Thermodynamics

The purpose of this section is *not* to give a thorough account of equilibrium thermodynamics, but rather to extract from this large field a couple of important relationships which govern the behavior of substances. Since these relationships must be fulfilled if a substance is at equilibrium, they provide useful tests to determine whether or not a substance is at equilibrium. A later section, *Irreversible Thermodynamics,* deals with nonequilibrium states. Irreversible thermodynamics is used increasingly to treat substances which do not fulfill the criteria for equilibrium (Katchalsky and Curran, 1965).

In defining whether a substance is or is not at equilibrium, the concept of an electrochemical potential has proven useful. This concept is an extension of the common electrical potential to include the differences in concentration or chemical activity of a substance. In measuring any potential, it is necessary to define a ground potential, and then to determine the work necessary to bring the substance from the ground state to the state of interest.

In bringing a coulomb of positive charge to a state at $V_i$ volts, where $V_i$ is the voltage of the inside of a cell, $V_i$ joules of energy will be required. The electrical work $W_e$ to bring a mole of a substance $S$ (with a valency $Z_s$) to a voltage $V_i$ will be simply

$$W_e = Z_s F V_i \qquad (2.1)$$

where the Faraday constant $F$ is simply a conversion factor with the dimensions of coulombs/mole (1 mole contains the molecular weight of a substance in grams). A quick check of the dimensions on the left- and right-hand sides of this equation will verify that the constant $F$ is needed.

In addition to electrical work, it will take chemical work to bring a substance from a unit concentration such as 1 molar (1 M = 1 mole/liter) to the concentration within a cell. This chemical work $W_c$ is given by

$$W_c = RT \ln \left[ S \right] \qquad (2.2)$$

where $R$ is the gas constant, $T$ is absolute temperature in degrees Kelvin, ln represents logarithms to the base $e$ (natural logarithms), and $[S]$ is the concentration of $S$. The product $RT$ can again be thought of as a conversion factor to give the dimensions of joules, although this factor obviously depends on temperature. The basis for the use of the logarithm here will be considered later (Chapter 4). Note that because the logarithm of $[S]$ is used, the chemical work will be negative (energy will be released) if the concentration of the substance inside the cell is less than 1 molar.

For a substance to be at equilibrium across a cell membrane, the work required to bring the substance to points just outside and just inside the membrane must be equal. *No net work is expended when a molecule of that substance crosses the membrane.* If $i$ and $o$ represent the inside and outside of the membrane respectively, then

$$Z_s F V_i + RT \ln \left[ S \right]_i = Z_s F V_o + RT \ln \left[ S \right]_o \qquad (2.3)$$

Rearranging Equation (2.3) and defining the membrane potential according to the usual convention, $V_m = V_i - V_o$, yields

$$V_m = \frac{RT}{Z_s F} \ln \frac{\left[ S \right]_o}{\left[ S \right]_i} \triangleq E_s \qquad (2.4)$$

Equation (2.4) is an important equation first derived by Nernst, and is therefore known as the Nernst equation. It is also used to define ( $\triangleq$ means equals by definition) an *equilibrium potential*, $E_s$, for the substance $S$. Generally, logarithms to the base 10 are used rather than natural logarithms, and the temperature is held constant. Then

$$E_s \cong \frac{60}{Z_s} \log \frac{\left[ S \right]_o}{\left[ S \right]_i} \qquad (2.5)$$

The equilibrium potential in Equation (2.5) is measured in mV and is only approximately equal ($\cong$) to the right-hand side. The exact value of the number on the right-hand side will depend on the absolute temperature, varying from 58 mV at room temperature (20°C) to 62 mV at body temperature (37°C). The Nernst equation is a necessary and sufficient condition for a substance to be at equilibrium. In other words, if one measures the concentration of a substance inside and outside a cell and computes the equilibrium potential according to Equation (2.4), then the substance is assumed to be at equilibrium if and only if the equilibrium potential agrees with the membrane potential to within the limits of experimental error.

Experimentally, the most difficult quantity to measure is often the membrane potential. An alternative method for determining if a substance is at equilibrium is to measure the concentrations of two substances $S_1$ and $S_2$, assuming that one of these substances is at equilibrium. For substances with concentrations $[S_1]$ and $[S_2]$ and valencies $Z_1$ and $Z_2$ respectively, Equation (2.4) gives

$$V_m = \frac{RT}{Z_1 F} \ln \frac{[S_1]_o}{[S_1]_i} = \frac{RT}{Z_2 F} \ln \frac{[S_2]_o}{[S_2]_i} \tag{2.6}$$

Canceling the common $RT/F$ and taking exponentials of each side gives

$$\left( \frac{[S_1]_o}{[S_1]_i} \right)^{1/Z_1} = \left( \frac{[S_2]_o}{[S_2]_i} \right)^{1/Z_2} \tag{2.7}$$

Equation (2.7) is the Donnan or the Gibbs–Donnan equilibrium condition. More commonly, it is written out in the form of the ratios expected for common ions, for example

$$\frac{[K^+]_o}{[K^+]_i} = \frac{[Na^+]_o}{[Na^+]_i} = \frac{[Cl^-]_i}{[Cl^-]_o} \tag{2.8}$$

If both potassium and sodium ions are at equilibrium, then the ratio of the external to the internal concentrations should be equal. This ratio should also be equal to the ratio of the internal to the external concentrations of chloride ions, because these ions have a valence of $-1$. Still another way of stating the Gibbs–Donnan condition, which is also readily derived from the chemical law of mass action, is that the products of monovalent cations and anions should be equal on both sides, e.g., $[K^+]_o[Cl^-]_o = [K^+]_i[Cl^-]_i$. This derivation and the expected ratios for divalent ions are left to the problems at the end of this chapter.

When the Nernst equation or the Gibbs–Donnan equilibrium condition has been applied to biological membranes, ions are often *not* at equilibrium. Typical concentrations are listed in many texts for some ions in nerve or muscle cells (e.g., Ruch and Patton, 1965; DeVoe, 1974). Only

$Cl^-$ appears to be at equilibrium in some, but not all, nerve and muscle cells (Eccles, 1964; Ferreira *et al.*, 1972). $K^+$ ions are somewhat more concentrated inside cell membranes than would be expected, while $Na^+$ ions are grossly out of equilibrium. Both electrical and chemical diffusion forces should be driving $Na^+$ ions into the cell, yet $Na^+$ ions remain at low concentration inside most nerve and muscle cells. These findings do not in any sense invalidate the usefulness of equilibrium thermodynamics. Instead, they indicate that application of the formulas of equilibrium thermodynamics may only be a first step in understanding the movement of substances across cell membranes.

Why is a particular substance not at equilibrium? Is the cell membrane permeable to the substance? If the substance cannot get through the membrane to come to equilibrium, this provides a ready explanation for the nonequilibrium state. However, permeability studies indicate that $Na^+$ ions and the other common ions, with the exception of some large organic ions, can permeate muscle membranes. If the membrane is permeable to $Na^+$ ions, then they must be continually "pumped out" of muscle cells to maintain their low concentration internally. This process, known as *active transport*, will be considered in detail in Chapter 3. Use of the equilibrium conditions derived in this section is the logical first step in determining experimentally the processes controlling movements of substances across membranes. In applying these methods experimentally, caution must be exercised in the use of the measured concentrations. Some substances may be largely bound or sequestered in an intracellular compartment so the electrochemical activity may differ considerably from the concentration. The distinction between activities and concentrations is discussed in detail in most physical chemistry and general physiology texts (e.g., Moore, 1962; Dowben, 1969), but will not be considered further here.

## Osmotic Equilibrium

In addition to the equilibrium of individual substances dissolved in a solvent such as water, one can test whether the total osmotic pressure is equal on two sides of a membrane. Osmotic pressure results from solute molecules striking a membrane, and higher osmotic pressure inside a cell will tend to increase the cell's volume by pushing the cell membrane outward. This increase occurs by inward movement of water molecules, which will be more concentrated in the external solution if the external concentration of solute molecules is less.

An interesting consequence of the conditions for osmotic equilibrium is that if a cell contains an impermeable substance which is not present externally, there will be a net osmotic pressure. The proof of this result is left as a problem (2.3) at the end of the chapter. Virtually all cells do contain considerable amounts of large impermeable molecules such as proteins, nucleic acids, etc. There is thus an inevitable tendency for cells to swell and for water to rush in until either: (1) the membrane is stretched to the point where there is an equal and opposite restoring force; (2) metabolic energy is continually supplied to oppose this tendency; or (3) the cell bursts.

Option (1) is available in plants which have stiff cell walls. and also applies to some animal tissues. However, the second option is widely used and involves the action of the sodium pump, as will be explained in the next chapter. If the source of energy for the sodium pump is removed, isolated cells such as red blood cells will swell until they burst.

The proof of the result that there is an inevitable osmotic pressure difference across cell membranes depends on the assumption of *approximate charge neutrality*. This assumption may seem inconsistent with the idea that the normal resting potential difference between the inside and outside of a cell requires a separation of charges across the cell membrane. However, a subsequent problem (2.4) indicates that the number of charges required to produce a normal resting potential is insignificant compared to the total concentrations of ions in a cell. Thus, the assumption of approximate charge neuturality is a good one.


## Irreversible Thermodynamics

There are clear limitations in describing the properties of nerve and muscle cells using equilibrium thermodynamics. Although this approach provides useful criteria for determining which substances are at equilibrium, application of these criteria indicates that many, if not most, of the important ions are *not* at equilibrium. The resting cell is clearly in a steady state with respect to many important variables, but metabolic energy is continually being supplied to maintain this steady state. This energy is dissipated in the form of heat and entropy, which is a measure of the randomness of a particular state (Moore, 1962). In other words, the ordered structure of the high-energy phosphate compound, adenosine triphosphate (ATP), is continuously broken down to a less ordered or more random state. A branch of thermodynamics has been developed to deal with just such limitations in which energy is being continuously supplied and irreversibly dissipated in the form of heat and entropy (Prigogine, 1967).

The theory of irreversible thermodynamics is not completely general, but has two major limitations: (1) only *steady state* situations can be treated, and (2) only deviations from equilibrium are considered which are small enough that *linear* approximations can be used. As mentioned above, the assumption of a steady state seems reasonable enough, but the assumption of linearity is less clear.

Many of the processes which will be considered are reasonably linear over quite wide regions. For example, Ohm's law states that the current $I$ flowing at any point in a medium will be proportional to the voltage gradient at that point (grad $V$)

$$I = - G \text{ grad } V \qquad (2.9)$$

where $G$ is the electrical conductivity of the medium. The current will flow in a direction opposite to the maximum gradient or rate of change. Thus, if the voltage were increasing along the $x$ axis, current would flow from high values of $x$ and $V$ to lower values. Hence, a negative sign is required in Equation (2.9).

Similarly, the diffusion of substances under concentration gradients is governed by Fick's first law

$$J = - D \text{ grad } C \qquad (2.10)$$

where $J$ is the flux of a substance produced by the gradient in concentration $C$, and $D$ is the diffusion constant relating the two. In general, the equations in irreversible thermodynamics have a form in which a flow is proportional to the force producing this flow. A coefficient, which depends on the particular phenomenon being studied (known as a *phenomenological coefficient*), relates the flow to the force.

If the forces and fluxes are defined appropriately, the increase in entropy is equal to the sum of the products of the various flows and forces acting on the system

$$T \frac{dS}{dt} = \sum_i J_i X_i \qquad (2.11)$$

where $T$ is the absolute temperature (in degrees Kelvin) of the system, $dS/dt$ is the rate of change of entropy $S$, $J_i$ is the flow of the $i$th substance produced by the force $X_i$. Equation (2.11) is a central result of irreversible thermodynamics, but will not be derived here. Note, however, that the quantity on the left-hand side of Equation (2.11) represents the rate at which energy is being dissipated in the form of entropy and is sometimes referred to as the *dissipation function* (Kedem and Katchalsky, 1958). By contrast, a reversible process proceeds at infinitely low rates and is opposed by exactly opposite forces, so that no entropy is generated. In real systems such as biological membranes, the dissipation function will generally be positive. An individual product may be negative (e.g., a substance may be "pumped" from a region of low concentration to a region of high concentration), but this process must be coupled to another process or to a chemical reaction which supplies enough energy to drive this reaction as well as producing an increase in entropy.

The idea of coupling between processes is an important concept in irreversible thermodynamics. There are many examples of coupling such as the piezoelectric crystal, in which mechanical forces that produce deformation of the crystal also generate voltages, and vice versa. This example involves coupling between mechanical and electrical forces, but similar coupling occurs in diffusional processes. Application of a pressure difference will lead directly to a volume change. As indicated in the previous section, a concentration difference will also lead to movements of water and hence produce volume changes. Thus (Kedem and Katchalsky, 1958)

$$J_v = L_p \Delta p + L_{pD} RT\Delta C \qquad (2.12)$$

where $J_v$ is the total flow of volume, $L_p$ is the phenomenological coefficient relating this flow to a pressure difference $\Delta p$, and $L_{pD}$ is a cross-coefficient (known as the *osmotic* coefficient) relating a concentration difference $\Delta C$ across the membrane to the volume flow. Volume flow can be measured in liters ($1\ L = 10^{-3}\ m^3$) per second (s). The forces generated by kinetic motion of solute molecules will depend on temperature and the constant $R$ is again required to make the units appropriate for inclusion as a force [e.g., in Equation (2.11) for the dissipation function].

In addition to mass or volume flow, the solute molecules will move relative to those of the solvent. If this differential flow is $J_D$, again measured in terms of the flow in $L/s$ of the partial volume occupied by solute molecules, then

$$J_D = L_{Dp} \Delta p + L_D RT\Delta C \qquad (2.13)$$

$L_D$ is the phenomenological coefficient relating a concentration difference to diffusional flow. However, separation of solute and solvent can also be produced by a pressure difference. Pressure tends to push a solvent through a membrane faster than the solute, a process known as *ultrafiltration*, and the cross-coefficient $L_{Dp}$ is referred to as the ultrafiltration coefficient.

Onsager first showed that for any system of cross-coupled flows the cross-coefficients must be equal (see Dowben, 1969, for a brief proof and discussion of the Onsager relation). In the present example, the ultrafiltration coefficient $L_{Dp}$ will automatically equal the osmotic coefficient $L_{pD}$ when the forces and flows are defined as indicated above. Thus, three distinct coefficients are required to fully describe the diffusion of a solute through a membrane. This result is intuitively acceptable since there are essentially three types of interaction taking place in such a diffusion process: (1) the interaction of the solute with the membrane, (2) the interaction of the solute with the solvent, and (3) the interaction of the solvent with the membrane. These interactions will not be related here to the coefficients $L_D, L_p$, and $L_{pD}$ (the interested reader can consult Kedem and Katchalsky, 1961). However, this discussion illustrates why numerous people were led astray in trying to measure "the permeability coefficient" without considering the linked flows of water which occur (several examples are discussed by W. D. Stein, 1967, p. 48).

For an *ideal semipermeable membrane* (one that distinguishes perfectly between solute and solvent), a single coefficient is sufficient. Such a membrane will not allow any solute molecules to permeate, so the total volume flow $J_v$ will simply be the flow of solvent. Since solute molecules do not move at all, their flow relative to the solvent will be

$$J_D = -J_v \qquad (2.14)$$

Equation (2.13) must hold under all pressure differences and all concentration differences, so from Equations (2.12) and (2.13) and Onsager's relation between the cross-coefficients,

$$L_p = L_D = -L_{pD} = -L_{Dp} \qquad (2.15)$$

The degree to which a membrane approaches an ideal semipermeable membrane can be measured by forcing a solution through a membrane into an identical solution by means of a pressure gradient. The ratio

$$\sigma = -J_D/J_v \qquad (2.16)$$

under these conditions ($\Delta C = 0$) is known as the *reflection coefficient*. For an ideal semipermeable membrane, it is clear from Equation (2.14) that $\sigma = 1$. At the opposite extreme is a completely nonselective membrane which permits the solute and solvent to pass through with equal ease. Then, $J_D = 0$ and $\sigma = 0$. As the size of a solute molecule is increased relative to the size of the pores through a membrane, the reflection coefficient will increase smoothly from 0 to 1. The size of the solute molecule at which this transition takes place can be used to measure the size of a membrane pore (Goldstein and Solomon, 1960).

## Problems

2.1. If the ratio of $[K^+]_i/[K^+]_o = 10$ at equilibrium, what should the ratio of $[Ca^{++}]_i/[Ca^{++}]_o$ be? In fact, the internal concentration of $Ca^{++}$ in muscle cells is normally less than $10^{-7}$ M compared to over $10^{-3}$ M outside. What possible explanations are there for this low concentration? The most probable explanation is discussed in Chapter 3.

2.2. The law of mass action states that for a chemical reaction $A + B = C + D$ there is an equilibrium constant $K$ given by

$$K = \frac{[C][D]}{[A][B]}$$

Show that if we consider the movement of an uncharged salt such as KCl across the membrane as a chemical reaction, that the Gibbs–Donnan equilibrium condition of Equation (2.7) can be derived. Hint: Let $A = K_i^+$, $B = Cl_i^-$, etc. What will the equilibrium constant be for this reaction?

2.3. Consider the situation below in which an impermeable anion is contained on the inside of a membrane:

| Outside | Inside |
|---|---|
| $K_o^+$ | $K_i^+$ |
| $Cl_o^-$ | $Cl_i^-$ |
| | $A^-$ |

Assuming that approximate charge neutrality holds on each side of the membrane (the number of positive charges equals the number of negative charges on each side of the membrane) and that the other ions are distributed at equilibrium, show that the permeable ions will distribute themselves in such a way that there will always be a net osmotic gradient across the membrane.

2.4. How many coulombs of electrical charge $Q$ are required to produce a membrane potential of 100 mV? Hint: Use the relation $Q = CV_m$, where $C$ is the electrical capacity of the cell membrane, which is typically in the range $1-10$ $\mu F/cm^2$. How many moles of monovalent ions will this require? (The Faraday constant $F = 96,500$ coulombs/mole.) What fraction of the total number of ions does this represent, if the concentration of monovalent cations in a spherical cell with a diameter of 10 $\mu m$ is 100–200 mM? Is this fraction sufficient to affect the result of Problem 2.3?

2.5. The flow of a solute is not normally measured in L/s but rather the number of molecules or moles/s permeating the membrane is measured using radioactive tracers. What is the relation between this flow and the volume flows defined in the text? [The answer is given by Equation (2.13) of W. D. Stein, 1967.]

2.6. Show that if a fluid is surrounded by a membrane which is sufficiently stiff that no volume changes are possible, a single permeability coefficient is sufficient to relate fluxes to flows.

# 3

# Carrier Transport

In this chapter we turn from more general thermodynamic considerations to specific models for the transport of substances across cell membranes with the help of special molecules such as *carriers*. The reason for carriers is obvious since many substances such as sugars and amino acids that are continually needed inside a cell and are continually being used up cannot cross a membrane without special help. In some examples the carrier merely functions as a catalyst or enzyme, and no energy is required. The process is known as *facilitated transport* and is treated in the first section of this chapter. In other examples, movement of substances is directly or indirectly linked to a chemical reaction which provides a source of energy. This process is known as *active transport* and is discussed in the second section. The coupling of flows to chemical reactions such as occurs in active transport can be included in the equations of irreversible thermodynamics simply by adding extra terms to the right-hand side of equations such as (2.12) (Kedem, 1961). However, this chapter is directed towards more specialized results, and the last section shows that the inclusion of active transport is sufficient to account qualitatively for the voltage and concentration differences across cell membranes and also for the regulation of cell volume.

## Facilitated Transport

Transport of a substance across a biological membrane usually involves three distinct steps (Tonomura and Yamada, 1973). The first step is the *recognition* of the substance to be transmitted by the carrier molecule. This step involves the formation of a complex between the substance and the carrier due to chemical or other affinity (see Chapter 1). Next comes the *translocation* or movement of the complex across the membrane. This

OUTSIDE        MEMBRANE          INSIDE        Fig. 3.1. A model of facilitated transport from the
                                              outside to the inside of a cell in which $[S]_o$ and $[S]_i$ are

$$S + \begin{matrix} | \\ | \\ | \end{matrix} \quad C \underset{d}{\overset{a}{\rightleftharpoons}} SC \xrightarrow{b} C \quad \begin{matrix} | \\ | \\ | \end{matrix} + S$$

the concentrations of the substances on the outside and
the inside of the membrane respectively. $C$ is the carrier whose total concentration in the membrane is $C_m$,
of which a concentration $[C]$ is unbound; $SC$ is the complex formed by binding of $S$ to $C$,
which has a concentration $[SC]$; $a$ is the rate constant for the *association*, of $S$ to $C$ at the
outer surface; $d$ is the rate constant for the *dissociation* of the complex at the outer surface; $b$
is the rate constant for the *translocation of the bound complex* to the inner surface.

step may involve crossing a substantial energy barrier and hence is often
rate limiting. The final step is the *release* of the transported substance on
the opposite side of the membrane.

A model for the use of carriers in facilitated transport which is
analogous to the action of enzymes is illustrated in Fig. 3.1. This model
assumes that (1) the internal concentration of the substance $[S]_i$ is low (it
is continually being used up so that the back reaction at the inner surface
is negligible), and (2) neither the release of the substance at the inner
surface nor the movement of the free carrier back across the membrane
limit the transport, as these processes are ignored. Then, the steady rate or
flow $J_s$ of the substance across the membrane will be simply

$$J_s = b[SC] \tag{3.1}$$

In the steady state, the rate of formation of the complex must equal its
rate of breakdown. Then

$$a[S]_o[C] = (b + d)[SC] \tag{3.2}$$

Formation at the outer surface depends on external concentration, the
amount of unbound carrier, and the rate constant $a$, whereas breakdown
occurs at both surfaces and depends on the amount of bound carrier. The
total concentration of carrier is

$$C_m = [SC] + [C] \tag{3.3}$$

Combining these three equations gives the rate of transport

$$J_s = \frac{abC_m[S]_o}{a[S]_o + b + d} \tag{3.4}$$

Equation (3.4) is a rectangular hyperbola which is analogous to the
*Michaelis–Menten equation* of enzyme kinetics (see, for example, Dowben,
1969).

In fact, Equation (3.4) can be rewritten in the form

$$J_s = \frac{M[S]_o}{[S]_o + K_m} \tag{3.5}$$

where $M = bC_m =$ the maximum possible flow and $K_m = (b + d)/a =$ the

*Michaelis constant*. This constant gives the external concentration at which there will be half-maximal flow. This same form of equation will arise in considering synaptic transmission (see Fig. 7.9C) and sensory processes (Chapter 9). Thus, rectangular hyperbolas are quite widely found and their occurrence does not necessarily imply the presence of a single enzymatic reaction.

This analysis can be extended to consider two-way movement of a single substance. If $J_s^{in}$ and $J_s^{out}$ represent the inward and outward fluxes of a substance $S$ and the two fluxes are independent, then the net movement will be

$$J_s = J_s^{out} - J_s^{in} = \frac{M[S]_i}{[S]_i + K_m} - \frac{M[S]_o}{[S]_o + K_m} \qquad (3.6)$$

Equation (3.6) is known as the *mobile carrier equation* (Widdas, 1952; see also Rosenberg and Wilbrandt, 1955), which is analogous to *product inhibition* in enzyme kinetics (Dowben, 1969). The conditions under which this equation is valid will be examined further in Problem 3.6.

A carrier will never be completely specific, and many other substances may combine with it. These other substances are often not transported across the membrane and serve to inhibit facilitated transport (LeFevre, 1959; Forsling and Widdas, 1968). Inhibitors may compete for the same site (*competitive inhibition*) or may attach at a different site (*noncompetitive inhibition*). These possibilities are considered in Problems 3.1 and 3.2. Further elaborations of this model and its analogy to enzyme kinetics could be pursued. However, the reactions within the membrane are somewhat more complex and are better described by Fig. 3.2.

A full analysis of the steady state fluxes of this model can be found in Britton (1964, 1966). Läuger (1972) also considered the transient behavior of this model and was able to determine each of the four rate constants experimentally. Only uncharged substances and carriers are considered in this chapter, so the rate constants for the movement of the carrier $C$ and complex $SC$ in the two directions are assumed equal. This assumption will be relaxed in Chapter 4. Also, the steady state fluxes will only be treated here under the simplifying assumption that the transport of the carrier across the membrane is rate limiting. Then, the reactions at the two interfaces will be approximately at equilibrium, with a dissociation con-

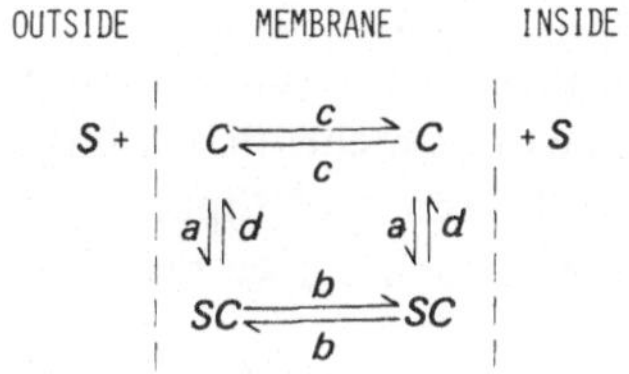

Fig. 3.2. A more accurate model of carrier diffusion (modified from Läuger, 1972), where $a$ is the rate constant for *association* of $S$ to $C$ at either edge of the membrane, $d$ is the rate constant for *dissociation* at either edge of the membrane, $c$ is the rate constant for diffusion of the *carrier* alone across the membrane in either direction, $b$ is the rate constant for diffusion of the *bound complex* across the membrane in either direction, $c$ is the rate constant for diffusion of the *bound complex* across the membrane in either direction, tion.

stant $K_d$, where

$$K_d = \frac{d}{a} = \frac{[C]_i[S]_i}{[SC]_i} = \frac{[C]_o[S]_o}{[SC]_o} \tag{3.7}$$

and $i$ and $o$ refer to the inside and outside edges of the membrane respectively. The total concentration of carrier $C_m$ is

$$C_m = [C]_i + [C]_o + [SC]_i + [SC]_o \tag{3.8}$$

Finally, a steady state situation is assumed, so no net movement of carrier occurs. Then,

$$[C]_i + r[SC]_i = [C]_o + r[SC]_o \tag{3.9}$$

where $r = b/c$ is the ratio for the rate of movement for the complex compared to that for the free carrier. If the movement of carrier is rate limiting, then the inward flux will depend on the amount of bound carrier at the outer edge of the membrane $[SC]_o$ and the rate it can be moved $b$. Thus,

$$J_s^{in} = b[SC]_o \tag{3.10}$$

Equations (3.7)–(3.9) can then be solved for $[SC]_o$ and then for $J_s^{in}$. The result is

$$J_s^{in} = \frac{bC_m[S]_o(K_d + r[S]_i)}{(K_d + [S]_o)(K_d + r[S]_i) + (K_d + [S]_i)(K_d + r[S]_o)} \tag{3.11}$$

This equation shows that the inward movement of a substance depends in a rather complex way on both its internal and external concentrations. There are, however, some simpler and interesting limits of this equation which are considered in Problems 3.4 and 3.5. Further analysis of the transport of charged substances can be found in Läuger (1972). W. D. Stein (1967) also considers the kinetics of linked movements of more than one substance. A particularly important case of linked transport is the active movement of $Na^+$ and $K^+$ ions, which we will now discuss.

## Active Transport

Active transport differs from facilitated transport in requiring metabolic energy from the high-energy phosphate compound ATP. Active transport is usually included as a type of carrier transport, but inclusion does not mean that a large intrinsic protein molecule actually carries ions in the sense of flipping from one side of the membrane to the other with ions attached (see below). A simplified reaction sequence for the $Na^+$ pump is given below (Hoffman, 1973)

$$ATP + C \xrightarrow{Na^+} ADP \cdot P \sim C \xrightarrow{K^+} ADP + P_i + C \qquad (3.12)$$

This sequence emphasizes once again the enzymic or catalytic nature of the process which involves the splitting of ATP. ATP can combine with the enzyme molecule to form a phosphorylated intermediate, which then splits and releases inorganic phosphate (adenosine diphosphate or ADP is also produced). $Na^+$ ions stimulate the phosphorylation. Translocation of the ions presumably occurs together with the dephosphorylation which is stimulated by $K^+$ ions. Often three $Na^+$ ions are pumped out and two $K^+$ ions are pumped in (reviewed by Thomas, 1972; De Weer and Geduldig, 1973). Thus, a net outward electric current is generated (*electrogenic pumping*) each time a molecule of ATP is split. The linkage of $Na^+$ efflux to $K^+$ influx is probably variable (see below) and other ions can be involved. However, the effects of the pump would hold even if the pump were electrically neutral (see the next section) and the same number of $Na^+$ and $K^+$ ions were transported in each cycle.

The molecular details of this transport process are still obscure, so the diagram in Fig. 3.3 merely indicates the ratios of substances involved in the reaction. Since the enzyme involved is a large protein molecule (Skou, 1974), transport cannot involve rotation or flip-flop of the whole molecule. Thermodynamically, flip-flop of large proteins requires a considerable amount of energy and must proceed at exeedingly low rates, if at all (Rothman and Lenard, 1977). Models of the process have been proposed based on the known subunit structure of the molecule (e.g., W. D. Stein *et al.*, 1974; Kyte, 1975), and visualization of the molecule by freeze fracture electron microscopy has been possible (Vogel *et al.*, 1977).

A further site is required for a cardiac glycoside such as ouabain, which inhibits transport, presumably by stabilizing the position of the carrier in the membrane. Under the assumption that one ouabain molecule inhibits transport of one carrier molecule, the number of ouabain molecules required to completely inhibit transport can be used to estimate the density of carrier molecules in the membrane. Baker and Willis (1972)

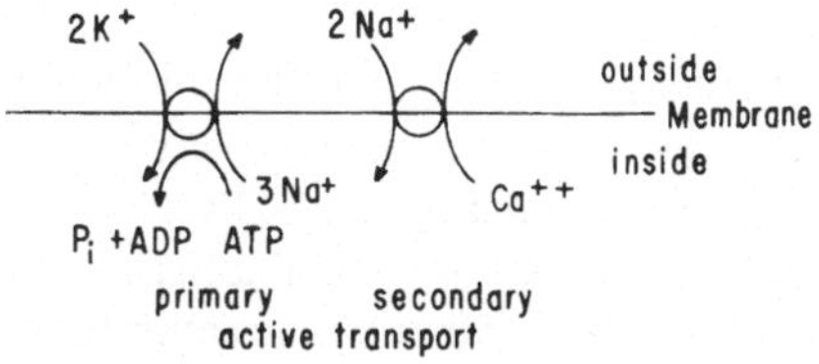

Fig. 3.3. A model of primary and secondary active transport. The primary active transport of $Na^+$ out of the cell provides potential energy which can be utilized to produce an outflow of $Ca^{++}$ which is coupled to an inflow of $Na^+$ down its electrochemical gradient. The outflow of $Ca^{++}$ represents an active transport *against* the electrochemical gradient of this ion.

found about $10^3$ carriers/$\mu m^2$ in a variety of cell membranes and estimated that 100 ions are pumped per site per second. For comparison, Läuger (1972) estimated that a single molecule of valinomycin can transport up to $10^4$ $K^+$ ions/s by facilitated transport, although this rate would only be reached at unphysiologically high concentrations. Finally, a molecule such as gramicidin A which serves as a channel for passive $K^+$ movements (see Chapter 1) can pass $10^7$ ions/s (Urry, 1978).

Some features of the carrier molecule are well established. The carrier is a membrane-bound enzyme (referred to as $Na^+$, $K^+$ activated ATPase) which can be isolated and purified. It has a large polypeptide component with a molecular weight near 100,000 daltons which appears to extend completely through the membrane (Nakao et al., 1973). A smaller glycoprotein component is anchored on the outside (Vogel et al., 1977). The native enzyme contains two of the large polypeptide components for a total molecular weight of 250,000 (reviewed by Wilson, 1978).

$Na^+$ ions and ATP bind preferentially on the inside of the membrane while $K^+$ ions and glycoside bind preferentially on the outside. However, the sites are not absolutely specific, and under certain conditions the molecule can produce $K^+$ exchange ($K_o^+ \rightleftharpoons K_i^+$) or $Na^+$ exchange ($Na_o^+ \rightleftharpoons Na_i^+$). This exchange diffusion does not involve the net breakdown of ATP (Garrahan and Glynn, 1967a). The idea of exchanging an ion of $Na^+$ (or $K^+$) on one side of the membrane for the same ion on the other side seems rather pointless, but simply indicates that the carrier sites are not completely specific for one ion. Indeed, under suitable conditions the pump can be reversed: with $Na^+$ ions moving inward and $K^+$ outward with a net formation of ATP (Garrahan and Glynn, 1967b). These data suggest that the sites on the inside and the outside function independently of one another in binding ions. However, all the ions are probably bound and then a single translocation process takes place (Skou, 1965; Garay and Garrahan, 1973), rather than a sequential process involving several stages.

An equation analogous to Equation (3.5) can be written (Garay and Garrahan, 1973; Nakao et al., 1973)

$$J = \frac{M}{(1 + K_{Na}/[Na^+])^3(1 + K_K/[K^+])^2} \tag{3.13}$$

where $K_{Na}$ and $K_K$ are the dissociation constants for the internal $Na^+$ and external $K^+$ sites respectively, and the exponents indicate the number of sites shown in Fig. 3.3. The efflux of $Na^+$ can be inhibited competitively by internal $K^+$ or other ions, although the affinity of the internal site for $Na^+$ is about 50 times greater than for $K^+$ (Garay and Garrahan, 1973). Similarly, $Na^+$ and other ions can compete for the external sites, although the affinity of the pump for internal $Na^+$ is about 160 times greater than for external $Na^+$.

With a model of this nature, complex interactions can be observed. For example, Hodgkin and Keynes (1955a) found that if $K^+$ ions are removed from the external solution: $Na^+$ efflux is reduced as well as the expected reduction in $K^+$ influx. Similarly, reduction of external $Na^+$ concentration leads to a reduction in $Ca^{++}$ efflux (Blaustein and Hodgkin, 1969), which suggests that the influx of $Na^+$ ions can be coupled to the efflux of $Ca^{++}$ ions. Since a primary action of the sodium pump is to extrude $Na^+$ and hence build up a concentration gradient for $Na^+$ ions, the inward movement will be down the concentration gradient. This "downhill" movement will generate enough energy so that for every two ions of $Na^+$ flowing in, one $Ca^{++}$ ion could be pumped out. This type of movement can account for the low internal $Ca^{++}$ concentration in many cells (see, for example, Blaustein *et al.*, 1978) and is referred to as *secondary active* transport.

Secondary active transport involves the extrusion of $Ca^{++}$ ions against an electrochemical gradient, but only indirectly involves the use of metabolic energy, as shown in Fig. 3.3. Other examples of secondary active transport are known in the transport of amino acids and sugars (W. D. Stein, 1967). Transport of $Ca^{++}$ ions is also directly linked to ATP consumption for storage in the sarcoplasmic reticulum of muscle (see Tada *et al.*, 1978; and Chapter 8) and in the presynaptic terminals of nerve (Blaustein *et al.*, 1978). Primary active transport of $Ca^{++}$ ions across the surface membranes of red blood cells is also well known (Schatzmann and Bürgin, 1978).

## Effects of the $Na^+$ Pump

Before proceeding in the next chapter to a more quantitative analysis of membrane potential, let us examine exactly how the action of the $Na^+$ pump leads to the regulation of membrane potential, concentration gradients, and cell size. Intially, let us consider a cell in which there is no $Na^+$ pump, no membrane potential and all ions are in equal concentration on the two sides of the membrane. If a $Na^+$ pump is now introduced, we can follow the changes that will take place very quickly. Woodbury (1965) introduced the unit of 1 *jiffy* to denote a short period of time: which for the values he used was about 20 ms. In this time he estimated that in a cell of typical size (10 $\mu$m diameter) 400,000 ions of $Na^+$ might be pumped out in exchange for 400,000 ions of $K^+$. Note that this assumes that the pump is electrically neutral and does not generate any electrical currents or potential differences directly. The extra effect of electrogenic pumping on the membrane potential will be considered in the next chapter.

However, even in the absence of electrogenic pumping, a membrane potential will soon develop due to the higher permeability of nerve and muscle membranes to $K^+$ ions. Thus, even if the $Na^+$ pump initially created the same concentration gradient for $Na^+$ and $K^+$ ions, more $K^+$ ions will diffuse back out through the membrane than $Na^+$ ions diffuse in. Woodbury estimated that perhaps 200 $K^+$ ions would diffuse out and only four $Na^+$ ions would diffuse in during the first jiffy. Thus, a net movement of 196 positive charges will flow out of the cell, which would tend to create a small potential difference with the inside of the cell, negative with respect to the outside. Other ions such as $Cl^-$ will tend to move in such a way as to reduce this potential, and Woodbury estimated that 192 $Cl^-$ ions might be attracted to the more positive outside of the cell in the first jiffy, leaving a net charge of only $-4$ inside the cell.

Thus, the action of this sophisticated molecular pump leads to the extrusion of common table salt (NaCl). This may seem frivolous until one realizes that the pump was developed by simple organisms living in seawater where the ability to control salt concentration would have considerable advantages. As soon as NaCl leaves the cell and creates an osmotic gradient, water will follow because of the high permeability of the cell membrane to water. Woodbury calculated that 70,000 molecules of $H_2O$ might leave in the first jiffy. Already in the first jiffy all the important actions of the sodium pump are evident. It will: (1) produce the normal concentration gradients with $Na^+$ concentration higher on the outside of the cell and $K^+$ concentration higher on the inside; (2) produce a potential difference directly if the pump is electrogenic, but also indirectly as a result of the differential permeability of the membrane to $Na^+$ and $K^+$ ions; (3) produce a redistribution of other ions such as $Cl^-$ as a result of the potential difference (with the normal polarity of the membrane being negative on the inside, $Cl^-$ ions will become more concentrated outside the cell); and (4) regulate cell volume. The $Na^+$ pump tends to cause the cell to shrink and concentrate impermeable substances internally, which will balance the tendency for cells which contain these substances to swell and burst [due to considerations of osmotic equilibrium (Chapter 2)].

These same processes will continue in later jiffies with certain quantitative differences. As the concentration gradients for $Na^+$ and $K^+$ ions become greater, the passive fluxes will become more comparable in size to the active fluxes produced by the pump. Similarly, the passive fluxes of $Na^+$ and $K^+$ ions will tend to become more equal as the potential difference develops, since the inward diffusion of $Na^+$ will be enhanced by the negativity inside the cell, while the outward diffusion of $K^+$ will be deterred by this potential difference. Eventually, the inflow and the outflow of each ion from active and passive movements will just balance, and no further net movement will occur. The voltage will then remain steady and the cell will remain constant in size. The next chapter will investigate some of the determinants of this steady state or resting condition of a cell, as it is sometimes called.

## Problems

3.1. For a *competitive inhibitor* of facilitated transport, the sequence of Fig. 3.1 becomes

| Outside | Membrane | Inside |
|---------|----------|--------|
| $S + C$ | $\rightleftharpoons SC$ | $\rightarrow C + S$ |
| $I + C$ | $\rightleftharpoons IC$ | |

where the complex $IC$ cannot cross the membrane. The total concentration of carrier is $C_m = [C] + [SC] + [IC]$. If the dissociation constant for the complex $IC$ is $K_i = [I][C]/[IC]$, show that the rate of transport is

$$J_s = \frac{M[S]_o}{[S]_o + K_m(1 + [I]/K_i)}$$

where $K_m$ and $M$ are as defined in the text. (A competitive inhibitor affects the Michaelis constant $K_m$, but not the maximum rate of transport $M$.)

3.2. For a *noncompetitive inhibitor* of facilitated transport, the reaction sequence of Fig. 3.1 becomes

| Outside | Membrane | Inside | |
|---------|----------|--------|-----|
| $S + C$ | $\rightleftharpoons SC$ | $\rightarrow C + S$ | (A) |
| $I + C$ | $\rightleftharpoons IC$ | | (B) |
| $S + IC$ | $\rightleftharpoons ISC$ | | (C) |
| $I + SC$ | $\rightleftharpoons ISC$ | | (D) |

Assume (1) the dissociation constants for $IC$ and $ISC$ (reactions B and D) are both $K_i$; (2) the inhibitor does not affect the affinity of the carrier for the substance $S$ (the dissociation constant is $K_m$ in A and C); (3) the complexes in B–D are not transported across the membrane; and (4) the transport of $SC$ across the membrane is rate limiting. Then, show that

$$J_s = \frac{M[S]_o}{([S]_o + K_m)(1 + [I]/K_i)}$$

where $K_m$ and $M$ are as defined in the text. (Under these assumptions a noncompetitive inhibitor affects the maximum rate of transport but not the Michaelis constant.)

3.3. Enzyme kinetics are often displayed on a Lineweaver–Burke plot in which $1/J_s$ is plotted against $1/[S]_o$. What will be the form of the equations in Problems 3.1 and 3.2 on this plot? What will be the effect of varying $[I]$ or $K_i$ on this plot?

3.4. Show from Equation (3.11) that (1) if the free carrier cannot move readily across the membrane ($b \gg c$), or (2) if $[S]_o$ and $[S]_i$ are both much larger than $K_d$, then little net steady state transport of a substance can occur.

3.5. Show that (1) if $b = c$ (the bound and free carrier move equally well across a membrane), or (2) if $c \gg b$ (the free carrier can cross the membrane much more readily), and the internal concentration is low ($[S]_i \ll [S]_o$), Equation (3.11) reduces to the form of the Michaelis–Menten equation (3.5).

3.6. Show from Equation (3.11) that the net flux (Britton, 1964) is

$$J_s = J_s^{out} - J_s^{in} = \frac{bC_m K_d ([S]_i - [S]_o)}{2K_d^2 + K_d([S]_i + [S]_o)(1 + r) + 2r[S]_i[S]_o}$$

Under what conditions will this equation reduce to the mobile carrier Equation (3.6) or to Fick's law [Equation (2.10)]?

# 4

# Membrane Permeability and Voltage

The previous chapters have discussed the various methods by which substances cross cell membranes: (1) diffusion through the lipid membrane itself, (2) diffusion through aqueous pores in the membrane, (3) facilitated transport with the aid of carriers, and (4) active transport with the use of metabolic energy. If the substances involved are charged, the ionic currents involved in any of these methods can generate voltages across the membrane, and these voltages were considered qualitatively at the end of the last chapter. The magnitude of the voltage expected when substances are distributed at equilibrium was also given in Chapter 2. However, many ionic substances are not at equilibrium, and in this chapter more realistic assumptions will be used in deriving the steady voltages normally found across cell membranes.

Consider the membrane illustrated in Fig. 4.1 with one boundary at $x = 0$ and the other boundary at $x = d$ (the thickness of the membrane is $d$). Note that the normal convention has been used in defining the *membrane potential* $V_m$ as the voltage *inside* the cell with respect to the voltage outside the cell. The voltage outside the cell is also conventionally taken to be the zero or reference potential.

For any value of $x$ there will be two forces acting on an ionic substance: diffusion forces and electrical forces. The net flux $J_s$ of an ionic substance $S$ with concentration $[S]$ will be

$$J_s = -D_s \frac{d[S]}{dx} - kD_s[S]\frac{dV}{dx} \qquad (4.1)$$

The two parts of this equation are easily understood: (1) the *net* movement due to diffusion will depend on the diffusion constant $D_s$ and concentra-

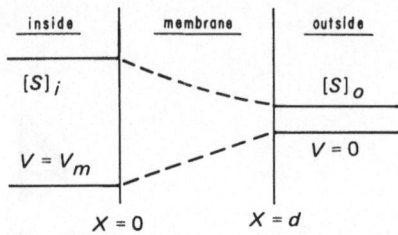

Fig. 4.1. Concentration and voltage gradients across a membrane. The values within the membrane are unknown (interrupted lines), but a linear change in voltage between the two sides (see the section *Constant Field Assumption* below) has been illustrated.

tion *gradients* (there will be no net movement due to diffusion if the concentrations at two points are equal). The first term on the right-hand side of Equation (4.1) is merely Fick's first law and the reason for the minus sign has been mentioned previously (Chapter 2). The more complex situation, where diffusion involves carrier molecules, will be considered later. (2) The net movement due to electrical forces is proportional to voltage gradients: Ohm's law states that the current flow between two points is proportional to the voltage difference. The constant of proportionality will again depend on $D_s$, but also on the concentration $[S]$ and on a further constant $k = Z_s F/(RT)$. This constant can be thought of as simply converting from units of volts, etc., to ones for flux, where $Z_s$ is the valence of the substance, $F$ is the Faraday which represents the number of coulombs/mole, $R$ is the gas constant, and $T$ the absolute temperature. A more detailed description is found for example in Ruch and Patton (1965, pp. 68–69).

Equation (4.1) can be rearranged if the rules for differentiation of a product are recalled, namely $d(AB) = A\,dB + B\,dA$. In particular,

$$\frac{d}{dx}\left([S]e^{kV}\right) = [S]e^{kV}k\frac{dV}{dx} + e^{kV}\frac{d[S]}{dx} \tag{4.2}$$

Comparing the right-hand sides of Equations (4.1) and (4.2), we find

$$J_s e^{kV} = -D_s \frac{d}{dx}\left([S]e^{kV}\right) \tag{4.3}$$

Integrating across the whole thickness of membrane from $x = 0$ to $x = d$ gives

$$\int_0^d J_s e^{kV}\,dx = \int_0^d -D_s \frac{d}{dx}\left([S]e^{kV}\right)dx \tag{4.4}$$

$D_s$ presumably does not depend on $x$ in this range, and $J_s$ must similarly be independent of $x$ in any steady state since charge cannot pile up at any point in the membrane indefinitely. Thus,

$$J_s \int_0^d e^{kV}\,dx = D_s\left([S]_i\, e^{kV_m} - [S]_o\right) \tag{4.5}$$

Rearranging this equation, we have

$$J_s = f(V)P_s([S]_i e^{kV_m} - [S]_o) \tag{4.6}$$

where $[S]_i$ and $[S]_o$ are the concentrations inside and outside the cell, $V_m$ is the transmembrane voltage or membrane potential, $P_s = D_s/d$ is the permeability of the membrane to the substance, and $f(V) = d/\int_0^d e^{kV} dx$. The last factor $f(V)$ depends on the whole profile of the voltage across the membrane. This can be solved for simple examples (such as the *constant field* assumption described later). For the moment, simply note that there will be one such function for all ionic species crossing the membrane. From the definition of this function, it will be dimensionless and will have a value of 1 if there is no potential difference across the membrane.

Equation (4.6) can be used to define the *permeability* of a substance *as the flux of a substance which crosses a unit area of membrane in unit time under a unit concentration gradient*. The permeability of an ion should be measured when there is no voltage difference across the membrane [and $f(V) = 1$]. The measurement should also be made when no volume changes are taking place, or otherwise three constants are required (see Chapter 2). The dimensions of the permeability constant can be determined from the definition. If flux is measured in moles per $cm^2$ of membrane area per second, and concentration is measured in moles per $cm^3$ of volume, the units of permeability will be (mol $cm^3/cm^2$ s mol) = cm/s. These units are the units of velocity, so the permeability constant can be thought of as a measure of how quickly a molecule moves across a membrane. The velocities measured will generally be much lower than the velocities in free solution. For an ion such as $K^+$, which permeates muscle membranes quite readily, the value is only about $10^{-7}$ (one ten-millionth) that in free solution (Ruch and Patton, 1965). The reason is that only a very small fraction of the membrane is occupied by pores, and the pores are small enough that the molecules which enter are somewhat restricted in their motion.

In four important examples the precise shape of the voltage profile within the membrane is unimportant [i.e., the function $f(V)$ defined earlier cancels out]. These examples will now be considered in successive sections of this chapter.

## Equilibrium (Nernst Equation)

At equilibrium, the net flux of an ion across the membrane is zero. For a substance which has a finite permeability, this can only be true if the quantity in the brackets on the right of Equation (4.6) is zero [$f(V)$ must be nonzero since $e^{kV}$ is a positive number at all points with a finite voltage, so the integral is greater than zero]. Thus, at equilibrium

$$[S]_o = [S]_i e^{kV_m} \tag{4.7}$$

The value of the voltage at which this is true is known as the *equilibrium potential* of a substance and will be denoted $E_s$. From Equation (4.7) we have

$$E_s = \frac{1}{k} \ln \frac{[S]_o}{[S]_i} = \frac{RT}{Z_s F} \ln \frac{[S]_o}{[S]_i} \cong \frac{60}{Z_s} \log_{10} \frac{[S]_o}{[S]_i} \tag{4.8}$$

The "constant" in the right-hand side of Equation (4.8) is normally about 60 mV, when logarithms to the base ten are used. This value will vary with absolute temperature, being about 58 mV at room temperature (20°C or 293°K) and 62 mV at body temperature (37°C or 310°K). Equation (4.8) can be recognized as the Nernst equation, which has been derived here under conditions relevant to cell membranes. Although less general than the usual thermodynamic treatment given in Chapter 2, the derivation illustrates the reason for the logarithmic factor which was merely assumed without proof in Chapter 2.

## Electrical Steady State

As indicated earlier, ions such as $Na^+$ and $K^+$ are *not* normally in equilibrium and a nonzero flux of these ions will occur. However, if the membrane potential is steady, the total flux for all ions must be zero in the absence of any externally applied current.

Only ions which are not in equilibrium must be considered, since ions in equilibrium contribute no net flux. If $Na^+$ and $K^+$ are the main ions not in equilibrium, the condition for an electrical steady state will be

$$J_K + J_{Na} = 0 \tag{4.9}$$

Substituting from Equation (4.6)

$$P_K([K^+]_o - [K^+]_i e^{kV_m}) + P_{Na}([Na^+]_o - [Na^+]_i e^{kV_m}) = 0 \tag{4.10}$$

Note that again the function $f(V)$ cancels out; i.e., the electrical steady state does not depend on the particular voltage profile inside the membrane itself. Rearranging and taking logarithms of both sides

$$V_m = \frac{RT}{F} \ln \frac{P_K[K^+]_o + P_{Na}[Na^+]_o}{P_K[K^+]_i + P_{Na}[Na^+]_i} \cong 60 \log_{10} \frac{[K^+]_o + q[Na^+]_o}{[K^+]_i + q[Na^+]_i} \tag{4.11}$$

where $q = P_{Na}/P_K$. If $q$ is sufficiently small, Equation (4.11) simplifies to the Nernst equation for $K^+$ ions, while if $q$ is sufficiently large, it simplifies

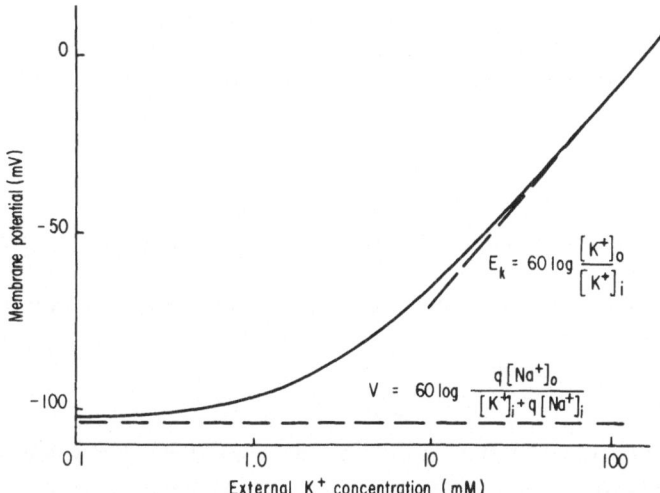

Fig. 4.2. Effect of varying external $K^+$ concentration on the membrane potential according to the Goldman equation (4.11). At high $[K^+]_o$, the potential approaches the equilibrium potential ($E_k$) for $K^+$, while at low $[K^+]_o$, it approaches a constant membrane potential ($V_m$), which can readily be calculated from Equation (4.11).

to the Nernst equation for $Na^+$ ions. In other words, *the steady state in membrane potential approaches the equilibrium potential of the most permeable ion*, and the greater the differences in permeability, the closer is the approach.

Equation (4.11) was first derived by Goldman (1943) using the assumption that the electric field across the membrane was constant (*constant-field assumption*). It will be called the *Goldman equation*, even though it is obvious from the present derivation that this relation is far more general than first assumed by Goldman. Equation (4.11) can be extended to include any other ionic species which has appreciable effect on membrane potential.

In Fig. 4.2, the predictions of the Goldman equation are plotted as a function of the logarithm of external $K^+$ concentration. At high external $K^+$ concentrations, the relation approaches the Nernst equation for $K^+$, while at low external potassium concentrations the membrane potential approaches a constant. The value of this constant potential can be used to determine the value of $q$ (Problem 4.1).

## Chemical Steady State

The electrical steady state described above is not a true steady state. Even though no net current flows, there is a net movement of particular ions which will change the concentration gradients slowly with time. To avoid changes, there must be an active flux equal and opposite to the passive fluxes. Thus, the condition for a true steady state, which will be called a chemical steady state, is that

$$J_K^a + J_K^p = 0$$
$$J_{Na}^a + J_{Na}^p = 0$$

(4.12)

where the superscripts $a$ and $p$ indicate the active and passive components respectively.

Generally, the active movements of $K^+$ and $Na^+$ ions are linked, as described in Chapter 3, so that for every potassium ion pumped out there are $n$ sodium ions pumped in. Then,

$$J_{Na}^a = -nJ_K^a$$

(4.13)

If $n = 1$, the pump is electrically neutral, while for other values of $n$ the pump generates an electric current (*electrogenic pump*). Combining Equations (4.12) and (4.13),

$$nJ_K^p + J_{Na}^p = 0$$

(4.14)

Substitution Equation (4.6) into (4.14) yields an equation identical to Equation (4.11) except that now $q = P_{Na}/(nP_K)$. Transiently, by a net outward movement of positive charge, the membrane potential will be *hyperpolarized* (i.e., the normal polarity will be increased so that the inside will be more negative with respect to the outside). In fact, one way of demonstrating that a pump is electrogenic is to show that the membrane potential becomes more negative than the electrochemical potentials of any important ions (see, for example, Thomas, 1972). The potential could not then arise from the passive movement of these ions. If more $Na^+$ is pumped that $K^+$ ($n > 1$) in the steady state, the constant $q$ defined above will be decreased relative to its value for an electrically neutral pump ($n = 1$). The steady-state potential may then approach more closely to the K equilibrium potential, but can not be more hyperpolarized than $E_K$. However, redistribution of both $Na^+$ and $K^+$ across the membrane can occur. To determine the net effect on $V_m$, equations for the movement of all permeable ions and water (volume changes) would have to be solved (see Jakobsson, 1980, *Am. J. Physiol.* **238**: C196–C206).

## Ussing's Flux-Ratio Test

Some years ago, Ussing (1949) developed a method of distinguishing among simple diffusion, facilitated transport, and active transport. This test involves measuring the ratio of the unidirectional fluxes in the two directions across the cell membrane by introducing radioactive tracers, first on one side of the membrane and then on the other side. The ratio of the two unidirectional fluxes will again be independent of the detailed profile of the voltage across the membrane. From Equation (4.6),

$$\frac{J_s^{out}}{J_s^{in}} = \frac{[S]_i e^{kV_m}}{[S]_o} = e^{k(V_m - E_s)} \tag{4.15}$$

The equation on the left of (4.15) follows from Equation (4.6) since the external concentration of $S$ can be set equal to zero when measuring the outward flux ($J_s^{out}$) with tracers. Similarly, in measuring the inward flux, the internal concentration can be set equal to zero. The equation on the right is then obtained from the definition of the equilibrium potential $E_s$ [Equation (4.8)].

To apply this ratio test, let the experimentally measured ratio be $m$ and let $U$ be the ratio predicted from Equation (4.15). For definiteness assume $U > 1$; i.e., the substance should have a net flow out of the cell. A number of possible results can emerge from this test:

(1) If $m = U$ to within experimental error, there is no reason to doubt that the movement of the substance is occurring by passive diffusion.

(2) If $1 < m < U$, the net movement is in the direction expected from the electrochemical gradient, but is not as large as expected. A smaller net movement could result from diffusion occurring by a process which saturates, such as facilitated transport (see Problem 4.4). Alternatively, active transport of some ions could be occurring in the opposite direction (against the electrochemical gradient).

(3) If $m < 1$, active transport must be occurring. Only by active transport can a net movement of a substance take place against an electrochemical gradient. To distinguish between facilitated transport and active transport in (2) above, the voltage or concentrations can be changed to see if the ratio becomes less than 1. Active transport should only be assumed if $m < 1$ under some conditions.

(4) Occasionally $m > U$. The movement is more strongly in the direction of the electrochemical gradient than expected, as can happen if the movements of ions are not independent. For example, a carrier molecule might transport more than one ion at a time. If some number $n$ ions are transported, Problem 4.4 indicates that ratios up to $U^n$ can be observed. A second possibility is that the movement of ions in and out might not be independent. To account for their experimental results, Hodgkin and Keynes (1955b) considered that ions might move in single file through a long, narrow pore. The flow of ions in the preferred direction down the electrochemical gradient would tend to prevent movement in the opposite direction. If $n$ ions were in the channel at any time, ratios up to $U^{(n+1)}$ could be observed (Hodgkin and Keynes, 1955b; Lea, 1963). Still other interactions that could produce values of $m > U$ are discussed by W. D. Stein (1967, pp. 95–97).

## Constant Field Assumption

The results for the steady state considered in previous sections do not depend on the voltage profile within the membrane and hence are fairly general. However, if the steady state voltage is modified by applying current, the resultant fluxes do depend on the profile, so more specific assumptions must be made. One such assumption is that the electric field is constant (i.e, the voltage changes linearly within the membrane as shown by the interrupted line in Fig. 4.1). Thus,

$$V = V_m(1 - x/d) \tag{4.16}$$

and

$$f(V) = d \Big/ \left( \int_0^d e^{kV} dx \right) = kV_m/(e^{kV_m} - 1) \tag{4.17}$$

Substituting in Equation (4.6) yields

$$J_s = kV_m P_s \frac{[S]_o - [S]_i e^{kV_m}}{1 - e^{kV_m}} \tag{4.18}$$

Equation (4.17) was also first derived by Goldman in 1943 and is often referred to as the *constant field equation*. This equation has rather simple limits

$$J_s = \begin{cases} kV_m P_s [S]_i, & V_m \gg 0 \tag{4.19} \\ kV_m P_s [S]_o, & V_m \ll 0 \tag{4.20} \end{cases}$$

When the voltage is very positive, the flux from Equation (4.19) is directly proportional to the internal concentration. However, the flux for very negative voltages [Equation (4.20)] is directly proportional to the external concentration (as well as to the permeability $P_s$ and the voltage $V_m$). If the internal concentration is greater than the external concentration, as is true for $K^+$ ions, these equations predict that ions should flow more easily in the outward direction. A plot of flux against voltage should have a curvature in the direction of outward fluxes (Fig. 4.3). This curvature is referred to as *outward rectification*, because the behavior is analogous to that of an electrical rectifier, such as a diode, which passes current largely in one direction.

The opposite curvature would be expected for cations such as $Na^+$ which are more concentrated outside. This is also shown in Fig. 4.3 and is referred to as *inward rectification*. Both curves cross the $x$ axis at the equilibrium potentials for the respective ions. As mentioned previously, the membrane potential is conventionally measured as the voltage inside a cell with respect to the voltage outside. If the inside is positive, cations will tend to flow outward. This flow will generate an outward current, so outward currents are taken to be positive by convention. The relation

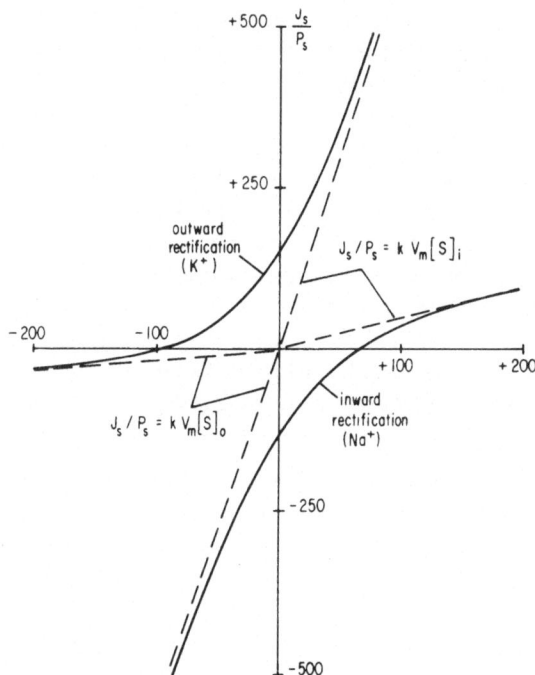

Fig. 4.3. Flux–voltage or current–voltage curves expected for $K^+$ and $Na^+$ ions in muscle cells from the *constant field equation* (4.18). The dashed lines give the predicted limits [from Equations (4.19) and (4.20)] for large positive and negative voltages in millivolts (mV). The ionic concentrations used in calculating these curves are taken from Ruch and Patton (1965). The ordinate gives the ratio of flux $J$ to permeability $P$ for a substance $S$. Flux can be converted to current using Equation (4.21). Experimentally measured current–voltage curves would have the shapes shown *if* permeability were independent of voltage. However, this condition is often not fulfilled, as will be considered in detail in the next chapter.

between flux and current is simply

$$I_s = Z_s F J_s \qquad (4.21)$$

where $I_s$ is the current carried by a substance $S$. The Faraday constant $F$ with the dimensions of coulombs/mole for a monovalent cation is again required to give currents in the usual dimensions. The valence $Z_s$ is required in Equation (4.21) so that the direction and magnitude of the currents will be correct when ions other than monovalent cations are considered.

Current–voltage curves can be measured for various ions (see Chapter 5). In fact, muscle cells show an inward rectification for $K^+$ ions. The results for muscle cells were referred to as *anomalous rectification* for many years (Katz, 1949) because they were in the opposite direction to that expected from the constant field equation. However, plausible explanations have been suggested for this finding (Adrian, 1969). One of these is contained in Problem 4.2, so the term inward rectification is preferable to anomalous rectification. The curvature for $Na^+$ ions is markedly different from the predictions of Fig. 4.3 in both nerve and muscle. The reasons for this result will be considered in the next chapter. Note, however, that some curvature or rectification is expected in the overall current–voltage curves (such as shown in Fig. 4.3) as long as there are concentration differences across the cell membrane. A curvature is predicted even though the initial equation in this chapter (based on Ohm's law) assumed that a linear current–voltage relation held at each point within the membrane.

Other factors can markedly affect the shape of the expected current–voltage curves. This subject has been well reviewed by Adrian (1969), who reached three conclusions:

(1) The presence of fixed charges within the membrane can reverse the direction of rectification from outward to inward, such as is found for $K^+$ ions in muscle. A proof of this result is included as Problem 4.2.

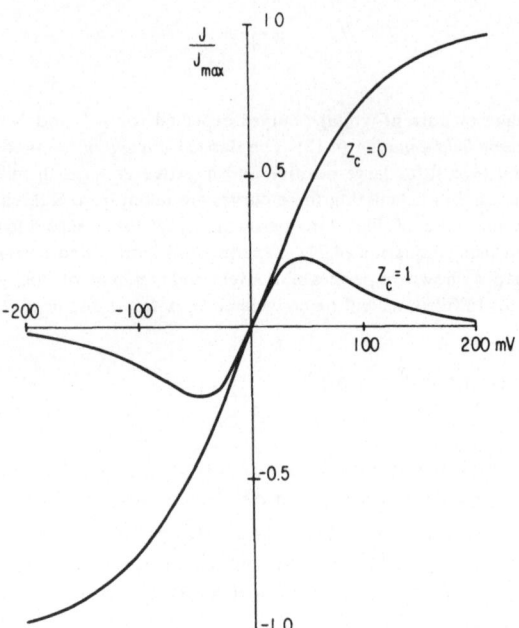

Fig. 4.4. Flux–voltage or current–voltage curves expected for transport of a monovalent cation when the carrier is uncharged ($Z_c = 0$) or charged ($Z_c = +1$). The numerical values were calculated from the results of Problems 4.5 and 4.6 assuming $[S]_o = [S]_i = cK_m/b$. If $[S]_o \neq [S]_i$, the curves would be asymmetrical but the current $J = 0$ would still be 0 when $V_m = E_s$ (Problem 4.7).

(2) If diffusion occurs by means of uncharged carriers or the complex *SC* is uncharged, the current–voltage relation will saturate for large positive and negative voltages. This result was considered by Adrian (1969) using the constant field assumption. It also follows from the carrier model of Fig. 3.2, in which movement from one side of the membrane to the other was considered as a single step over an energy barrier (see Problems 4.3–4.5). Current–voltage relations predicted from these problems are plotted in Fig. 4.4, and may be compared with Adrian [(1969), Fig. 6].

(3) If the carriers and the complex are both charged and have the same sign, the current may reach a maximum and then decline toward zero despite further increases in voltage (i.e., the current–voltage curves may have regions with a negative slope). This result is also the subject of a problem (4.5) and the predicted relationship is plotted in Fig. 4.4. The current–voltage curves are symmetrical about the origin when the internal and external ionic concentrations are equal and there are no fixed charges, but are generally asymmetrical. Readers interested in pursuing this topic further should consult Adrian's (1969) review.

## Problems

4.1. At what value of $[K^+]_o$ will the two straight lines in Fig. 4.2 cross? If $Na^+$ and $K^+$ are the main ions determining the relation between membrane potential and $[K^+]_o$, show how this intersection can be used to determine the constant $q$.

4.2. Assume that the voltage $V$ across a membrane is linear, but that fixed charge at the inner boundary causes a step of size $W$ as illustrated below.

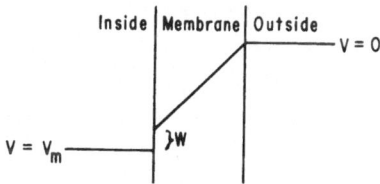

Show that:
  (1) The flux $J_s$ of a substance $S$ will be

$$J_s = k(V_m + W)P_s \frac{([S]_o - [S]_i e^{kV_m})}{1 - e^{k(V_m + W)}}$$

  (2) The flux will be zero when $V_m = E_s$ [$E_s$ is the equilibrium potential defined by Equation (2.4)].
  (3) The relation between flux and voltage is (a) outwardly rectifying when $W + E_s < 0$; (b) linear when $W + E_s = 0$; (c) inwardly rectifying when $W + E_s > 0$. (Hint: determine the slopes for large positive and negative values of $V_m$.)

4.3. If a carrier $C$ is uncharged but the complex $SC$ in Fig. 3.2 has a positive charge (i.e., $S$ is a monovalent cation), the movement of the complex across a membrane will be influenced by the membrane potential. If $b_o$ and $b_i$ represent the outward and inward rate constants for movement of this complex under a voltage gradient, show that

$$J_s^{\text{in}} = b_i[SC]_o = \frac{b_i C_m[S]_o(K_d + r_o[S]_i)}{(K_d + [S]_o)(K_d + r_o[S]_i) + (K_d + [S]_i)(K_d + r_i[S]_o)}$$

where $r_o = b_o/c$ and $r_i = b_i/c$.

4.4. If the outward and inward movements of the carrier complex in Problem 4.3 are independent, then from Ussing's flux-ratio test, the rate constants $b_o$ and $b_i$ will be related by $b_o = b_i e^{kV_m}$. Show that with this assumption the unidirectional fluxes of the substance $S$ will be $1 < J_s^{\text{out}}/J_s^{\text{in}} < e^{k(V_m - E_s)}$ when $V_m > E_s$. If the carrier transports some number $n$ ions ($n > 1$), show that $1 < m < U^n$, where $m$ and $U$ are as defined in the text.

4.5. In the carrier model described in Problems 4.3 and 4.4, $b_o = be^{kV_m/2}$ and $b_i = be^{-kV_m/2}$ if the membrane is symmetrical. Under this assumption, show that the net flux saturates at both large positive and large negative values of $V_m$. The net flux for this example is shown in Fig. 4.4. Will saturation also occur (1) for asymmetrical membranes; (2) if the carrier $C$ is negatively charged but the complex $SC$ is uncharged?

4.6. If the carrier $C$ in Fig. 3.2 has a single positive charge and the complex $SC$ has a double positive charge (i.e., both $C$ and $S$ are monovalent cations), show that the net flux will be

$$J_s = J_s^{\text{out}} - J_s^{\text{in}} = \frac{K_d C_m(b_o c_i[S]_i - b_i c_o[S]_o)}{(K_d + [S]_o)(c_o K_d + b_o[S]_i) + (K_d + [S]_i)(c_i K_d + b_i[S]_o)}$$

Show that if the membrane is symmetrical, the net flux will decline toward zero both at large positive and large negative membrane potentials. This example is also illustrated in Fig. 4.4. What is the physical reason for the difference between this result and that in Problem 4.5?

4.7. Assume in the carrier models described above that the carrier $C$ and complex $SC$ have charges $Z_C$ and $Z_{SC}$ and that the membrane is symmetrical. Show that

$$J_s = \frac{2C_m b([S]_o[S]_i)^{1/2}\sinh[k(V_m - E_s)/2]}{(K_m + [S]_o)(e^{k_c V_m/2} + \lambda[S]_i e^{k_{sc} V_m/2}) + (K_m + [S]_i)(e^{-k_c V_m/2} + \lambda[S]_o e^{-k_{sc} V_m/2})}$$

where $\lambda = b/(cK_m)$, $k_c = Z_c F/(RT)$, and $k_{sc} = Z_{sc}F/(RT)$. Note that regardless of the charge of the carrier, the net flux $J_s$ will always be zero when $V_m = E_s$.

# 5

# *Ionic Currents in Nerve and Muscle*

Considerable advances in understanding the ionic currents in nerve and muscle cells followed rapidly after the introduction of glass capillary microelectrodes by Ling and Gerard (1949). These electrodes made it possible to record routinely the membrane potentials within living cells. In fact, two microelectrodes are often inserted into a cell, the second electrode being used to control either the current or the voltage across the cell membrane (Moore and Cole, 1963). These two possibilities are illustrated in Fig. 5.1. On the left-hand side of this figure is shown the arrangement for passing constant currents of variable magnitudes into a cell. The magnitude is controlled schematically by a battery and a variable resistor which limits the current. The second microelectrode measures membrane potential. The cell body is assumed to be small enough that it is *isopotential* (i.e., there are no significant voltage gradients within the cell body). This assumption can be checked theoretically (Clark and Plonsey, 1966) or experimentally by measuring the voltage at different points. The spread of voltage in cells which are not isopotential will be considered in Chapter 6. When small currents are passed through the membrane, the voltage changes smoothly to a new value. From the steady state values measured in this way, a current–voltage curve such as shown in Fig. 4.3 or 4.4 can be measured. However, with currents that exceed a certain *threshold* voltage, a characteristic sequence of voltage changes, known as an *action potential*, takes place. The current–voltage curve can then no longer be measured with steady currents.

## *Voltage Clamp Technique*

An alternative method known as the *voltage clamp* (Cole, 1949) is preferable. This method is illustrated on the right-hand side of Fig. 5.1.

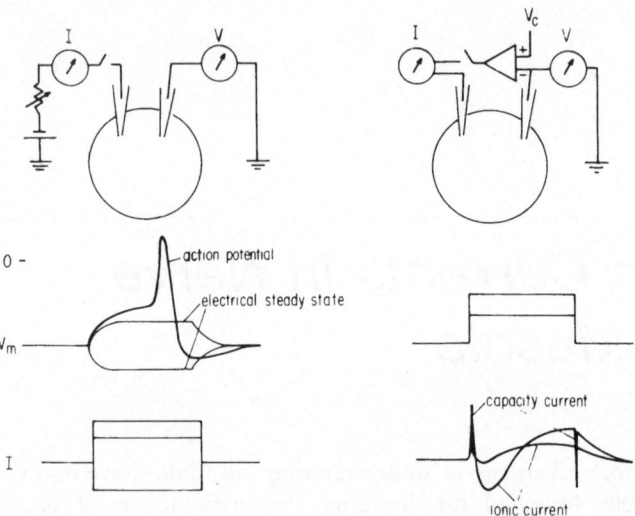

Fig. 5.1. The use of two microelectrodes to control the current $I$ (left) or voltage $V$ (right) across a cell membrane. The membrane potential $V_m$ is normally negative and changes to less negative values (depolarization) or more negative values (hyperpolarization), which are maintained with time as long as small, constant currents are passed in one direction or the other. Larger depolarizing currents generate one or more action potentials. On the right a high-gain differential DC amplifier (operational amplifier) is used to control the membrane potential to some desired level $V_C$, when the switch is closed. In order to change the voltage in a nearly stepwise fashion, the membrane currents shown on the right are required. These currents can be subdivided into capacity and ionic components. See further explanation in the text.

The battery and resistor have been replaced by a voltage source which can be varied to set the control value $V_c$ for the membrane potential. This voltage is fed into the positive side of a high-gain DC amplifier while the measured membrane potential is fed into the negative side. When the switch is closed, the membrane potential will be brought automatically to the desired control voltage. If the membrane potential is less than the control voltage, the amplifier will supply positive current as a result of amplifying the positive difference between the voltage at its two terminals. If the membrane potential is more than the control voltage, the amplifier will supply negative current by amplifying the negative difference. In both instances, the membrane potential will be driven toward the desired potential. The rate and closeness of the approach to the desired voltage will depend on the amplification and the ability of the microelectrode to pass current across the membrane. The circuit illustrated is a simple example of a feedback circuit, and there can be problems of instability or oscillations in such circuits. However, assuming that the circuit has been properly stabilized (Moore and Cole, 1963), the voltage will change abruptly to the desired value, and the currents required to produce this change are shown in the lower part of Fig. 5.1. According to Kirchhoff's law, current may be neither created nor destroyed, so the current passed by the microelectrode must just equal the current flowing out through the membrane at other points in the cell. Cell membranes have the nice property that they generally seal around microelectrodes, so that little or no current is lost through leakage at the points where the microelectrodes penetrate the cells.

The records on the right-hand side of Fig. 5.1 do not seem appreciably simpler than the ones on the left-hand side of this figure. However, they do have definite advantages for analyzing the nature of the ionic currents. The first advantage arises from the fact that the total membrane current $I_m$ can be divided into a *capacitative* current and an *ionic* current $I_i$, where

$$I_m = C \, dV_m/dt + I_i \tag{5.1}$$

Cell membranes contain a capacitance $C$ by virtue of their structure. Note that the letter $C$ is used here and in subsequent chapters to represent *capacity*, rather than a *carrier* molecule as was done in previous chapters. The polar phospholipid groups at the surface of the membrane can store charge, while the hydrocarbon chains in the center of the membrane act as an effective insulator or dielectric. Since the structures of most cell membranes have the similarities indicated by the unit membrane concept (Chapter 1), the capacitance of most nerve cells is similar and equal to about 1 $\mu$F/cm$^2$. The effective capacitance of muscle membranes is higher because of their more complex structure, which includes specialized tubular systems (see Chapter 8). Furthermore, the capacitance does not appear to change appreciably during normal function (Cole and Curtis, 1939). Current only flows through a capacitor when the voltage is changing and the current depends on (1) its capacity $C$ or ability to store charge, and (2) the rate at which the voltage is changing ($dV_m/dt$). Hence, by changing the voltage in a nearly step-wise fashion, the capacity current is of a known amount and is confined to the sharp peak at the beginning of the records. The ionic currents can then be studied in isolation from the later portions of the record.

If the ionic current were a simple function of voltage which did not change with time, then the separation of capacitative and ionic currents would not be a particular advantage. For example, if a membrane obeyed Ohm's law

$$I_i = (V_m - V_r)/R \tag{5.2}$$

where $V_r$ is the *resting potential* (when $I_i = 0$) and the resistance $R$ of the membrane to current flow is a constant, then Equation (5.1) could be solved to give

$$V_m = V_r + IR(1 - e^{-t/\tau}) \tag{5.3}$$

where $\tau = RC$. This equation holds reasonably well for small currents and forms the basis for the smooth exponential approach to a new steady state shown in the left-hand side of Fig. 5.1. It can be used to measure the resting membrane resistance and capacitance. However, most membrane currents show nonlinearities with larger deviations, and many change markedly with time, as illustrated in the right-hand side of Fig. 5.1.

A second important advantage of the voltage clamp technique in applications to biological membranes is that by holding the voltage constant, this time dependence can be studied separately from the voltage dependence. In their Nobel prize work, Hodgkin and Huxley (1952) exploited these advantages fully in deriving a semiempirical set of differential equations describing the ionic currents in squid nerve fibers. These fibers are large enough (up to 1 mm in diameter) that a wire can be passed down the long axis of the fiber. Such a wire can pass much greater currents than a microelectrode and hence is more effective in clamping the voltage of squid nerve fibers. This analysis, which proceeded in several stages, required certain assumptions which have proven more or less valid in various types of excitable cells. These assmptions will be listed here before describing Hodgkin and Huxley's analysis. The validity of the assumptions will be considered in later sections. The physical interpretation of these results at the molecular level is still uncertain, and the equations could be consistent with more than one interpretation. However, in their original papers, Hodgkin and Huxley (1952) discussed a possible interpretation which will be helpful to use in this analysis.

## Assumptions of the Hodgkin–Huxley Equations

Six assumptions were made in the original derivation of the equations: (1) The ionic current is carried entirely by ions moving down their respective electrochemical gradients. This assumption required either that the $Na^+ - K^+$ pump should be electrically neutral or that the current which it generates should be negligibly small. (2) $Na^+$ and $K^+$ ions flow through separate channels in the membrane and there is no direct interaction between them. The flow through any channel was expressed as the product of the ionic conductance of the channel and the electrochemical force driving ions through the channel. (3) Each kind of channel can be in one of two states, *open* or *closed*. Only in the open state can ions pass through the channel. (4) Each channel is controlled by one or more independent *gates*. Gates consist of charged groups on the proteins or phospholipids which open or close, depending on the electrical field. (5) The condition for a channel to be in the closed state is that at least *one* of the gates is closed. The condition for a channel to be open is that *all* of the gates are open. (6) Each gate obeys a first-order reaction with voltage-dependent rate coefficients. This reaction is a change in the orientation of the charged group.

## Form of the Equations

Assumptions (1) and (2) imply that the total ionic current can be subdivided into components due to each ion, namely

$$I_i = I_K + I_{Na} + I_L \tag{5.4}$$

$Na^+$ and $K^+$ ions proved to be the most important contributors to the ionic current, and the contributions of other ions were lumped into a single term, which was referred to as *leakage current*. This division into separate ionic currents was made experimentally as shown on the left of Fig. 5.2. If the currents are independent, removal of $Na^+$ from the external medium will permit one to measure the potassium currents in isolation except for the small leakage current. The $Na^+$ currents are obtained by subtracting the current in $Na^+$-free solution from the total current in normal solution. Fig. 5.2 shows that the $K^+$ current is well maintained while the $Na^+$ current is only transient. Hodgkin and Huxley used less drastic changes than total removal of $Na^+$ in their experiments and tested the assumption of independence (see Problem 5.2).

The assumption that these currents are carried by $Na^+$ and $K^+$ ions was examined further by varying the voltage in successive steps and

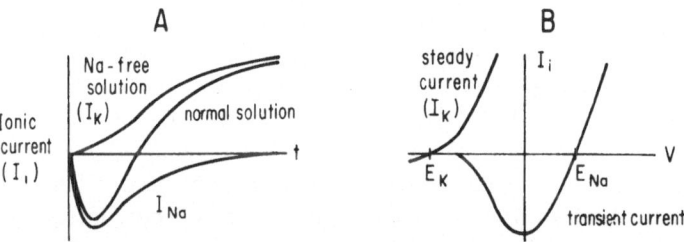

Fig. 5.2. Separation of ionic currents into components carried by $Na^+$ and $K^+$ ions. (A) After stepping the voltage from near the resting potential to near 0 volts in a normal solution, a negative (inward) ionic current is followed in time by a delayed positive (outward) current. In $Na^+$-free solution only the outward current is observed and is mainly carried by $K^+$ ions ($I_K$). If the movements of different ions are independent, the difference between the two curves gives the $Na^+$ currents ($I_{Na}$). The extent of the current axes is about $\pm 1$ mA per $cm^2$ of membrane and the time scale covers about 4 ms. (Modified from Hodgkin, 1958.) (B) The early transient currents reverse in sign at voltage steps to near the calculated $E_{Na}$, while the steady currents in $Na^+$-free solution reverse near the value of $E_K$. This provides further evidence for the role of $Na^+$ and $K^+$ ions in producing these currents. The extent of the voltage axis is about $\pm 100$ mV. (Modified from Hodgkin et al., 1952.)

plotting the magnitude of the maintained current and the peak of the transient current. As shown on the right of Fig. 5.2, these currents reverse close to the presumed equilibrium potentials for $K^+$ and $Na^+$ ions. Assumption (2) also implies a specific form for each of the terms in Equation (5.3), for example

$$I_K = G_K(V_m - E_K) \qquad (5.5)$$

This equation is of the same form as Ohm's law [see Equation (5.2)] where $G_K$ is the conductance of the membrane to $K^+$ ions and the force driving them through the channel is the difference between the membrane potential $V_m$ and the potential at which they are in equilibrium ($E_K$). Equation (5.5) does not imply that $G_K$ is a constant, but from Assumptions (3) to (5) $G_K$ should have a maximum value $\overline{G}_K$ when all the channels are open. When a fraction $n$ of the gates are open, and there are $c$ identical gates controlling each channel, then the conductance will be

$$G_K = \overline{G}_K n^c \qquad (5.6)$$

For example, if $n = 1/2$ at a given voltage and each of $c = 4$ gates must be open for ions to pass through, then the conductance would only be $1/16$ of its maximum value. Finally, Assumption (6) implies that if the voltage is suddenly changed, the fraction $n$ changes according to a first-order equation, namely

$$dn/dt = (n_\infty - n)/\tau_n \qquad (5.7)$$

The solution of Equation (5.6) will be an exponential which can be specified by its steady state value $n_\infty$ and the time constant $\tau_n$ with which it approaches this steady state value [see Equation (5.2)]. If $c = 1$, $G_K$ would be directly proportional to $n$, and the conductance would change exponentially. If $c$ is greater than 1, the change in conductance will be sigmoid (see Fig. 5.3A). By comparing the experimental curves with the predicted shapes, Hodgkin and Huxley (1952) found that a value of $c = 4$ provided the best fit. In a similar study on amphibian myelinated nerve, Frankenhaeuser (1963) found that a value of $c = 2$ gave the best fit. These results suggest that there are four gates controlling each $K^+$ pore in squid nerve, but only two in myelinated nerve. However, the situation is probably more complex, since Cole and Moore (1960) found the currents produced in squid nerve following steps from very hyperpolarized voltages were so delayed that a value of $c \sim 25$ was required to fit the data.

Once $c$ has been determined, values of the steady state parameter $n_\infty$ and the time constant $\tau_n$ can be determined as a function of the control voltage for the membrane potential. As seen in Fig. 5.3B, $n_\infty$ is a monotonically increasing function of voltage. The variable $n$ is referred to as the *potassium activation variable* since an increasingly large fraction of the gates are open or activated for $K^+$ ions to pass through as the nerve is

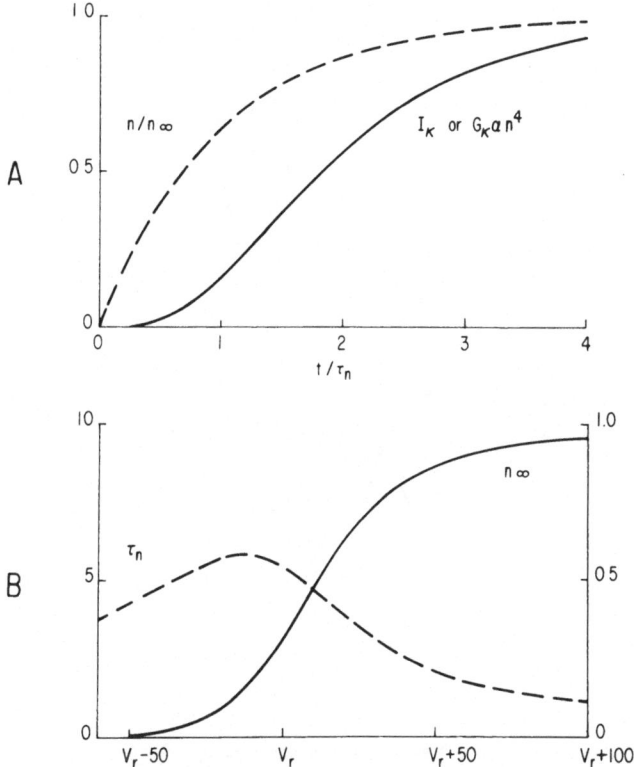

Fig. 5.3. (A) Changes in current $I$ or conductance $G$ for $K^+$ ions following a step change in voltage, and the inferred changes in potassium activation variable $n$. The axes have been normalized in terms of the steady state value $n_\infty$ and the time constant $\tau_n$ for the exponential approach to the steady state. (B) Variation of $\tau_n$ and $n_\infty$ with voltage changes from the resting potential $V_r$, based on the experiments using squid axons (Hodgkin and Huxley, 1952). The units of $\tau_n$ are ms. (Modified from Hille, 1970.)

depolarized. The time constant $\tau_n$ is longest over the range of voltages where $n_\infty$ changes and becomes shorter whenever most of the gates are switched to either one or the other state. The implications of this finding will be discussed later (see also Problem 5.3).

To describe the $Na^+$ currents which are transient, two variables are required, one which increases with voltage ($m$) and one which decreases with voltage ($h$). The equation for $Na^+$ conductance used by Hodgkin and Huxley is of the form

$$G_{Na} = \overline{G}_{Na}m^3h \tag{5.8}$$

where $m$ and $h$ are normalized variables which follow equations similar to Equation (5.7) for $n$. A plot of $m_\infty$ and $h_\infty$ against voltage is shown in Fig. 5.4. The variable $m$ is referred to as the *sodium activation variable* (Fig.

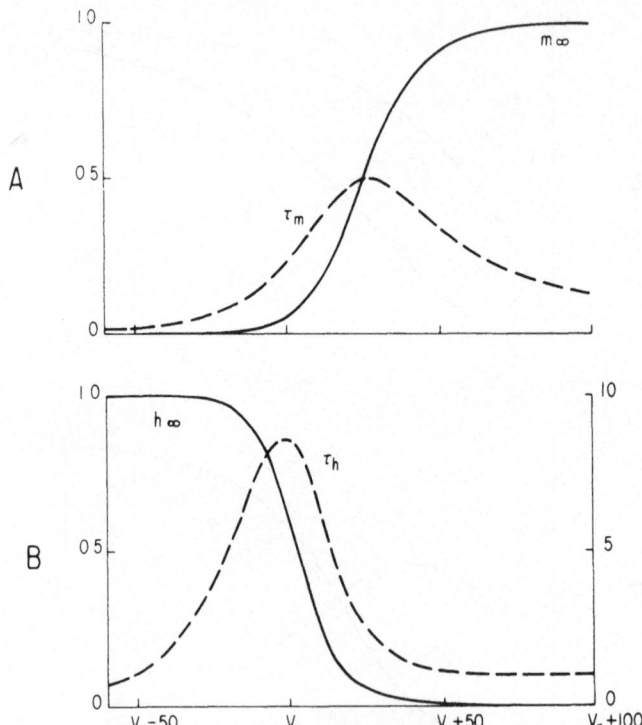

Fig. 5.4. Variation of (A) the Na$^+$ activation variable $m$ and (B) the Na$^+$ inactivation variable $h$ with voltage changes from the resting potential $V_r$, based on experiments using squid axon (Hodgkin and Huxley, 1952). The steady state values of $m_\infty$ and $h_\infty$ vary between 0 and 1. The time scale in (B) (0–10 ms) is an order of magnitude greater than in (A). (Modified from Hille, 1970.)

5.4A) since an increasing fraction of its gates will be open as the nerve is depolarized. The variable $h$ is referred to as the *sodium inactivation variable* (Fig. 5.4B) since it has just the opposite voltage dependence. The time constant $\tau_m$ is considerably smaller at most potentials than $\tau_h$ (note the change in scale from 1 ms in Fig. 5.4A to 10 ms in Fig. 5.4B). Thus, the Na$^+$ conductance will be activated first and then inactivated, which accounts for the transient nature of the Na$^+$ current. Problem 5.4 is included at the end of the chapter so that the reader can test his knowledge of which variables are responsible for some commonly observed properties of nerve fibers.

## Structure of the Channels

Although the physical basis of these equations has not been fully established, and there are still conflicting theories (Ling, 1962; Nachmanson, 1970), the form of the equations proposed by Hodgkin and Huxley in 1952 has stood the test of time remarkably well. The separation of $Na^+$ channels and $K^+$ channels has been strengthened by the discovery of drugs (reviewed by Hille, 1970) which have quite selective action on the $Na^+$ channel (*tetrodotoxin* or TTX and *saxitoxin* or STX) and on the $K^+$ channel (*tetraethyl ammonium* or TEA). The time constant and extent of inactivation for the $Na^+$ currents can also be affected fairly selectively by drugs or the enzyme *pronase* (Ulbricht, 1969; Armstrong *et al.*, 1973). Selective application of these agents has made it possible to estimate the number of $Na^+$ pores in a square $\mu$m of membrane. Estimates vary from the order of 10–20 in squid nerve up to several thousand at the nodes of Ranvier of mammalian nerves (Ritchie, 1979). If the pores are of the order of 0.4 nm in diameter (see below), only a very small fraction of the membrane is required to generate the $Na^+$ and $K^+$ currents underlying the action potential in squid axon, but a larger fraction of mammalian nodes will be needed. The number of $Na^+$ pores overlaps the estimate for the number of $Na^+$ pump sites mentioned in Chapter 3. However, each pump site can handle only 100 ions/s while a $Na^+$ pore can pass 1000 ions during an impulse lasting less than 1 ms (Hille, 1970). Thus, the transient currents during an action potential are much larger than any steady currents which may be produced by an electrogenic pump.

The selectivity of these channels has also been investigated by replacing the normal solutions with ones containing a variety of other ions. The $Na^+$ channel is actually 10% more permeable to $Li^+$ ions than $Na^+$ ions, so it might be referred to as the transient or fast channel. However, since $Na^+$ is the major ion going through this channel *in vivo*, we will retain the term *Na⁺ channel*. $K^+$ ions go through this channel as well, but at a much slower rate (only 8% of the rate for $Na^+$ ions). The order of permeabilities for the alkali earth metals is $Li^+ > Na^+ > K^+ > Rb^+ > Cs^+$. The potassium channel is most permeable to $K^+$ ions, but $Rb^+$ ions will go through fairly well. The order of permeability for this slower channel is $K^+ > Rb^+ > Na^+$ (Chandler and Meves, 1965).

Eisenman (1962) developed a theory of ion selectivity which was originally applied to glass electrodes but has since been used in relation to nerve membranes. The differences in ion selectivity could be accounted for in Eisenman's theory if there were a high anionic field strength due to negative charge around the $Na^+$ pore, and a weaker field around the $K^+$ pore. From his study on the pH dependence of the permeability through the $Na^+$ pore, Hille (1971) suggested that there is at least one carboxyl group at a margin of the pore. At normal pH these groups will be dissociated and the oxygen molecules will carry a negative charge. Hille (1971) has also investigated the permeation of a large number of molecules through this pore, and concluded that the minimum pore size is $0.3 \times 0.5$ nm (Fig. 5.5A). By utilization of hydrogen bonds, $Na^+$ ions with one water

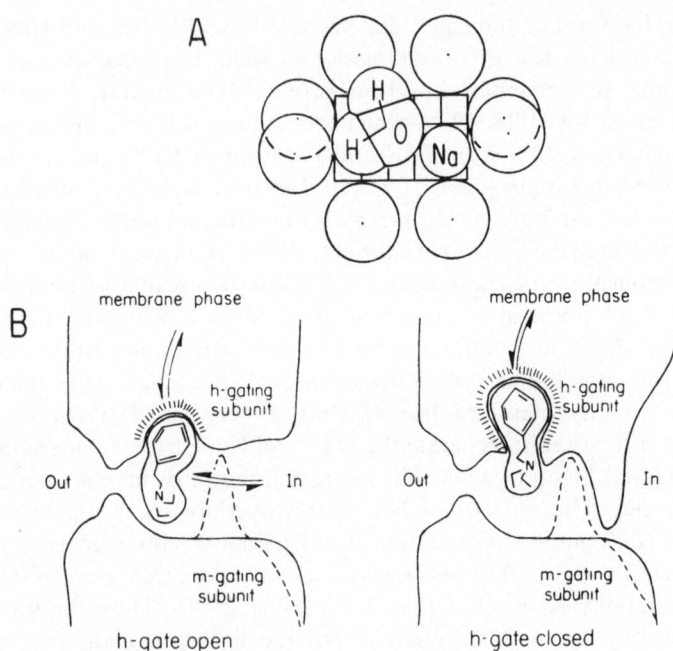

Fig. 5.5. (A) Possible molecular model for a filter to distinguish Na$^+$ selectively from other ions (from Hille, 1971). The 0.3 × 0.5 nm pore is lined by oxygen molecules which allow a Na$^+$ ion with one water of hydration to pass through. Passage results from the formation of a hydrogen bond to an oxygen molecule. (B) A longitudinal diagram of a Na$^+$ pore with a selectivity filter at the outer edge, a binding site for local anesthetics, and subunits for the *m* and *h* gates (from Hille, 1977).

of hydration could just fit through a channel of this size. Organic molecules of similar size which can utilize hydrogen bonds also permeate through this channel, but those which cannot form hydrogen bonds do not permeate. This provides further evidence for the pores being lined with carboxyl or other groups which can accept hydrogen bonds. TTX fits snugly into Hille's model of the Na$^+$ pore, and hence the action of this drug in blocking Na$^+$ permeability can be explained (Hille, 1975). TTX only acts on the outside of nerve membranes, so the groups described by Hille are presumably located near the outside of the membrane.

The nature of the charged groups which act as gates for the channels has not been determined. However, by removing Na$^+$ and other ions which normally pass through the Na$^+$ channel, Armstrong and Bezanilla (1973) and Keynes and Rojas (1974) have measured a residual current, which they call a *gating current* (reviewed by Hille, 1976). Its kinetics and other properties are approximately those expected for the movement of the charged *m* gates within the membrane which were described by Hodgkin and Huxley (1952).

Tsien and Noble (1969) have also given a plausible explanation of the form of the curves for the steady state values and time constants of the Hodgkin–Huxley variables based on transition state theory. Assume that a molecule has two states, $A$ and $B$, corresponding to the gates being active and blocked respectively, and that there are rate constants $\alpha$ and $\beta$ for the transitions between states, i.e., the activation or opening and the blocking or closing of the gates. Then, if $E$ is the voltage at which the two states are equally likely, the ratio of the two rate constants will be

$$\frac{\alpha}{\beta} = \exp\left[ \frac{ZF}{RT}(V - E) \right] \tag{5.9}$$

Equation (5.9) is identical in form to Ussing's ratio for the unidirectional fluxes through a membrane [Equation (4.15)]. The same arguments given in Chapter 4 could be used to justify this equation in the present context. The form of this equation is also identical with the Maxwell–Boltzmann distribution in statistical physics (Morse, 1962).

If for a given voltage the rate of the gating molecules moving from $B$ to $A$ is twice as high as from $A$ to $B$, eventually two-thirds of the molecules will end up in form $A$. As shown in Problem 5.3 the steady state values of an activation variable such as $m$ should be

$$m_\infty = \left\{ 1 + \exp\left[ -\frac{ZF}{RT}(V - E_m) \right] \right\}^{-1} \tag{5.10}$$

Alternatively, Equation (5.9) can be rearranged to give

$$\exp\left[ \frac{ZF}{RT}(V - E_m) \right] = \frac{m_\infty}{1 - m_\infty} \tag{5.11}$$

Then, from a semilogarithmic plot, the effective valency $Z$ of the charged group acting as a gate can be determined. Keynes and Rojas (1974) tested Equation (5.11) and found that there appeared to be $3m$ gates/pore, each with a valency of $Z = 1.3$, or a total charge near 4. Transition state theory can also be used to predict the time constants (see Problem 5.3), the activation energy, and the entropy changes involved in these changes (Tsien and Noble, 1969).

Work is also proceeding on the chemical nature of the inactivation variable and on the mode of action of local anesthetics on the Na channel (Hille, 1977). Recent models for the inactivation process implicate a mobile protein molecule attached to the inner surface of the membrane at one end and free to plug the $Na^+$ channel at the other (Armstrong and Bezanilla, 1977). Inactivation can be removed by arginine-specific (Eaton *et al.*, 1978) and tyrosine-specific (Brodwick and Eaton, 1978) reagents applied to the inside of a squid axon, suggesting an important role for these amino acids in the protein molecule. Local anesthetics in their uncharged form can dissolve in the membrane and block the central region of a sodium pore (organic molecule in Fig. 5.5B). Alternatively, in their charged form they can enter the pore from the inside during activity when the $m$ and $h$ gates are open (reviewed by Ritchie, 1979). A molecular picture of the sodium channel must include (1) a selectivity filter near the outer edge of the pore which distinguishes $Na^+$ from other ions, (2) an $m$-gating or activation subunit (dashed lines), (3) an $h$-gating or inactivation subunit (solid lines near the inner edge of the membrane), and (4) one or more sites for binding local anesthetics. Since the pore is so complex, interactions between different subunits might be expected, and possible interactions between a local anesthetic and the inactivation subunit are illustrated in Fig. 5.5. Further progress can no doubt be expected in clarifying the nature of both the $Na^+$ and $K^+$ channels.

## Role of Ca$^{++}$ Ions

$Ca^{++}$ is the most carefully controlled ion in the human body, and has long been known to affect the excitability of nerve and muscle (see, for example, Brink, 1954). Yet $Ca^{++}$ ions were not directly involved in the original formulation of Hodgkin and Huxley (1952) except insofar as they entered into the term for the leakage current. More recently, $Ca^{++}$ currents have been demonstrated in many tissues, including squid axons (Baker *et al.*, 1971). The currents vary in magnitude from being quite small in squid axons (permeability about 1% of that for $Na^+$) to being much larger than the $Na^+$ currents, for example in crustacean muscles (Fatt and Ginsborg, 1958; Hagiwara and Naka, 1964). These $Ca^{++}$ currents are not blocked by tetrodotoxin, but generally are abolished by $Mn^{++}$ and so are presumably moving through a distinct channel in the membrane (Hagiwara and Nakajima, 1966). As mentioned in Chapter 3, $Ca^{++}$ ions are generally maintained at a low internal ionic level like $Na^+$ ions, so $Ca^{++}$ ions will normally provide an inward current. This inward movement of $Ca^{++}$ appears to be particularly important in the release of transmitter by nerves (Chapter 7) and in producing contractions of muscles (Chapter 8).

In addition, Frankenhaeuser and Hodgkin (1957) showed that the concentration of $Ca^{++}$ had an important indirect effect in shifting the activation and inactivation curves of the variables governing the $Na^+$ and $K^+$ channels. Figure 5.6 illustrates the effect of changing the extracellular $Ca^{++}$ concentration on the $Na^+$ activation or $m$ variable for a squid axon. A ten-fold increase shifts the value of $E_m$ in Equation (5.10) by 15–27 mV in different experiments. If a certain fraction of $Na^+$ current must be activated to trigger an action potential, the threshold will be increased or the excitability decreased by an increase in $Ca^{++}$ concentration. Conversely, a decrease in $Ca^{++}$ concentration will shift the threshold closer to

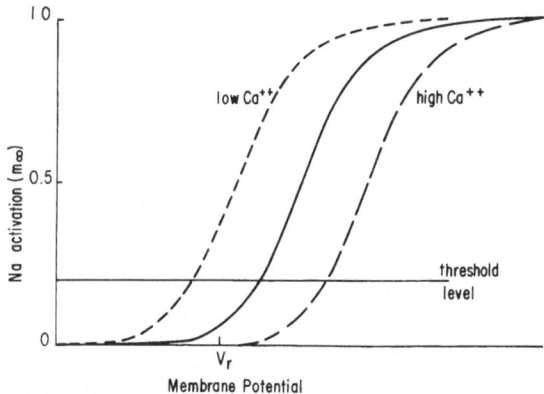

Fig. 5.6. Increasing $Ca^{++}$ in the external medium shifts the activation curves for $Na^+$ (and $K^+$) ions to the right. If a threshold level of $Na^+$ activation is required for generating an action potential, high $Ca^{++}$ will increase the deviation from the resting potential ($V_r$) required to reach this level. Conversely, with low $Ca^{++}$ concentrations, sufficient activation may occur even at the resting potential that action potentials are generated spontaneously each time the membrane potential returns to $V_r$ following the after-hyperpolarization.

the resting potential, and if the $Ca^{++}$ concentration is lowered sufficiently, spontaneous rhythmic activity will result (Huxley, 1959).

One possible explanation is that $Ca^{++}$ ions act by binding to the membrane, perhaps to phosphate groups or charged protein molecules as illustrated in Fig. 5.7. If $Ca^{++}$ ions bind to the outside of the membrane, they would increase the electric field (voltage gradient) *locally* near the membrane (solid line in Fig. 5.7A) over that expected with a constant field (interrupted line). This binding would not necessarily change the membrane potential difference between inside and outside, which would still be determined by the thermodynamic considerations discussed in Chapters 2 and 3. However, the charged groups within the membrane would behave as if the membrane potential were greater, and depolarization of the membrane to threshold would be more difficult. McLaughlin *et al.* (1971) obtained support for a role of $Ca^{++}$ binding by showing that $Ca^{++}$ ions only affected the local surface charge of artificial membranes made out of negatively charged phospholipids (such as phosphatidylserine) and not out of neutral lipids (such as phosphatidylethanolamine). However, in natural membranes most of the phosphatidylserine is on the inside, not the outside of the membrane (Rothman and Lenard, 1977). Changes in surface charge are also seen with higher concentrations of ions such as $Ba^{++}$ and $Sr^{++}$, which don't bind to phospholipid groups. The presence of these ions may partially screen the negative charge within the membrane. Some ions (e.g., $Ca^{++}$ and $Mg^{++}$ ions) may both bind and screen charges, whereas $UO_2^{++}$ ions merely bind charges.

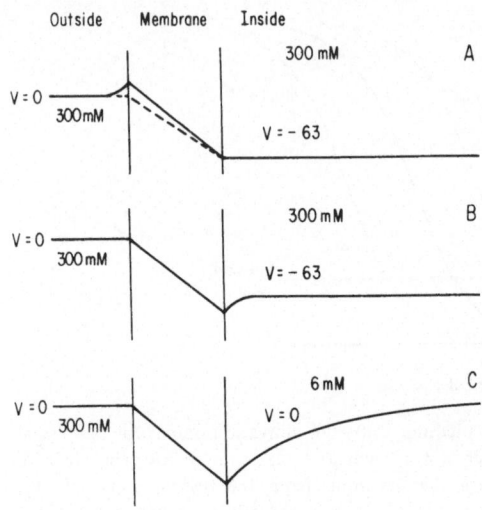

Fig. 5.7. Effects of fixed charge on the voltage distribution across a membrane. (Modified from Chandler *et al.*, 1965.) (A) The binding of $Ca^{++}$ ions to the external surface of the membrane can increase the voltage gradient or electric field within the membrane (solid line) above that expected from the constant field theory (interrupted line). (B) At the low levels of $Ca^{++}$ ions internally, the negatively charged phosphate groups within the membrane can cause a similar effect. (C) Reducing the concentration of monovalent cations in the internal solution (from 300 mM to 6 mM) enormously enhances the effect by removing the *screening* of charge normally provided by the monovalent cations. *Gates* within the membrane for the generation of action potentials would behave as if there were a normal resting potential, even though the net potential difference between the internal and external solutions is zero.

This idea also explains one of the surprising results when squid axons are perfused internally. *Internal perfusion* of a squid giant axon is a procedure in which the normal axoplasm is squeezed out with a rubber roller, the axon is cannulated, and then perfused with the solution of interest. If virtually all the $K^+$ ions are replaced by sucrose, action potentials are still seen (Narahashi, 1963), but they are slowed enormously because of the relatively small amount of $K^+$ or other ions to carry outward current. Furthermore, action potentials can still be obtained, even in the absence of a resting potential. A possible explanation is shown in Fig. 5.7C, where the removal of $K^+$ and other ions from the inside removes the screening that even monovalent ions normally give. Thus, the effect of negatively charged groups within the membranes (Fig. 5.7B) would be enormously enhanced. The membrane gates would behave as if there were a substantial voltage gradient across the membrane, even though no net potential difference exists between the internal and external solutions (Fig. 5.7C).

Since some $Ca^{++}$ ions normally enter cells during an action potential, these ions could modify later conductance changes. Thus, using an agent such as EGTA to bind internal $Ca^{++}$ and reduce its concentration depresses the $K^+$ conductance changes underlying the after-hyperpolarization in motoneurons (Krnjevic *et al.*, 1978a), while agents such as DNP, which release intracellular $Ca^{++}$, increase $K^+$ conductance (Krnjevic *et al.*, 1978b). The effect of intracellular $Ca^{++}$ ions in modulating $K^+$ currents has been demonstrated in many types of nerve cells (Meech, 1978).

An important conclusion from these studies is that even if an ion such as $Ca^{++}$ does not go through a channel, it can still have an important function in regulating membrane excitability. One must interpret cautiously the effects of removing ions (Fig. 5.2) and check that the actions cannot be accounted for by the binding or screening of charge and the associated changes in the local potential fields. The effects of removing an ion can now in many cases be confirmed by pharmacological agents known to have specific actions on various types of channels.

## Cardiac Muscle

The voltage clamp technique has been used on a variety of tissues, and equations analogous to the Hodgkin–Huxley equations have been derived which permit the action potentials to be reconstructed for other invertebrate cells (Connor and Stevens, 1971), myelinated vertebrate nerve fibers (Frankenhaeuser and Huxley, 1964), skeletal muscle (Adrian *et al.,* 1970; Adrian and Peachey, 1973), and cardiac muscle (McAllister *et al.,* 1975). In many ways, cardiac muscle is the most strikingly different of these tissues from the squid axon and hence can be used to illustrate the similarities and differences between different excitable cells. Figure 5.8 shows the form of an action potential in a squid axon and a cardiac Purkinje fiber. Cardiac Purkinje fibers (named after the Czech anatomist

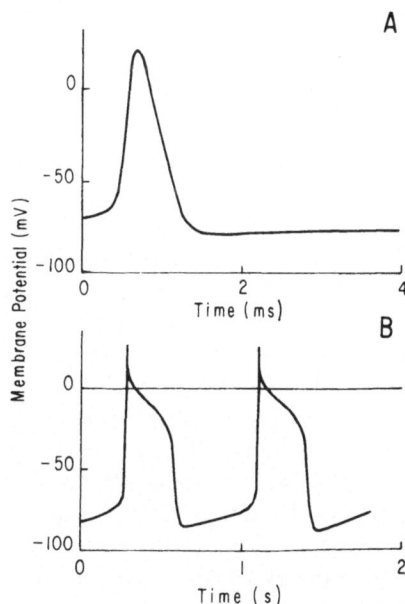

Fig. 5.8. Computed action potentials from (A) squid nerve at 6°C (Hodgkin, 1958) and (B) cardiac Purkinje fibers at 37°C (Noble, 1962). Note the greatly lengthened time scale, and the repetitive nature of the action potentials from the Purkinje fiber.

Purkinje who first described them) form a conduction system running from the atria to the ventricles in mammalian hearts. They are responsible for the nearly synchronous contraction of the ventricles, but Purkinje fibers don't contract appreciably themselves. Thus, voltage clamp studies can be carried out with two microelectrodes as indicated in Fig. 5.1, without the electrodes being dislodged by muscular contractions.

The obvious differences from a squid axon are: (1) Action potentials in a Purkinje fiber have a much greater duration, lasting hundreds rather than about 1 ms. The greater duration is characterized by a long *plateau potential* near 0 volts which follows a briefer spike-like depolarization. (2) Cardiac action potentials have an inherent rhythmicity which continues even after the heart has been dissected out of the body. This rhythmicity is evident as a slow depolarization, known as the *pacemaker potential*, which eventually reaches the threshold level and generates the next action potential. (3) Another marked difference which is not shown in Fig. 5.8 is in the pattern of conductance changes which take place during the action potential. During a nerve action potential, the membrane conductance increases on the order of 100 times above its resting value. In contrast, the conductance of the Purkinje fiber during the plateau phase is only about half as great as during the early part of the plateau potential when it is maximally polarized.

A composite picture can now be built up from various voltage clamp studies (see also the review by Trautwein, 1973) which qualitatively accounts for the three differences outlined above. Some points are still controversial, but there is not space to go into details here. Instead, I have generally given the view of Denis Noble and his group in Oxford, who have contributed greatly to our understanding of cardiac muscle. Several groups have also clamped atrial and ventricular muscle (Rougier *et al.*, 1968; Brown and Noble, 1969; Reuter and Beeler, 1969a,b), and many of the major factors are present in other types of heart muscle, although there are differences in these tissues from Purkinje fibers.

The major ion responsible for the rapid depolarization of the action potential is the $Na^+$ ion (Draper and Weidmann, 1951) as in squid nerve. This current can be blocked by tetrodotoxin (Rougier *et al.*, 1969). In addition, there is an inward current due to $Ca^{++}$ and perhaps $Na^+$ (see the discussion in Trautwein, 1973) which begins more slowly and may be important in maintaining the plateau. The inward movement of $Ca^{++}$ ions could also be important in the contraction of heart muscle. The $Ca^{++}$ current is considerably larger relative to the $Na^+$ current in heart muscle (10–15%) than in the squid axon (1%; Baker *et al.*, 1971).

The $Na^+$ currents are not completely inactivated during the plateau phase of the action potentials and represent a second factor in maintaining the plateau. This result is qualitatively similar to that in the squid axon, where the Hodgkin–Huxley equations do not predict a complete inactiva-

Fig. 5.9. Instantaneous current–voltage relationships for $K^+$ currents across membranes may show outward rectification according to the constant field equation (myelinated nerve), linearity according to Ohm's law (squid nerve), or inward rectification (cardiac Purkinje fibers). One component of the $K^+$ currents in these fibers ($K_2$) has a negative slope region (Noble and Tsien, 1969) in which the current declines as the voltage is increased.

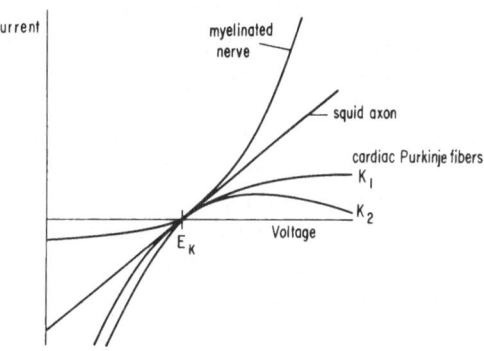

tion of the $Na^+$ currents (Noble, 1962). More recently, Chandler and Meves (1970) have shown that there are components of the $Na^+$ currents even in squid axons which are only inactivated over a much longer period of time than used by Hodgkin and Huxley in their initial studies.

The $Ca^{++}$ currents and the residual $Na^+$ currents would not be sufficient to maintain the plateau depolarization if it were not for a third factor, the *inward rectification* of cardiac muscle. If the potassium current is measured as soon as possible after changing the voltage to a new level (for technical reasons this may take as much as 10 ms), the results shown in Fig. 5.9 are obtained. These curves are known as *instantaneous current–voltage curves* because they are measured before any time-dependent changes in conductance have taken place. In contrast to the observed results, the constant field equation (see Chapter 4) predicts a curvature in the opposite direction (outward rectification). An outward rectification has been observed for myelinated nerve and indeed the predictions of the constant field equation were well obeyed by the instantaneous $K^+$ currents in myelinated nerve (Frankenhaeuser, 1962). A third possibility, which was found by Hodgkin and Huxley (1952), is that the instantaneous current–voltage curve is linear [see Fig. 5.9 and Equation (5.5)].

The reasons for the differences between the potassium currents in these tissues are not known with certainty, but the differences could arise from variations in the fixed charges on the membrane (Adrian, 1969; see also Frankenhaeuser, 1962, and Problem 4.2). Whatever the basis, the functional implications are clear; much less outward $K^+$ current will flow when cardiac muscle is depolarized during the action potential than is true of either squid axon or myelinated nerve. The small outward potassium currents which flow despite quite large depolarizations also account for the decreased conductance of the cardiac membrane. The small outward current plus increased $Ca^{++}$ currents and the residual $Na^+$ currents, which are inward, stabilize the membrane potential near zero during the plateau.

There are several currents involving $K^+$ ions in cardiac muscle, including an inward rectifier which does not change with time (referred to as $I_{K_1}$ by Noble, 1962). As in squid nerve, there are also time-dependent increases in potassium currents, referred to as *delayed rectification*. The delayed rectifier in cardiac muscle (Noble and Tsien, 1969) has several quantitative differences from that in squid axon. First, it is much slower in cardiac muscle, turning on with time constants of the order of 100 ms rather than 1 ms. This slow current is mainly responsible for the long duration of cardiac action potentials mentioned previously. Second, the delayed rectifier is much smaller in magnitude than in squid axon and shows some degree of inward rectification when fully activated. Finally, this current is partially carried by some other ions, as indicated by the fact that its current–voltage curve crosses the voltage axis at a more depolarized value than $E_K$. The other ions responsible are not known and hence this current has been referred to as $I_{x_1}$. (There is also a current known as $I_{x_2}$ with a fairly linear current–voltage curve which is not too important for the generation of single action potentials and will not be considered here.)

One final component of $K^+$ current known as $I_{K_2}$ has the unusual property that the current becomes less as the voltage is depolarized (Fig. 5.9). Since the slope of the current–voltage curve is a measure of conductance, this property is referred to as *negative slope conductance*. This current is also turned on by the depolarization of the action potential and generates substantial currents, *once the membrane is repolarized*. As this current slowly turns off, the inward currents due to other ions are able to depolarize the membrane gradually. From the current–voltage curve, a further decrease in this current will occur as the membrane is depolarized, which will permit further depolarization. The unusual shape of this current–voltage curve is thus directly responsible for the pacemaker potential and the rhythmicity of cardiac muscle. The mechanism for the negative slope conductance of this current is unknown. However, as discussed by Adrian (1969; see also Chapter 4), a negative slope can occur if currents are carried by charged carrier molecules which tend to remain on one side of the membrane as the potential becomes more and more depolarized.

In summary, the membrane currents are much more complex in cardiac muscle than in the squid axon. Each component is of the general form indicated by

$$I_y = f(V - E_y)a^c i \tag{5.12}$$

where $I_y$ represents the $y$th component of current, which has an equilibrium potential $E_y$. As the membrane potential is varied from the equilibrium potential, the instanteous current–voltage curve varies according to some function $f(V - E_y)$ which may show outward rectification, linearity, or inward rectification, even to the extreme of having regions of negative slope conductance. Finally, there are normalized variables analogous to those described by Hodgkin and Huxley which are responsible for the activation ($a$) and inactivation ($i$) of this current with time at a given voltage. The current may vary as the $c$th power of the variable if there is more than one identical gate controlling the passage of ions through the channel.

Thus, the differences serve to generalize rather than contradict the basic assumptions in the formulation of ionic currents developed by Hodgkin and Huxley (1952). Functionally, the complexity of the currents in cardiac muscle is understandable since the extra components nicely account for the special properties of cardiac muscle. First, the long plateau phase is a result of inward $Ca^{++}$ currents and residual inward $Na^+$ currents. These currents are able to maintain the depolarization of the membrane because the marked inward rectification of the $K_1$ channel only permits the generation of relatively weak outward currents. Second, the membrane is eventually repolarized by a delayed rectifier ($X_1$) as in nerve. Finally, another potassium current ($K_2$), which declines slowly once the membrane is repolarized and which shows negative slope conductance, is responsible for the slow depolarization of the membrane to threshold. If the membrane potential is determined by a balance of inward and outward currents, a depolarization can clearly be caused either by a *decline* in an outward current or an *increase* in an inward current. These two possibilities will occur again when we consider synaptic transmission (Chapter 7).

## Problems

5.1. If a nerve cell behaves according to Equations (5.1)–(5.3) up to a sharp threshold voltage, show that the current $I$ and duration $t$ of square stimulus pulses which just excite the nerve are related by

$$I = I_{Rh}/(1 - e^{-t/\tau})$$

where $I_{Rh}$ is the *rheobasic* or minimum current needed to excite the nerve using very long pulses ($t \gg \tau$). The equation above was first considered for the *strength–duration relation* of nerve by Lapicque (1907). More complex formulations have been considered by Noble and Stein (1966, and their references to earlier work). A general result, which should be proven for this simple model, is that the charge required to excite a nerve by brief stimuli ($t \ll \tau$) is constant.

5.2. If movement of a substance $s$ through a membrane in one direction is independent of the movement of the substance in the opposite direction, show using the results of Chapter 4 that the ratio $r$ of the current ($I_s$) of the substance at two external concentrations ($s_1$ and $s_2$) will be

$$r = \frac{I_{s_2}}{I_{s_1}} = \frac{1 - \exp\left[-k(V_m - E_{s_2})\right]}{1 - \exp\left[-k(V_m - E_{s_1})\right]}$$

where $E_{s_1}$, for example, is the equilibrium potential for the substance at an external concentration $s_1$. If the movement of one substance is also independent of the movement of other substances and $\Delta I$ is the change in total current in going from an external solution of concentration $s_1$ to one of concentration $s_2$, show that

$$I_{s_1} = \frac{\Delta I}{r - 1}$$

5.3. Hodgkin and Huxley (1952) described the change in activation variables such as $m$ in terms of rate constants $\alpha$ for opening and $\beta$ for closing a gate. If the fraction of open gates follows a first-order equation,

$$\frac{dm}{dt} = \alpha - (\alpha + \beta)m$$

show that

$$m_\infty = \frac{\alpha}{\alpha + \beta} \; ; \qquad \tau_m = (\alpha + \beta)^{-1}$$

If $\alpha$ and $\beta$ are related according to Equation (5.9), show that Equation (5.10) will result. Write an equation which determines the voltage at which $\tau$ is maximum.

5.4. Which Hodgkin–Huxley variables determine the following well-known properties of nerve? (a) The presence of a critical threshold voltage beyond which the nerve produces a rapid depolarization; (b) the rapid repolarization once the peak depolarization is reached; (c) the *refractory period* of a nerve. Shortly after generating an impulse, a second impulse cannot be generated (*absolute* refractory period) or is more difficult to generate (*relative* refractory period); (d) the *accommodation* of a nerve to a slowly increasing stimulus. Accomodation can be observed as an increase in the threshold voltage for initiation of an action potential when such a stimulus is applied.

5.5. One of the most satisfactory descriptions of accommodation (see Problem 5.3) prior to the work of Hodgkin and Huxley (1952) was provided by the two time-constant theories of Monnier (1934), Hill (1936), and Rashevsky (1938). They assumed that the threshold $U$ increased with a second time constant $\mu$ ($\mu \geqslant \tau$) according to the equation

$$\frac{dU}{dt} = \frac{V - U}{\mu}$$

Assume that initially $V = 0$ and $U = 0$. Prove:

(a) If the change in $V$ for a constant current $I$ follows Equation (5.3), then

$$U = IR\left[ 1 + \left( \frac{\tau}{\mu - \tau} \right)e^{-t/\tau} - \left( \frac{\mu}{\mu - \tau} \right)e^{-t/\mu} \right]$$

(b) If the condition for a nerve to fire is $V - U = U_0$, then

$$I = \frac{U_0(\mu - \tau)}{\mu R(e^{-t/\mu} - e^{-t/\tau})}$$

(c) The latest time at which excitation can take place (*maximum utilization time*) is given by

$$t = \frac{\tau\mu}{\mu - \tau} \ln\left( \frac{\mu}{\tau} \right)$$

(Hint: at this time $V - U$ reaches its maximum value.)

# 6

# Cable Theory and Extracellular Recording

In the previous chapters, the cell membrane was assumed to be *isopotential*; i.e., all points within the cell were at a value $V_m$ which changed only with time. However, experimental precautions have to be taken to ensure that this assumption is valid. In voltage clamping a squid axon, a long thin electrode is passed down the long axis of the fiber so that current is passed uniformly across the membrane and the membrane is therefore held at the same potential all along its length. In work on Purkinje fibers, the preparations are generally cut to a short length and then tied off or allowed to seal at the ends. The short length again ensures that the voltage is relatively uniform throughout. In studies on myelinated nerve, the nodal membrane, where appreciable ionic currents flow, is so small that it again is isopotential. Finally, in work on other tissues, artificial nodes are often created by isolating small gaps of membrane with a solution of sucrose. To voltage clamp a membrane using *sucrose gaps*, three gaps are created. The first is depolarized by applying a potassium solution of the same concentration as found internally. The voltage in the second is then measured with respect to the first and current is passed through the third node to change this potential as desired.

The reasons for developing these methods is that normally cells are not isopotential. The very reason for having action potentials is to ensure that a signal passes reliably from one region of a fiber to another without the voltage decrements which would normally take place. The spread of current from one region of a cell to another can be analyzed for simple geometric configurations. The most extensive analysis has been carried out on cylindrical geometries where the nerve fiber is assumed to be a uniform, cable-like structure. Elements of this *cable theory* are presented below.

Under some conditions, the more complex geometry of a dendritic tree can be considered as an equivalent cable (Rall, 1962). Spherical geometries have also been analyzed (see Jack *et al.*, 1975, pp. 132–134), and finally the properties of a sphere attached to a dendritic tree or equivalent cable have been extensively studied (Rall, 1969; Jack and Redman, 1971; Butz and Cowan, 1974). This whole field has been reviewed in an extensive monograph (Jack *et al.*, 1975).

My purpose here is simply to outline the introductory concepts in this field, and derive some important relationships. Applications to conduction and excitation of myelinated and unmyelinated nerve, and to skeletal muscle, will be emphasized, rather than mathematical techniques. Nonetheless, some standard methods for the solution of differential equations such as the use of Laplace transforms are required. Readers who are not familiar with Laplace transforms may wish to read an introductory account of these methods such as is found in mathematics (Sokolnikoff and Redheffer, 1958) or engineering texts (D'Azzo and Houpis, 1966) before studying the details in the following sections. Laplace transforms will be used again in later chapters.

## Assumptions and Definitions

The analysis that follows makes certain basic assumptions. These are listed below and are used to define the notation required. Common units for the quantities defined are also given in parentheses. These units are internally consistent in that for example the product of a resistance and a capacitance gives a time constant in the right units.

(1) An axon is a uniform, cable-like structure, as shown in Fig. 6.1, with a conduction core of intracellular fluid with diameter $a$ (cm) and a surface membrane.

(2) The electrical properties of the intracellular fluid are completely defined by its resistivity $\rho$ (K$\Omega$ cm).

(3) The membrane consists of a simple capacity $C$ for a unit area of membrane ($\mu$F/cm$^2$) and a resistance $R$ for a unit area of membrane (K$\Omega$

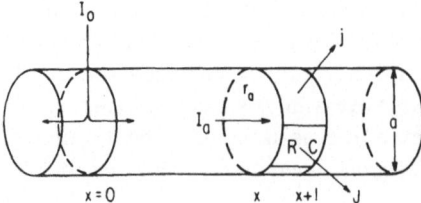

Fig. 6.1. Diagram of an axon considered as a cylindrical cable indicating the notation used, which is explained in the text.

cm$^2$). The current density through the unit area of membrane will be $J$ ($\mu A/cm^2$). Note that this usage of the letter $J$ differs from previous chapters in which it represented an ionic flux with different units.

(4) The axon is thin enough and the membrane resistance is high enough for intracellular current flow to be almost entirely axial. This assumption has been verified in detail by Clark and Plonsey (1966). If true, the voltage $V$ (mV) will be a function only of distance $x$ (cm) along the cable and of time $t$ (ms). We will measure the voltage as a *deviation* from the quiescent or resting potential inside the axon.

(5) The extracellular fluid is large enough to have a negligible resistance and to be virtually isopotential. There are of course small voltage changes in the extracellular space and these form the basis for extracellular recording methods using fine electrodes or gross electrodes; e.g., the electrocardiogram (ECG or EKG), electromyogram (EMG), and electroencephalogram (EEG). The theoretical basis for these methods is described in other places (Chung, 1974; Kiloh *et al.*, 1972; Rosenfalck, 1969; Stein and Pearson, 1971), and will only be considered briefly in the last section of this chapter. In the rest of the chapter, the fiber is assumed to be immersed in a large volume of fluid, and any extracellular potential changes are negligibly small compared to the potential changes within the fiber.

## Derivation of the Basic Cable Equation

Figure 6.1 illustrates the quantities which have been defined above. In addition, it shows a current $I_o$ ($\mu A$) being injected at a point in the cable. This point can serve as a reference for the coordinate system by setting $x = 0$ there. The current divides and there will be an axial current $I_a(\mu A)$ flowing along the long axis of the cable. If there is a resistance $r_a$ per unit length of cable (K$\Omega$/cm), then the voltage drop across a small length $\Delta x$ of cable will be $\Delta V = -I_a r_a \Delta x$ according to Ohm's law, or as $\Delta x \rightarrow 0$

$$\partial V/\partial x = -I_a r_a \qquad (6.1)$$

Partial differentials are used here because $V$ is a function both of $x$ and $t$.

Except for points at which external current is injected, the change in $I_a$ only depends on the current density through the membrane since current flows in closed loops according to Kirchhoff's law and is neither created nor destroyed. If the current density through the membrane is $j$ per unit length of cable ($\mu A/cm$), then the change in axial current over a short distance $\Delta x$ will be $\Delta I_a = -j\,\Delta x$, or again taking the limit as $\Delta x \rightarrow 0$,

$$\partial I_a/\partial x = -j \qquad (6.2)$$

Differentiating Equation (6.1) and substituting from Equation (6.2), we find

$$\partial^2 V / \partial x^2 = r_a j \tag{6.3}$$

Equations (6.1) and (6.3) imply that the axial current in a fiber (and also outside a fiber under some conditions; see the section *Extracellular Recording* below) varies as the first spatial derivative of the membrane potential fluctuations, while the membrane current density varies as the second spatial derivative. In Equations (6.1)–(6.3), some quantities in small letters have been used, which refer to values *per unit length of cable*, rather than the more basic quantities defined in the previous section, which relate to a *unit membrane area*. By considering the unit length of cable between $x$ and $x + 1$ in Fig. 6.1, it is easily seen that $j = J\pi a$ and $r_a = 4\rho/\pi a^2$ since the circumference of the cable will simply be $\pi a$ and its cross-sectional area will be $\pi a^2/4$; $\rho$ is the resistivity of the fluid inside the cable. The membrane current density will also have several components, a capacitive component charging up the membrane capacity, an ionic component $J_i$, and a third component $J_e$ from external sources such as stimulating electrodes or synaptic connections from other fibers. These components, like $J$, will have the units $\mu A/cm^2$. Substituting these quantities into Equation (6.3) and rearranging gives the basic partial differential equation for a cable

$$\frac{a}{4\rho} \frac{\partial^2 V}{\partial x^2} = C \frac{\partial V}{\partial t} + J_i + J_e \tag{6.4}$$

Each of the terms of this equation can be easily understood. The first term on the right, $C\,\partial V/\partial t$, is the capacitative current which was discussed previously in Chapter 5. The second and third terms are simply the ionic currents and the externally applied currents mentioned above. The new term on the left-hand side of this equation represents the spatial changes in voltage because of the cable structure chosen. Even without solving this equation, several important results can be determined for both unmyelinated and myelinated fibers. These will be given in the next two sections.

## Unmyelinated Fibers

The diameter $a$ of the cable enters directly into Equation (6.4) only once. The dependence on cable diameter can be completely eliminated if a new parameter $X = x/a^{1/2}$ is introduced. Then, it is easily shown that $\partial V/\partial X = a^{1/2}\partial V/\partial x$, or $\partial^2 V/\partial X^2 = a\partial^2 V/\partial x^2$, so

$$\frac{1}{4\rho} \frac{\partial^2 V}{\partial X^2} = C \frac{\partial V}{\partial t} + J_i + J_e \tag{6.5}$$

Thus, we can consider equivalent lengths of cables which have identical membrane properties, but different diameters, for example 1 and 4 in

Fig. 6.2. Scaling factors in unmyelinated nerve. In a fiber four times as large, subthreshold or suprathreshold responses will spread twice as far in the same time (conduction velocity is twice as great). The area of membrane which must be depolarized by membrane currents to produce this voltage pattern is $4 \times 2 = 8$ times as great. A distant electrode will produce axial current densities proportional to the cross-sectional area ($4^2 = 16$ times as great in the larger fiber). The implications of these geometric relationships are discussed in the text.

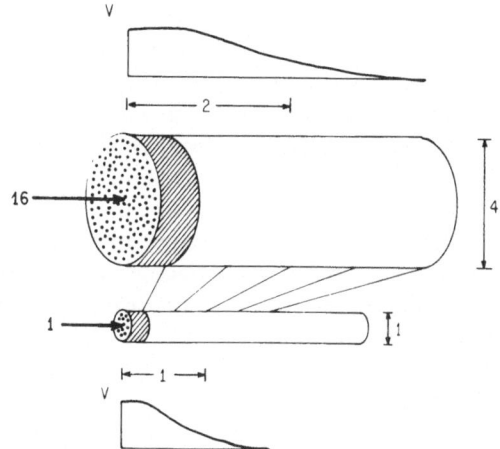

Fig. 6.2. From this figure and Equation (6.5), three important results follow immediately:

(1) If the currents are suitably scaled (*suitable* scaling of currents will be considered below), and a solution for $V$ as a function of $x$ and $t$ is derived for the cable of diameter $a$, then this solution will apply equally well to the cable of diameter $4a$. However, the pattern of voltage spread will extend a distance $(4)^{1/2} = 2$ times as far in the larger cable, since the same solution holds for $X$, but $x$ in centimeters is $X(a)^{1/2}$. If any response, whether suprathreshold or subthreshold, spreads along a cable with a conduction velocity $v$, then the same response will spread along a cable four times as large with twice this velocity. Thus, *conduction velocity in unmyelinated nerve fibers will increase as the square root of diameter* if the membrane properties of small and large fibers are identical. This result would apply to muscle fibers as well, except that they have a more complex cable structure (see below), so certain additional assumptions are required.

(2) To reach threshold for initiating an action potential in a fiber with a diameter four times as large, four times as much current must be applied so that the density through a unit membrane area will be identical. In addition, the equivalent length of cable that must be excited will be $4^{1/2} = 2$ times as long. Thus, eight times as much current must be applied to a nerve fiber with identical membrane properties, but four times as large a diameter. *The threshold for initiating an action potential through direct application of current across a membrane will increase as the 3/2 power of fiber diameter.* The synaptic currents that normally excite nerve fibers are applied directly across the membrane and this result forms the basis of the *size principle* which will be discussed in Chapter 10.

(3) If a stimulus is applied externally to a mixed nerve between two distant electrodes, the amount of current that will flow down the long axis (see arrows in Fig. 6.2) will vary as the cross-sectional area. In a fiber with

four times the diameter, sixteen times as much current will flow. Since eight times as much current will be required to excite the fiber [as indicated in (2) above], the threshold to an external electrical stimulus would be half as great. Thus, *the threshold for initiating an action potential with a distant external electrode will vary inversely with the square root of fiber diameter.* This result explains the common observation that in stimulating a nerve through the skin, the largest fibers are excited first and fibers of decreasing size are excited in turn as the stimulus is increased. This result also applies to a whole nerve, which is picked up into a nonconducting medium for stimulation, to the extent that the current density is uniform throughout the nerve at the point of stimulation. Obviously, nerve fibers near the outside of the nerve, and hence closer to the stimulating electrodes, will tend to be stimulated somewhat more readily.

## Myelinated Fibers

The three results derived in the previous section are of great practical importance in neurophysiology. They follow in a straightforward fashion from the cable properties of unmyelinated nerve, but they can be extended to myelinated nerve as well. In myelinated nerve, a short region capable of generating action potentials (node) alternates with a region incapable of generating action potentials (internode). Although the internodal region is formed by the fusion of many layers of a Schwann cell membrane, a unit area of the total membrane can be approximated by a capacitor and a resistor which are independent of voltage (see Fig. 6.3). Thus, current will vary linearly with voltage, in agreement with Ohm's law. This simplification of a *linear cable* is also important in other tissues (see Problem 8.2 on skeletal muscle membrane). Even for membranes which show complex, nonlinear, time-dependent current–voltage curves as in Chapter 5, the

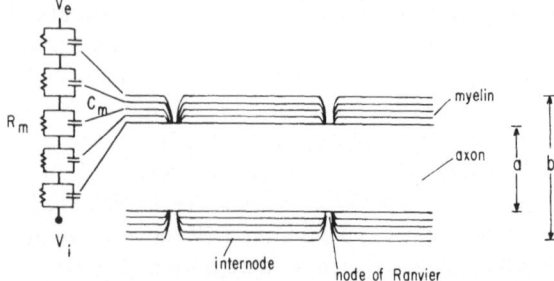

Fig. 6.3. Schematic representation of a myelinated nerve giving the terms commonly used and a simple electrical circuit. The circuit consists of resistors ($R_m$) and capacitors ($C_m$) in series between the external and the internal portions of the fiber, which are at voltages $V_e$ and $V_i$ respectively.

membrane will behave linearly for small deviations in voltage. Therefore, the properties of linear cables will be analyzed in some detail.

In dealing with linear cables, it is useful to define two further quantities: a *time constant* $\tau = RC$, and a *space constant* $\lambda = (Ra/4\rho)^{1/2}$. The full significance of these quantities will only become apparent later, but note for the moment that the units of $\tau$ are in milliseconds and the units of $\lambda$ are in centimeters. Thus, the names, *time constant* and *space constant* are appropriate. For a linear cable $J_i = V/R$ and substitution in Equation (6.4) will then give the basic equation for a linear cable:

$$\lambda^2 \frac{\partial^2 V}{\partial x^2} = \tau \frac{\partial V}{\partial t} + V \tag{6.6}$$

The following interesting problem was considered by Rushton (1951). If the total diameter of a nerve axon *plus* its myelin sheath is $b$, what fraction of that total diameter should be composed of an axon (of diameter $a$) and what fraction should be composed of myelin in order that the spread of voltage along the fiber be optimal? From its structure, myelin can be considered as a series of membranes which are packed so tightly that the fluid has been extruded from between each layer (see Chapter 1). No current can then flow axially between the layers, and the circuit diagram shown in Fig. 6.3 will apply to a unit membrane area of myelin. The time constant is then independent of the thickness of the myelin. For example, the effective resistance of $n$ layers of membrane in series will be $nR_m$ while the capacitance of $n$ layers of myelin will be $C_m/n$, where $R_m$ and $C_m$ are the resistance and capacitance of a single layer of myelin of unit area. Actually, each layer of myelin will be slightly larger than the one inside it, but as shown in Problem 6.2, the product of the resistance and capacitance, which determines the time constant, will still be independent of the number of layers.

The effective transverse resistance $r$ of a *unit length of cable*, assuming each layer has a resistance $R_m$ for a *unit membrane area* and there are $n$ layers per cm packed from an inner radius $a/2$ to an outer radius $b/2$, is obtained by integrating between these limits:

$$r = \int_{a/2}^{b/2} \frac{nR_m}{2\pi z} \, dz = \frac{nR_m}{2\pi} \ln(b/a) \tag{6.7}$$

The effective transverse resistance $R$ *for a unit membrane area* in an axon of diameter $a$ will be $R = \pi a r$. After substitution of these quantities into the definition of $\lambda$ and some algebraic manipulation, it follows that

$$\lambda^2 = \frac{nR_m}{8\rho} a^2 \ln(b/a) \tag{6.8}$$

For a given external diameter $b$, $\lambda$ will be maximum when $d\lambda^2/da = 0$. Differentiating Equation (6.8) and rearranging

$$\ln(a/b) = -\frac{1}{2} \tag{6.9}$$

The natural logarithm of $-1/2$ is just over 0.6, so the space constant will be optimized when the total diameter is divided into an axon 60% and a sheath 40% of the total diameter. If this optimal ratio is obeyed, then from Equation (6.8) the space constant will increase linearly with diameter. Since the time constant is not affected (see above), in the same period of time voltage will spread twice as far along the internodes of a cable twice the size.

Now consider what area of nodal membrane would be required to generate sufficient current for an identical action potential to be produced in two fibers of differing diameter $b$ and internodal length $l$. The axial resistance of the internode will increase as the length $l$ and inversely as the cross-sectional area. Thus, to get the same voltage drop will require an axial current

$$I_a \propto a^2/l \tag{6.10}$$

If the *internodal* length $l$ increases linearly with diameter $a$, the required current would be proportional to $a$. This current would automatically be generated if the *nodal* lengths and membrane properties were identical in both fibers, since both the membrane area and nodal current would then increase linearly with $a$. In short, if the following three conditions are obeyed, the spread of charge along a myelinated fiber will be optimized:

(1) the ratio of internal to external diameter is constant and approximately equal to 0.6;
(2) nodal length and membrane properties are independent of fiber diameter;
(3) internodal length increases linearly with fiber diameter.

Then, three results follow corresponding to those for unmyelinated nerve derived earlier: (1) An action potential will propagate with a conduction velocity varying directly as the diameter of the fiber; (2) the threshold current needed for initiating an action potential through direct application of current across a membrane will increase linearly with fiber diameter; (3) the threshold for initiating an action potential with a distant external electrode will vary as the reciprocal of fiber diameter.

## Comparison with Experimental Data

The approach in the previous section differs from that in many other sections of the book. Following Rushton's (1951) approach to the problem, the optimal design for a myelinated nerve fiber was attacked as a straight-forward engineering problem without consideration of biological data. A condition [Equation (6.9)] was derived which would give an optimal design for the spread of voltage along a nerve fiber. Then, when Rushton (1951) considered available evidence, he found that "he'd been scooped." The nervous system has been using this design for millions of years. The available evidence indicated that the ratio of internal to external diameter of myelinated fibers was relatively constant and that the internodal length increased reasonably linearly with fiber diameter. Little direct evidence was available about the properties of the nodal membranes from fibers of different size. Membrane properties are not completely independent of size. However, before considering the discrepancies, a few of the interest-ing results will be examined that emerge from assuming approximate independence:

(1) If the conduction velocity of unmyelinated nerve varies as the square root of diameter, and the conduction velocity of myelinated nerve varies linearly with diameter, then by the very nature of the two curves, they must cross as indicated in Fig. 6.4. This figure, taken from Waxman and Swadlow (1977), shows that, depending on the assumptions made, the intersection will occur somewhere between 0.2 and 1 $\mu$m. These values are of interest since the largest unmyelinated fibers (found in peripheral nerves) are about 1 $\mu$m in diameter, while the smallest myelinated fibers (found in the central nervous system) are about 0.2 $\mu$m.

(2) There are constraints against having very small myelinated fibers. Hille (1970) calculated that if a node of Ranvier in a 1-$\mu$m fiber had the same properties as those in larger fibers, the opening of only two $Na^+$ pores would be sufficient to depolarize the nerve to threshold. Such a fiber would be quite unreliable, so vertebrate fibers probably become myelinated as soon as they reach a diameter where there is an advantage in

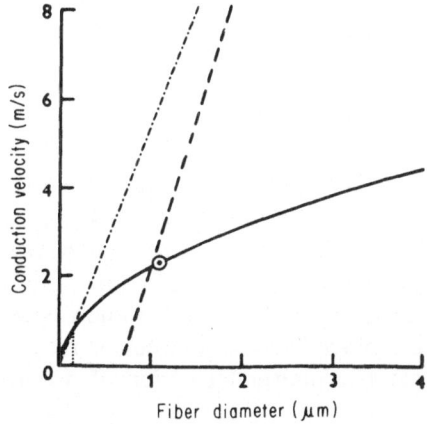

Fig. 6.4. Predicted relationship between conduction velocity and fiber diameter for unmyelinated fibers (solid curve) and myelinated fibers (dashed lines). Depend-ing on the assumptions made, the lines cross at between 0.2 and 1 $\mu$m, indicating that above this diameter myelinated nerve fibers will conduct faster than un-myelinated fibers. The circle gives the conduction velocity of the largest unmy-elinated fibers measured by Gasser (1950). (From Waxman and Swadlow, 1977.)

conduction velocity *and* a myelinated nerve can achieve a reasonable degree of reliability.

(3) Once a nerve becomes myelinated, the number of nodes is set for life. As the animal grows, the fiber diameter grows and the length of the internodal region grows. If these two increase proportionally, the conduction velocity will also increase at the same rate and the time for impulses to go to or from the central nervous system will remain unchanged. Thus, *the pattern of nerve impulses that produces a coordinated sequence of movements in a young animal will produce a similar coordinated sequence in the adult.* The advantages of such a system are obvious, since the basic patterns of locomotion, breathing, etc., would not have to be continually modified as an animal grows.

Having stated the advantages, one must point out that Rushton's predictions are not obeyed exactly in either myelinated or unmyelinated nerve (Paintal, 1978). In carefully reinvestigating the relation between conduction velocity and fiber diameter, which is usually taken to be linear (Hursh, 1939), the data were better fitted by a power function with an exponent of 1.5 (Coppin and Jack, 1972) or by two straight line segments with differing slope (Boyd and Kalu, 1979). The conduction velocity of unmyelinated fibers has long been known to increase somewhat faster than the square root relation (Pumphrey and Young, 1938), and Pearson *et al.* (1970) found better fits in insect nerves with power functions having an exponent of 0.7 or 0.8. The reasons for these discrepancies or their effects on the relations stated above are not certain. However, the derivation assumes that membrane properties are independent of fiber diameter. Then, the same time course of action potential will result, only differing in conduction velocity. Yet the duration of the action potential in small nerve fibers (below 5 $\mu$m) is substantially longer in both myelinated and unmyelinated nerve (Paintal, 1966, 1967; Pearson *et al.*, 1970). The differences in membrane properties underlying the different time course of action potential may well account for the observed deviations in conduction velocity (Pearson *et al.*, 1970).

## Linear Cables

The results in the previous section were all obtained from analyzing the form of the cable equations without obtaining complete solutions. Solutions for tissues with nonlinear current–voltage relations must often be obtained by numerical computation for the special values of interest. If a linear current–voltage relation is assumed, then general solutions can be obtained. This assumption is reasonably valid in a number of situations: (1) The internodal region of myelinated nerve behaves as a linear cable transmitting the voltage changes at one node to the next node. (2) Dendritic regions of cells, which have synaptic or sensory inputs impinging on them, often cannot generate action potentials. This does not apply to all dendrites (see, for example, Llinás and Nicholson, 1971), but linear models are applicable to many. (3) The current–voltage curves of many cells, although grossly nonlinear for large voltage changes, can be treated by linear analysis for sufficiently small fluctuations. For example, a single packet of synaptic transmitter released at a neuromuscular junction produces voltage changes which are often less than 1 mV in a muscle cell. Although the muscle cell will give a propagated action potential if a sufficient number are released synchronously, the response is essentially linear to single packets which are being released continuously.

In solving the equation for a linear cable [Equation (6.6)], it is convenient to define two dimensionless quantities: $X = x/\lambda$ and $T = t/\tau$. Then, this equation becomes simply

$$\partial^2 V/\partial X^2 - \partial V/\partial T - V = 0 \qquad (6.11)$$

By using Laplace transforms, this equation can be converted from a partial differential equation to an ordinary differential equation. This simplification results from the algebraic relation between the Laplace transform of a variable and the Laplace transform of its derivatives (see, for example, Sokolnikoff and Redheffer, 1958). If $V^*$ is the Laplace transform of $V$, the the Laplace transform of $\partial V/\partial T$ is $sV^* - V(0, X)$, where $V(0, X)$ gives the initial conditions at $T = 0$. In a quiescent cable $V(0, X) = 0$ for all $X$, since we are measuring voltage deviations from the resting potential. The Laplace variable $s$ replaces the time variable $t$, and Equation (6.7) becomes

$$\partial^2 V^*/\partial X^2 - (s + 1)V^* = 0 \qquad (6.12)$$

This ordinary differential equation is well known and has exponential solutions with two constants $A^*$ and $B^*$ in general:

$$V^* = A^*\exp\left[-X(s + 1)^{1/2}\right] + B^*\exp\left[X(s + 1)^{1/2}\right] \qquad (6.13)$$

No voltage change would be expected in a linear cable a long distance away, so $V^*$ must be zero when $X = \infty$. Hence, $B^* = 0$ for very long cables and

$$V^* = A^*\exp\left[-X(s + 1)^{1/2}\right] \qquad (6.14)$$

$A^*$ can be determined from Equation (6.1)

$$I_a = -(1/r_a\lambda)(\partial V/\partial X) \qquad (6.15)$$

and $I_a^* = -(1/r_a\lambda)(\partial V^*/\partial X)$, where $I_a^*$ is the Laplace transform of $I_a$.

Differentiating Equation (6.14) gives

$$\partial V^*/\partial X = -A^*(s + 1)^{1/2}\exp\left[-X(s + 1)^{1/2}\right] \qquad (6.16)$$

From Equations (6.15) and (6.16)

$$I_a^* = \left[ A^*(s+1)^{1/2}/(r_a\lambda) \right] \exp\left[ -X(s+1)^{1/2} \right] \qquad (6.17)$$

At $X = 0$, $I_a = I/2$ ($I$ = electrode current). Hence

$$I^* = 2A^*(s+1)^{1/2}/r_a\lambda$$

so that

$$A^* = r_a I^*\lambda/2(s+1)^{1/2}$$

Substituting into Equation (6.14) gives $V^*$ as a function of $s$.

$$V^* = I^*\left[ r_a\lambda/2(s+1)^{1/2} \right]\exp\left[ -X(s+1)^{1/2} \right] \qquad (6.18)$$

The *transfer function* $Z^*$ for this cable is given by

$$Z^* = V^*/I^* = \left[ r_a\lambda/2(s+1)^{1/2} \right]\exp\left[ -X(s+1)^{1/2} \right] \qquad (6.19)$$

The inverse transform can be obtained from a table of Laplace transforms (e.g., Selby, 1975, pairs 84 and 11). This gives what is known as the *impulse response* and physically corresponds to the response when a unit charge is applied at $t = 0$. The action of a short-acting transmitter such as ACh at a muscle end plate corresponds closely to this situation. The result is

$$V = \frac{r_a\lambda}{2(\pi T)^{1/2}}\, e^{-T}e^{-X^2/4T} \qquad (6.20)$$

where $V$ is the change in voltage after placing a unit charge on the membrane. The physical interpretation of this expression can be understood by noting that the last factor on the right has a close resemblance to the Gaussian or normal distribution which plays a central role in statistics. This distribution is characterized by its standard deviation $\sigma$ and is

$$\Phi(y) = (2\pi\sigma^2)^{-1/2}e^{-y^2/\sigma^2}$$

Remembering $X = x/\lambda$ and $T = t/\tau$, one can show by algebraic manipulation that

$$V = \frac{2^{1/2}R}{\pi a}\, e^{-t/\tau}\Phi(x) \qquad (6.21)$$

where in our example $\sigma = 2\lambda(t/\tau)^{1/2}$ and $V$ is the change in voltage after placing a unit charge on the membrane. Thus, the total charge on the membrane decays exponentially with a time constant $\tau$. At the same time the charge is being redistributed along the cable according to a Gaussian distribution, whose standard deviation increases as the square root of $t$. This redistribution is illustrated in Fig. 6.5A. Alternatively, one can show (see for example Fig. 3.20 in Jack et al., 1975) that the overall decay of voltage approaches an exponential with time constant $\tau$. Close to the point

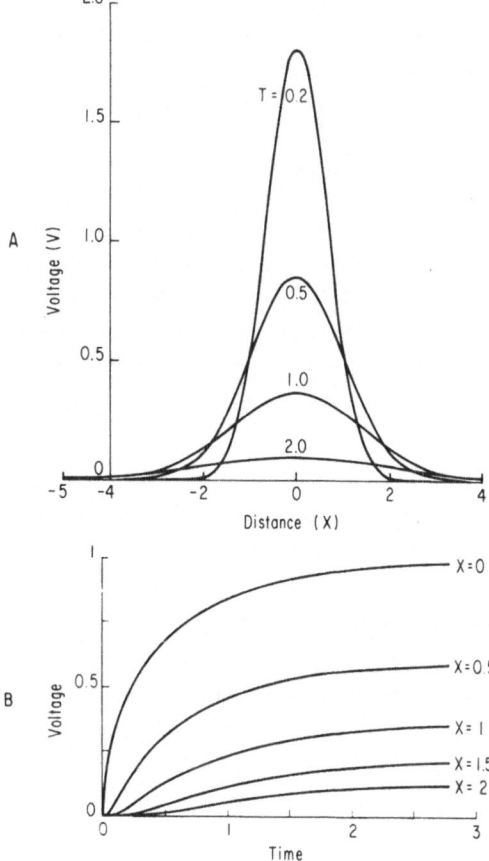

Fig. 6.5. (A) Spread of charge applied instantaneously at $T = 0$ at a point $X = 0$ in a linear cable [from Equation (6.21)]. (B) Spread of charge applied as a constant current $I$ at a point $X = 0$ [from Equation (6.22) and Jack et al., 1975]. Note that in (A), the total charge decays exponentially while spreading along the cable. In (B), the voltage builds up, but more slowly and to a lesser extent at points more distant from the point of application.

of application of current, the initial decay is faster than exponential, while further away the intial decay is slower than exponential.

From the transfer function, the response of a cable to any current waveform can be obtained. For example, if a constant current $I$ is applied beginning at time $t = 0$, the Laplace transform of $I$ is $I^* = I/s$. From Equation (6.19)

$$V^* = Z^* I^* = \left\{ (r_a I \lambda) / \left[ 2s(s + 1)^{1/2} \right] \right\} \exp\left[ -X(s + 1)^{1/2} \right]$$

The inverse Laplace transform of this expression is

$$V = \frac{r_a I \lambda}{4} e^{-X} \left\{ \left[ 1 - \operatorname{erf}\left( \frac{X}{2T^{1/2}} - T^{1/2} \right) \right] \right.$$

$$\left. - e^{X} \left[ 1 - \operatorname{erf}\left( \frac{X}{2T^{1/2}} + T^{1/2} \right) \right] \right\} \tag{6.22}$$

where the error function $\text{erf}(y) = 2\pi^{-1/2}\int_0^y e^{-z^2}\,dz$, is closely related to the Gaussian function mentioned previously. Hodgkin and Rushton (1946) first derived Equation (6.22) which predicts that the voltage near the electrode will rise rapidly to 85% of its steady state value (Fig. 6.5B) at a time $t = \tau$ after the current is switched on. At a distance from the electrode the voltage will rise much more slowly along a sigmoid curve towards a lower steady state. Thus, $V$ is a rather complicated function of $x$ and $t$, but some fairly simple and useful relationships emerge from the equation. One of these can help in understanding the meaning of the cable constant $\lambda$ and will be considered here. Another simple relationship is given as Problem 6.4.

When $T \to \infty$, Equation (6.22) becomes

$$V = \frac{r_a\lambda I}{4}\, 2\exp(-X) = \frac{r_a\lambda I}{2}\exp(-x/\lambda) \qquad (6.23)$$

since $\text{erf}(\infty) = 1$ and $\text{erf}(-\infty) = -1$. Thus, $V$ decays exponentially, reaching $1/e$ of its initial value when $x = \lambda$, which is the rationale for referring to $\lambda$ as the *space constant*.

## Extracellular Recording

The final topic in this chapter concerns the signals recorded extracellularly from nerve or muscle cells. Despite the advances in intracellular recording methods discussed in Chapter 5, extracellular recordings are still

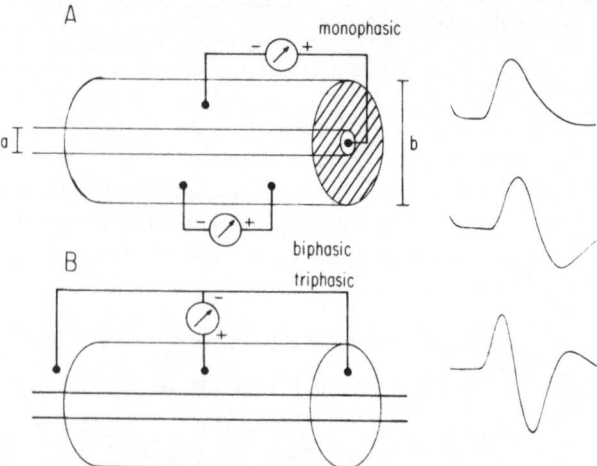

Fig. 6.6. Methods for recording monophasic, biphasic and triphasic potentials from an axon of diameter $a$ in a restricted extracellular space of diameter $b$. Typical action potentials recorded extracellularly in this way (from Stein et al., 1977) have one, two, or three phases for each phase of the intracellularly recorded action potentials.

extremely important experimentally and clinically. The basis for these methods is considered in detail in other places (Chung, 1974; Kiloh *et al.*, 1972; Rosenfalck, 1969), but the form and amplitude of neural signals expected in various experimental situations remain a source of confusion to many students and research workers. Two limiting cases will be considered here which illustrate the principles involved and which extend the ideas that have already been discussed in relation to cable theory.

The first example concerns a fiber in a restricted extracellular space, as shown in Fig. 6.6A. This example applies to the traditional method of recording in which a cut nerve is pulled up into a nonconducting medium such as paraffin oil. Restricting the extracellular space increases the voltage recorded extracellularly since an equal and opposite current must flow outside a fiber to that flowing internally from Kirchhoff's law. Increasing the extracellular resistance increases the voltage drop produced by these external currents. Then, if $r_a$ and $r_e$ are the resistances per unit length of the axon and the extracellular space respectively, there is a set of equations corresponding to Equation (6.1),

$$\frac{\partial V_i}{\partial x} = -I_a r_a; \qquad \frac{\partial V_e}{\partial x} = I_a r_e \qquad (6.24)$$

$$\frac{\partial V}{\partial x} = \frac{\partial(V_i - V_e)}{\partial x} = -I_a(r_a + r_e) \qquad (6.25)$$

where $I_a$ is the current flowing down the axon and $V_i$, $V_e$, and $V$ are the intracellular potential, the extracellular potential, and the transmembrane potential respectively. From Equation (6.25), it is easily shown (Problem 6.5) that the cable equation (6.6) still holds, although the space constant $\lambda$ decreases as the extracellular resistance increases.

From Equations (6.24) and (6.25) it also follows easily that

$$\frac{\partial V_e}{\partial x} = \frac{-r_e}{r_a + r_e} \frac{\partial V}{\partial x} \qquad (6.26)$$

Equation (6.26) indicates that a fraction of the change in membrane potential is recorded in the extracellular space and the axoplasmic resistances act like a simple electrical voltage divider (Skilling, 1959). If the extracellular resistance $r_e$ is made sufficiently high, the fraction approaches 1 and the entire membrane potential change is recordable extracellularly. The principle is utilized in the sucrose gap technique mentioned previously in Chapter 5, where a nonconductor such as sucrose continuously perfuses a length of axon to maintain a very high external resistivity. In this way the membrane potential of axons can be monitored continuously without having to impale the fiber with microelectrodes.

The form of the potentials recorded depends on the configuration of the electrodes used. If a nerve is cut and placed in a restricted extracellular space which is sealed at one end, the difference between the potential at the sealed end ($x = l$) and at some point $x$ in the restricted space becomes from Equation (6.24)

$$V_1(x) = V_e(l) - V_e(x) = \int_x^l \frac{dV_e}{dx} \, dx = \frac{-r_e}{r_a + r_e} \int_x^l \frac{dV}{dx} \, dx$$

$$= \frac{r_e}{r_e + r_a} \left[ V(x) - V(l) \right] \tag{6.27}$$

Equation (6.27) indicates that the potential $V_1$ recorded with this configuration will give a faithful replica of the change in membrane potential at the point $x$ with respect to the constant potential at the cut end, only scaled down by the factor $r_e/(r_a + r_e)$. This type of recording is also referred to as *monophasic* recording, since each phase of the action potential will produce a single phase at the extracellular electrode.

If both electrodes are on intact nerves some distance $\delta$ apart, the potential difference recorded $V_2$ is

$$V_2(x, \delta) = V_e(x + \delta) - V_e(x) = V_1(x) - V_1(x + \delta) \tag{6.28}$$

The potential recorded will be the first difference of the monophasic potentials, so that for every phase of the intracellular potential, two phases will generally be recorded extracellularly. Hence, this method is known as *biphasic recording*.

Still another common recording configuration is shown in Fig. 6.6B. Here a potential at the center of the extracellular space is recorded with respect to the potential at the two ends which are shorted together, which gives the average of the two voltages. It can be shown from Equation (6.24), with a little manipulation, that the potential recorded $V_3$ is given by

$$V_3(x, \delta) = V_e(x) - \frac{1}{2} \left[ V_e(x - \delta) + V_e(x + \delta) \right]$$

$$= \frac{1}{2} \left[ V_2(x - \delta, \delta) - V_2(x, \delta) \right] \tag{6.29}$$

The potentials recorded are therefore the first difference of the biphasic potentials. They are often referred to as *triphasic potentials* in that a single phase of the intracellular action potential can give rise to three phases with this recording configuration (see Fig. 6.6). Triphasic potentials can also be recorded from an intact nerve which passes through a restricted space, as shown in Fig. 6.6B. The potentials will again correspond to the second difference of the intracellular action potential (Stein and Pearson, 1971). For close spacing of the electrodes, Equations (6.28) and (6.29) become proportional to the first and second differentials of the membrane potential. Biphasic recordings can then be used to measure the axial current flowing within a nerve fiber [cf. Equation (6.1)] and triphasic recordings can be used to measure the membrane currents without penetrating the fiber [cf. Equation (6.3)].

All the results up to this point assume that the extracellular space is so restricted that there are no significant voltage drops in the radial direction. A condition for testing this assumption is given in Problem 6.1. At the other extreme, neural signals may be recorded in a large extracellular space. The voltages expected in this situation have been studied for many years (e.g., Lorente de Nó, 1947) and will now be considered briefly.

If there is a point source of positive charge, as shown in Fig. 6.7A, current flow will be directed radially away from that point and the surfaces

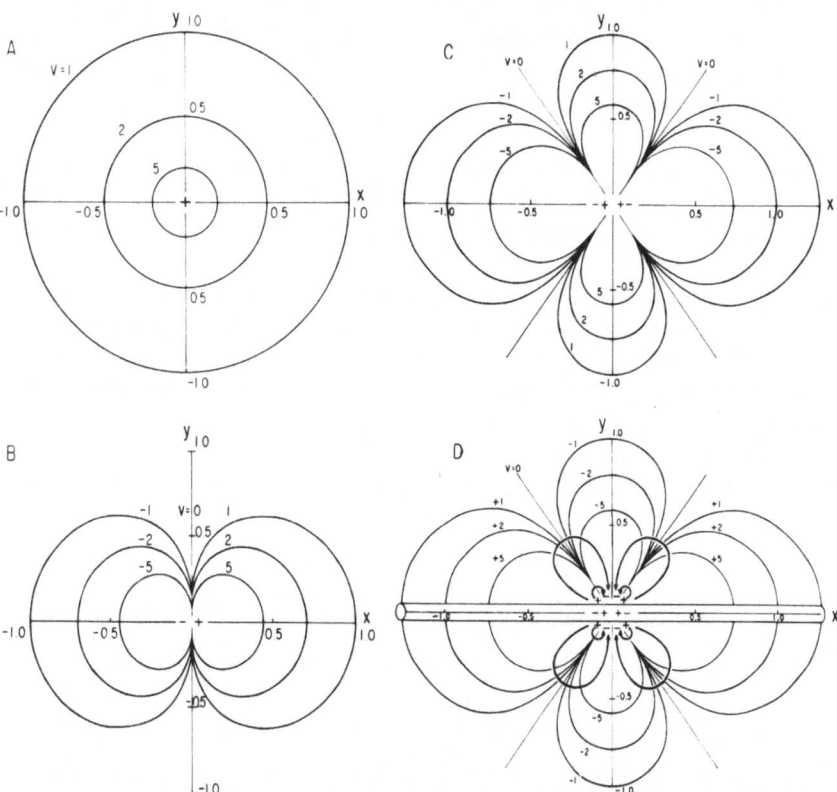

Fig. 6.7. Lines of equal potential ($V = 0$, $\pm 1$, $\pm 2$, $\pm 5$) are shown in Cartesian coordinates for (A) a point positive charge, (B) a dipole and (C) two oppositely directed dipoles. The potentials were calculated from (A) Equation (6.30), (B) Equation (6.31), and (C) Equation (6.32), assuming (A) $q = 1$, (B) $q\delta = 1$, and (C) $q\delta^2 = 1$. Part (D) most closely approximates the usual situation of an action potential conducting along a cable-like nerve fiber. A few lines of current flow are shown diagrammatically (arrows).

of equal potential will be spheres centered about that point. In a two-dimensional representation such as Fig. 6.7A, there will be circular equipotential lines. From electrostatic theory, the potential recorded varies inversely with distance, i.e.,

$$V_1 = \frac{q}{\left(x^2 + y^2\right)^{1/2}} \tag{6.30}$$

where $q =$ the amount of charge, $(x^2 + y^2)^{1/2}$ is the distance from the charge which is centered on ordinary Cartesian coordinates, and the symbol $V_1$ is again used because all the potentials have a single polarity.

A dipole is a close apposition of equal positive and negative point sources of charge, as shown in Fig. 6.7B. The lines of current flow radiating from the positive charge now curve around to return to the negative charge. The current flow is identical above and below the dipole (the horizontal or $x$ axis is an axis of rotational symmetry), but each current line extends further in the horizontal direction than in the vertical or $y$ direction (horizontal eccentricity). The equipotential lines shown in Fig. 6.7B also curve around so as to remain perpendicular to the lines of current flow. By definition, no current can flow in a medium of finite resistivity along an equipotential line since this would produce a voltage drop according to Ohm's law. As shown in Fig. 6.7B each equipotential line is confined to half the plane, but a line of opposite polarity is found in the other half plane. The $y$ axis is the line of zero potential (in a three-dimensional representation it is actually a plane, perpendicular to the plane of the paper, which passes through the $y$ axis) and there is a mirror symmetry about this axis (plane).

The potential $V_2$ is the difference between the potentials produced by the two sources. If they are separated along the $x$ axis by a distance $\delta$ we have

$$V_2 = V_1(x, y) - V_1(x + \delta, y)$$

Using the usual definition of a partial differential, it follows easily that

$$V_2 = -\delta\left(\frac{\partial V_1}{\partial x}\right) = \frac{xq\delta}{\left(x^2 + y^2\right)^{3/2}} \tag{6.31}$$

An action potential can be considered as two oppositely directed dipoles, since at the peak of the action potential, the inside of the nerve fiber becomes positive, while elsewhere it is negative. As shown in Fig. 6.7C, there is still a rotational symmetry about the $x$ axis and around the $y$ axis. The lines of current flow now curve around so as to remain in a single quadrant, while the equipotential lines are confined to a quadrant of the plane represented by the paper.

The potential $V_3$ is the difference between the potentials produced by two dipoles, which if they are again separated by a distance $\delta$ would give

$$V_3 = V_2(x, y) - V_2(x - \delta, y)$$

$$= \delta\left(\frac{\partial V_2}{\partial x}\right) = \frac{(y^2 - 2x^2)q\delta^2}{\left(x^2 + y^2\right)^{5/2}} \tag{6.32}$$

Equation (6.32) and Fig. 6.7C actually predict a pattern of voltage changes opposite to that observed experimentally, because they neglect the fact that the nerve membrane itself serves as a dipole. Since the membrane can store charge (and act as a capacitor), there will be an equal and opposite charge on the outside of the membrane to that found inside. Equipotential lines, as observed on the outside of the membrane, are shown in Fig. 6.7D.

Now consider the potential changes at various distances from the $x$ axis, if the charges move at constant velocity along the $x$ axis. The potential changes expected from Fig. 6.7A are shown in Fig. 6.8A. The potential changes are monophasic and symmetrical, reaching a maximum when the charge is closest to the point of recording. It follows immediately from Equation (6.30) that the maximum voltage falls off inversely with the distance, so that at a distance twice as great (interrupted lines) the potential change is only half as large. In addition (Fig. 6.7A), potential lines will be crossed less frequently at a greater distance from the point source, so that the potential changes will be more spread out in time. In fact, the waveforms at twice the distance will be just twice as spread out in space or time (assuming constant velocity of movement), and so they will contain correspondingly lower frequency components.

Fig. 6.8B shows the potential changes expected for a dipole moving along the $x$ axis. The changes are now biphasic, as first positive and then negative equipotentials are crossed (or vice versa, depending on the direction of motion). The time between the two peaks increases linearly with distance away from the dipole [see Problem 6.7 or Pollak (1971)], so the waveform is again more spread out in time at greater distances. The maximum voltage also declines with distance, but for a dipole the voltage varies inversely as the square of the distance (Problem 6.7), rather than the first power.

For the double dipole of Fig. 6.7D, the potential changes are shown in Fig. 6.8C. They are now triphasic and symmetrical in shape. As might be expected from the above comments, the waveforms are again more spread out at greater distances and the amplitudes decline even more sharply with distance. Hence, to record a measurable signal extracellularly in a large volume of fluid, a microelectrode should be used to minimize the distance from the nerve fiber. Note that the waveform will be independent of the

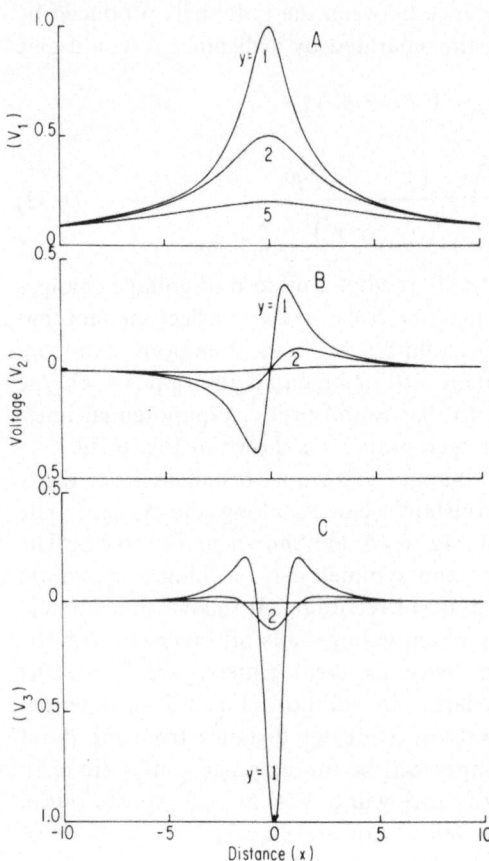

Fig. 6.8. Voltages recorded at different distances ($y$) from a nerve fiber as (A) a charge, (B) a dipole, or (C) a double dipole passes at constant velocity along the $x$ axis. The same conditions and assumptions have been made as in Fig. 6.7.

direction of movement and will be positive–negative–positive for the potential distribution shown in Fig. 6.7D.

These general considerations have to be refined in several ways in considering nerve fibers:

(1) The arguments above assume implicitly that the medium in which the charges are embedded is electrically *isotropic*; i.e., the resistance to current flow in all directions is the same. The differences are not too great for unmyelinated nerves, but Tasaki (1964) found that the radial or transverse resistivity of a myelinated nerve is fifty times the longitudinal resistivity. Presumably, the myelin makes it difficult for currents to flow in the radial direction except at the nodes, but currents can flow more easily along the inside of fibers in the longitudinal direction. Based on the work of Plonsey (1974), Marks and Loeb (1976) suggested that the anisotropy will convert the circles of Fig. 6.7A into ellipses in which the ratio of the long or $x$ axis to the short or $y$ axis is proportional to the square root of the

ratio of the longitudinal resistivity to the transverse resistivity. The same sort of reduction in the spread of currents in the $y$ direction will apply to the other recordings of Fig. 6.7.

(2) The action potential recorded intracellularly from a nerve fiber is usually asymmetrical in that the falling phase is generally about three times the duration of the rising phase (Marks and Loeb, 1976; Table 1). Thus, for a wave traveling at a constant velocity the dipole of the falling phase will be three times as spread out in space as that of the rising edge. Only when the distance from the fiber is large compared to the spatial extent of the action potential in space will the triphasic waveforms be symmetrical. As an electrode approaches a fiber more closely, the dipole of the falling phase will become spread out. Portions of the dipole will be at different angles from one another and hence will generate different voltages. Cancellation of positive and negative contributions may occur, and in general the third phase will be less prominent than the first two phases.

(3) The membrane of the nerve fiber imposes certain boundary conditions on the flow of current. Since the ionic channels are oriented perpendicularly to the surface of the membrane, current flow must be perpendicular to this surface. (The capacitance of the membrane, arising from its molecular structure as a bilayer, also requires that capacitative current flow perpendicularly to the surface.) Thus (Fig. 6.7D), current will initially exit or enter the fiber perpendicularly before spreading out in a curved pattern. The intracellular current flow will also be curved, but Clark and Plonsey (1966) showed that the radial changes in voltage were negligibly small compared to the longitudinal ones when reasonable assumptions were made about the diameter, conduction velocity, etc., of nerve fibers.

(4) The amplitudes of the potentials fall off steeply with distance so that at the skin only potentials generated from large structures, e.g., the heart (electrocardiogram), muscles (electromyogram), or large portions of the brain (electroencephalogram) are readily detected. Even these potentials will not be exactly as shown in Fig. 6.7 for an infinite field, but will be affected by the "boundary conditions" imposed by the body's surface. Accurate solutions for the complex geometry of the human body are generally difficult to compute and a given pattern of potentials at the surface can in general arise from more than one distribution of charges. Even when recording internally, monophasic, biphasic, and triphasic potentials can be recorded from a restricted extracellular space (Fig. 6.6), from an infinite field (Fig. 6.8), or for conditions between these two extremes. To interpret these records properly requires a consideration of the geometry in which a fiber is situated, as well as the ionic mechanisms and cable properties of the fibers. However, the ideas developed here are sufficient for many experimental situations without having to resort to detailed calculations.

## Problems

6.1. If an axon of diameter $a$ is placed in a restricted extracellular space of diameter $b$, as shown in Fig. 6.6, show that the voltage gradients in the radial direction can be neglected compared to the longitudinal gradients if

$$(b^2 - a^2)\ln(b/a) \ll 8l^2$$

where $l$ is the length of axon over which membrane currents spread. In a linear cable the length $l$ might be chosen as the space constant $\lambda$. (Hint: Calculate the longitudinal and radial resistance of the fluid in an extracellular space of length $l$. The inequality follows from assuming that the radial resistance, and hence the voltage drop produced by a given current in the radial direction will be small compared to that in the longitudinal direction.) Stein and Pearson (1971) give further details.

6.2. Show that the time constant $\tau = R_m C_m$ of a layer of myelin is independent of the distance of the layer from the center of the fiber.

6.3. (a) Show, starting from Equation (6.20), that the time at which the voltage reaches its maximum value varies nearly linearly with the distance from the point at which the charge is applied. [Hint: The maximum value occurs at the time when $\partial V/\partial T = 0$. The exact relation was verified by Fatt and Katz (1951).]
(b) If there were a linear relation between distance and the time to peak voltage, how would the peak amplitude depend on $X$?
(c) If the peak voltage is 1 mV at $X = 1$, what will the amplitude be at $X = 3$? How would this compare with the noise level in microelectrode recording systems which is generally about 100 $\mu V$?

6.4. Show from Equation (6.23), that the input resistance $= V_{x=0}/I$ is proportional to the square root of the change in $R$. This result is important in determining changes in $R$ with intracellular electrode techniques (Fatt and Katz, 1951).

6.5. Derive the cable equation (6.6) from Equation (6.25). Show that $\lambda$ becomes

$$\lambda = \left\{ \frac{Ra(b^2 - a^2)}{4[\rho_a(b^2 - a^2) + \rho_e a^2]} \right\}^{1/2}$$

where $\rho_a$ and $\rho_e$ are the resistivity of the axoplasm and the extracellular space respectively.

6.6. Show from Equations (6.28) and (6.29) that for large values of $\delta$, the peak-to-peak values of the recorded biphasic and triphasic action potentials approach two and one and a half times the value recorded monophasically.

6.7. Show from Equation (6.31) that the distance between the two peaks recorded when an action potential moves along the $x$ axis varies directly with the distance $y$. (Hint: The peaks will occur when $\partial V_2/\partial x = 0$.) Show also that the magnitude of the peaks varies inversely with the square of the distance $y$.

# 7

# Synaptic and Neuromuscular Transmission

The last chapter dealt with the spread of voltage along a nerve fiber. Eventually, the signal reaches the end of the fiber and must be transmitted across the gap or *synapse* that separates one nerve cell from another. Nerve cells may also make synaptic connections with effector cells such as muscle cells to produce contraction, or with endocrine cells to release hormones. In line with the theme of this book, much of the discussion will be related to the synaptic connection between nerve and muscle, which is often referred to as neuromuscular transmission. Because of the accessibility of these synapses, much of our knowledge of synaptic transmission has come from studying neuromuscular synapses. In fact, the neuromuscular junction is often taken as a model of all synapses, but in recent years it has become clear that there is a great variety of synaptic connections. In addition to the example of a nerve axon releasing a chemical *transmitter* which can diffuse across the synaptic cleft and excite or inhibit another nerve cell or muscle cell, numerous examples of electrical connections have been found. A model of such an electrical connection is shown in Fig. 7.1, and is often referred to as a *gap junction* because of the continuity of an anatomical pathway from one cell to another.

It is somewhat ironic that in the last century a controversy centered around the question of whether nerve cells were in fact distinct from one another, as other types of cells were thought to be (see Introduction in Eccles, 1964). However, evidence has accumulated over the last few years that most types of cells other than nerves are in fact electrically connected to one another. The connections or gap junctions (Fig. 7.1) are large enough to allow quite large molecules up to a molecular weight of 1,000 and a diameter of 2 nm to pass from one cell to another (Payton *et al.*, 1969). The membrane surrounding these junctions prevents the molecules

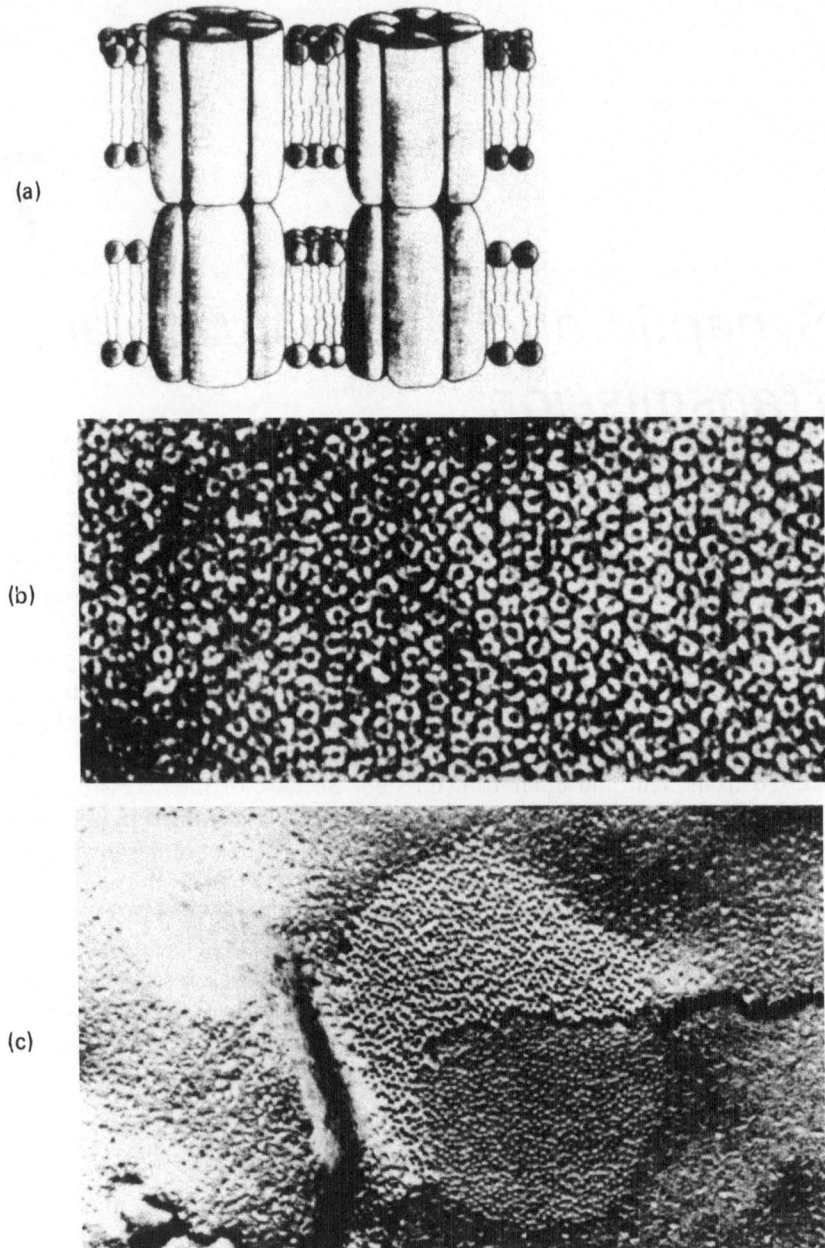

(a)

(b)

(c)

Fig. 7.1. Structure of a gap junction. (A) A molecular model with arrays of six connexin molecules surrounding a 2-nm central core. Two such arrays form a continuous path from one cell to another. (From Goodenough, 1976.) (B) An electron micrograph at 450,000 × shows a lattice of such cylindrical arrays with a center-to-center spacing of 8–9 nm. (C) A freeze-fracture picture at 110,000 × shows an aggregation of particles on one half-membrane face and pits on another half-membrane face to form the complex known as a gap junction. (From Staehelin and Hull, 1978.)

from escaping into the extracellular space and is distinct from ordinary cell membranes, for example in containing a specialized protein called *connexin* (Goodenough, 1976). This protein has a molecular weight of 18,000. These molecules can aggregate in a six-fold symmetry to form a cylinder with an external diameter of 7 nm. Two such cylinders would provide a continuous path from one cell to the other with the required 2 nm central core. An actual gap junction is shown at high power in Fig. 7.1B and in freeze fracture at lower power in Fig. 7.1C. This electrical and chemical continuity may be important in regulating the normal growth and development of cells, and is notably absent in certain types of tumor cells (Loewenstein, 1974). Electrical or gap junctions are also absent from the majority of nerve and skeletal muscle cells, although numerous examples of electrical connections have been described in lower animals (Bennett, 1974) and to a lesser extent even in mammals (Sotelo *et al.*, 1974).

A diagram of a neuromuscular junction is shown in Fig. 7.2 in which the *presynaptic* (nerve) fiber is separated by a space of about 20 nm from the *postsynaptic* (muscle) fiber. The large size of the muscle cell with respect to the fine nerve terminal would make it impossible for the nerve to depolarize the muscle significantly by direct spread of current, even if there were gap junctions. However, by releasing vesicles of a specialized synaptic transmitter, which can diffuse across and produce large permeability changes in the postsynaptic fiber, the weak nerve signals can be amplified sufficiently to produce action potentials reliably in many muscle fibers. An action potential leads to a brief contraction or twitch in the muscle fiber, as will be described in the next chapter. A principle first suggested by Sir Henry Dale (1935) holds that a single nerve cell only possesses the synthetic machinery to produce one chemical transmitter. Although it has been shown that the same transmitter may have different actions on different postsynaptic cells (Kandel *et al.*, 1967), clear examples of a mature nerve cell violating *Dale's principle* by producing two transmitters are controversial (cf., Burnstock, 1976; Osborne, 1979).

The presence of vesicles concentrated on one (presynaptic) side of a synapse and membrane specialization and thickening on the other (postsynaptic) side are some of the anatomical correlates of the *chemical synapse* in which transmission involves a specific transmitter chemical as mediator. Other types of connections, such as *tight junctions*, have also been described (see Chapter 1) in which the 20 nm gap between two cells is replaced by a much closer apposition of membranes (Porter and Bonneville, 1968). In absorptive cells and myelin, these junctions are thought to form a seal which prevents diffusion of material through extracellular space, rather than having a synaptic function. In addition, membrane thickenings occur frequently without the presence of synaptic vesicles, which are thought to be points of adhesion rather than synapses (Peters *et al.*, 1970).

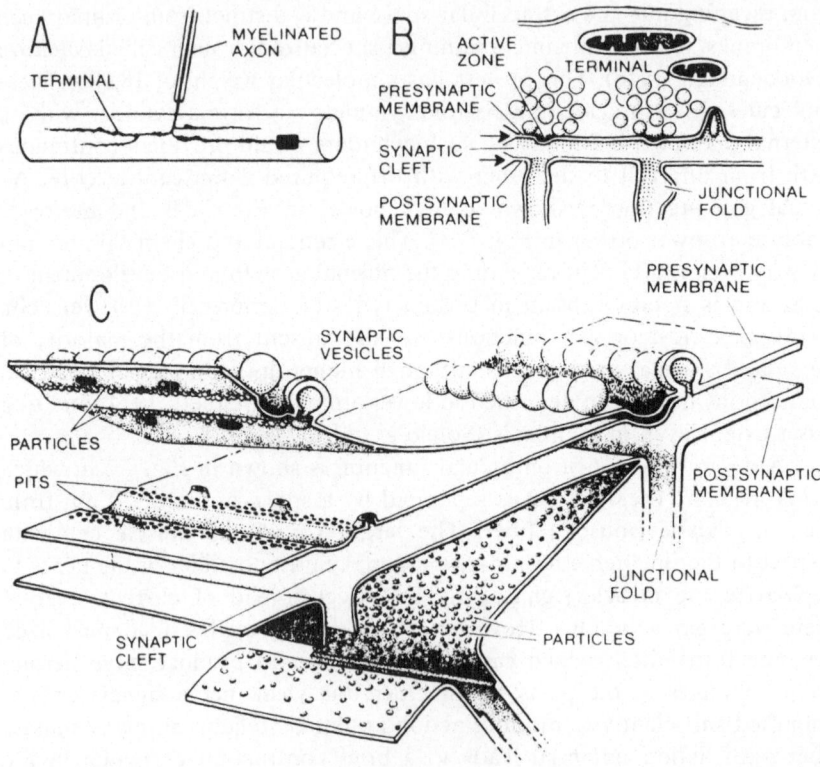

Fig. 7.2. Structure of a neuromuscular junction. (A) Schematic representation of a single myelinated motor nerve fiber connecting with a muscle fiber to form a neuromuscular junction. (B) At higher power, a number of vesicles are shown near an active release zone opposite a junctional fold in the postsynaptic membrane across the synaptic cleft. (C) The plasma membranes are split as they would be by freeze fracturing to show the particles and pits in the active zone adjacent to where synaptic vesicles can empty their contents into the synaptic cleft. The particles in the postsynaptic membrane are thought to be ACh receptors. (From Kuffler and Nicholls, 1976.)

Even within the class of chemical synapses defined anatomically or electrophysiologically, a great variety of connections has been described. As well as the "classical" examples of an axon synapsing with muscle cells or the dendrites and somas of nerve cells, connections between one axon and another and between dendrites have been described (Fig. 7.3). In fact, of the nine possible connections between dendrites, axons and somas as pre- and postsynaptic elements in chemical synapses, virtually all occur in nature (Shepherd, 1979).

*Reciprocal synapses* have also been found in which one cell is presynaptic to another, which in turn is presynaptic to the first cell at a nearby synaptic connection (Peters *et al.,* 1970). Finally, chemical synapses involving more than two cells have been described. Some of these are illustrated in the schematic diagrams of the synaptic connections in the retina (Fig. 7.3A) and the mammalian olfactory bulb (Fig. 7.3B). The first example involving synapses between three cells to be studied electrophysiologically was *presynaptic inhibition,* in which one axon inhibits the release of transmitter from a second axon and thereby reduces synaptic transmission to a third cell. This mechanism is particularly common in the initial relay stages of sensory pathways and will be treated in a later section. To discuss the wide variety of synapses and their various roles would take us too far afield. Fortunately, the subject has been summarized in a very readable book (Shepherd, 1979).

This chapter will be concerned more with the mechanisms which take place at a single synapse, such as the neuromuscular junction. Even with this limitation, a variety of processes on a broad time scale ranging from milliseconds to months will be considered. First, presynaptic events will be discussed; then, postsynaptic mechanisms will be considered, and finally some of the slower processes which affect synaptic transmission will be introduced.

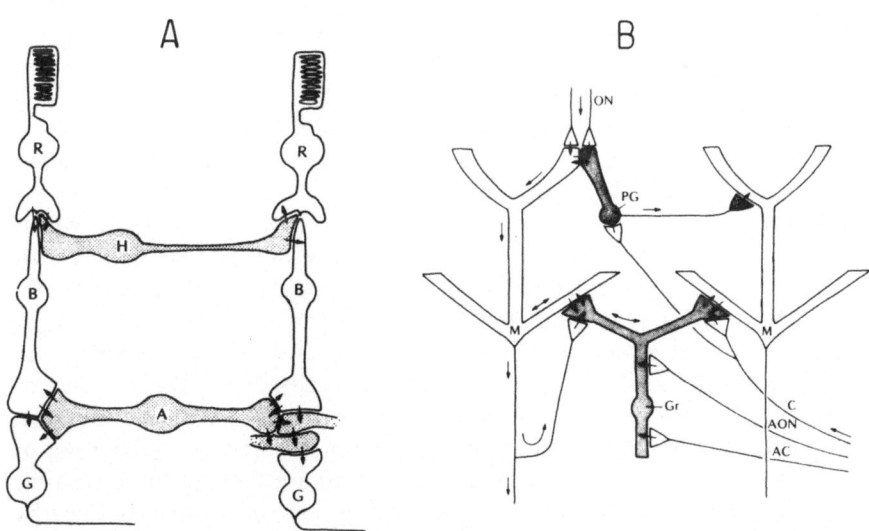

Fig. 7.3. **(A)** Schematic representation of the complex pattern of synaptic connections observed in the retina. R = rod, H = horizontal cell, B = bipolar cell, A = amacrine cell, and G = ganglion cell. **(B)** Circuit diagram for the mammalian olfactory bulb. ON = olfactory neuron, PG = periglomerular cell, M = mitral cell, Gr = granule cell; C, AON and AC are different types of axonal connections. (From Shepherd, 1979.)

## Presynaptic Mechanisms

Since their discovery in early electron micrographs (De Robertis and Bennett, 1955), synaptic vesicles have been central to understanding presynaptic mechanisms. Vesicles containing ACh, the transmitter at the vertebrate neuromuscular junctions, are 40–50 nm in diameter and have a clear center surrounded by a unit membrane (Robertson, 1956). Other transmitters are stored in somewhat different-sized vesicles and may possess dense cores or other distinctive morphological features (Bloom, 1972). However, it is not generally possible on histological evidence alone to specify which transmitter is being used as a given synapse. Nor is it possible on histological grounds alone to be certain whether a given synapse is excitatory or inhibitory, although certain morphological features (Gray, 1959) and shapes of vesicles (Uchizono, 1965) are commonly associated with the two different functions. Biochemical methods are widely used to separate a fraction of brain tissue called *synaptosomes* containing large numbers of synaptic vesicles, and to characterize the chemical transmitters they contain (Whittaker, 1970). These methods suggest that there are several thousand ACh molecules contained in one vesicle. By direct application of ACh from a micropipette at synaptic junctions and careful measurements of the amount of ACh needed to mimic the effect of the release of single vesicles, Kuffler and Yoshikami (1975b) produced an upper estimate of 10,000 molecules/vesicle. This estimate seems quite reasonable, since about 6,000 molecules would produce an isotonic solution of ACh (Problem 7.1).

In recent years electron micrographs have been published which support the suggestion from physiological studies that vesicles can empty their contents into the synaptic clefts (Fig. 7.4). In fact, Heuser and Reese (1973) have also popularized the idea of a life cycle of vesicles in which they fuse with the surface membrane and later reform and are refilled with ACh for reuse. Heuser and Reese described a population of coated vesicles (Fig. 7.4D) which they feel is in the process of recycling. The coating might contain substances such as mucopolysaccharides which are found on the outside of cell membranes and are taken internally during the recycling of the membrane. The method of freeze cleavage (see Fig. 7.2) has also shown details of the internal structure of synaptic membranes. Regular rows of particles have been observed (Fig. 7.4E) which are thought to be sites for releasing synaptic vesicles. The coated vesicles reform at other, but nearby points in the membrane.

There is also good evidence for the recycling of choline, which results from the breakdown of ACh by an enzyme cholinesterase at the postsynaptic membrane (reviewed by Hubbard, 1973). The reuptake of choline, and to some extent the resynthesis of ACh, can be blocked pharmacologically by hemicholinium ($HCh_3$). The acetyl portion of the molecule is

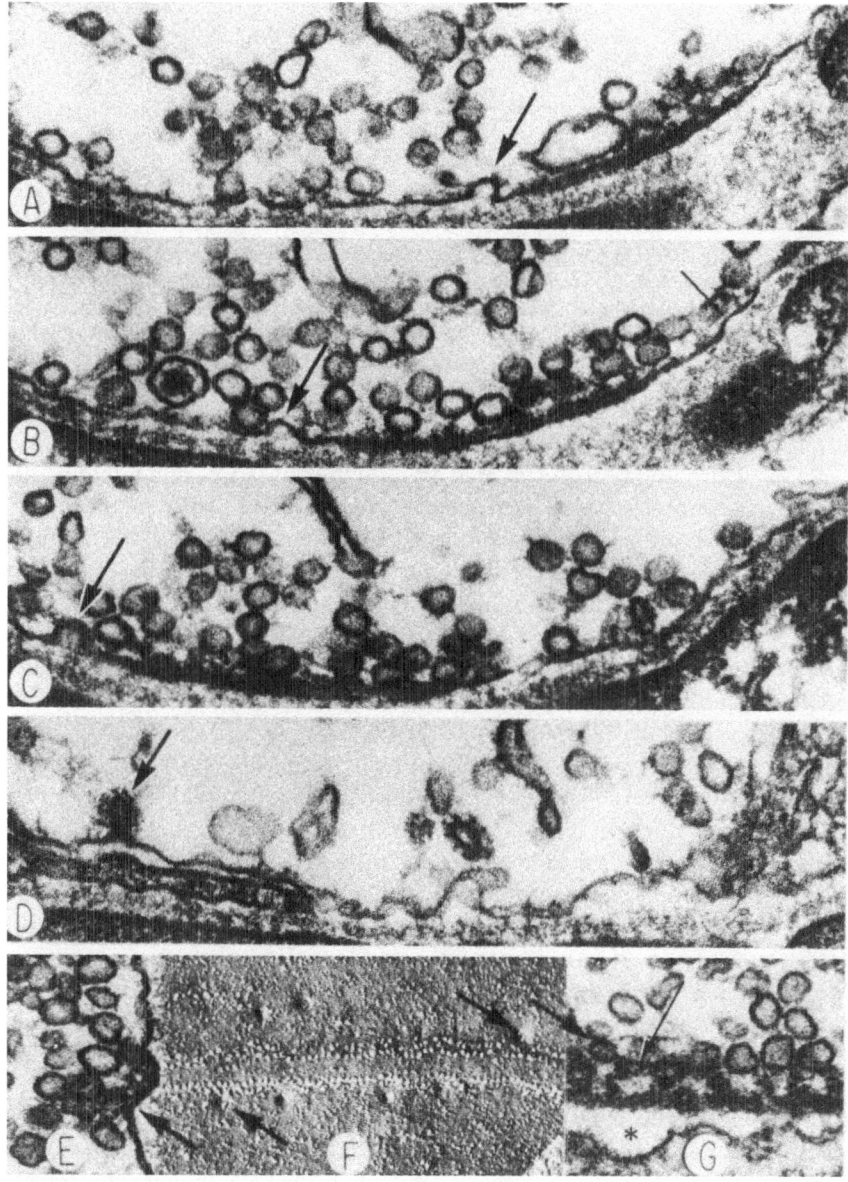

Fig. 7.4. Presumed recycling of synaptic membrane. (A) Vesicle fuses with the membrane at the active zone and opens its contents into the cleft (see arrows). (B) Vesicle becomes absorbed as part of the cell membrane. At a different location (C) membrane separates from the cell membrane and (D) becomes a coated vesicle which can be filled with ACh and reused. Combination of freeze fracture (F) and transmission electron micrographs (E and G) in a different plane of section to show an active zone with particles and vesicles in the process of fusion (compare with Fig. 7.2). (From Heuser *et al.*, 1974.)

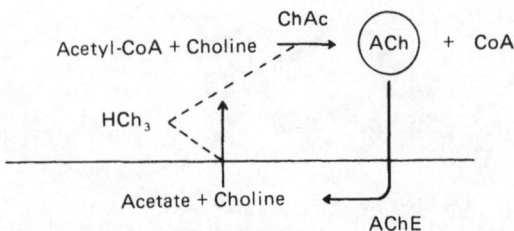

Fig. 7.5. Simplified view of the cycling of ACh at the neuromuscular junction. ACh is formed in a reaction between Ac-CoA and choline which is catalyzed by the enzyme ChAc. The ACh is packaged in vesicles and released, whereupon it is broken down by a second enzyme, AChE. The choline portion can be taken back into the presynaptic cell and reused. The reuptake and also the synthesis of ACh can be blocked by the drug $HCh_3$.

found in combination with coenzyme A (acetyl-CoA in Fig. 7.5). The reaction with choline to form ACh is catalyzed by an enzyme choline-o-acetyl transferase (ChAc). The acetyl-CoA is located between the inner and outer membranes of mitochondria, which are found in relatively large numbers at synaptic terminals and are the site of many processes requiring energy. Thus choline and the enzyme ChAc are found in the cytoplasm. ATP is also found in association with synaptic vesicles, and the incorporation and concentration of ACh in the vesicles may therefore be an active process requiring energy.

The picture that emerges from these studies is of a highly organized and ecologically sound factory for producing, packaging, and recycling synaptic transmitter. The process is decentralized in that the synaptic terminals can operate to some extent independently of the distant cell body. Overall control is under the executive direction of the cell body, which must supply some of the raw materials and energy supplies. For example, the mitochondria, together presumably with the acetyl-CoA, are transported down the axon using the fast axoplasmic transport system. Slow transport and regrowth take place at a rate on the order of 1 mm/day, whereas the fast transport system operates at 400–500 mm/day (Ochs, 1974). Nonetheless, communication to the cell body and back in axons up to a meter in length can take several days, which is the rationale for some degree of decentralization. The fast transport system involves the system of long microtubules. From direct light microscopy of living axons, together with electron microscopy, mitochondria are thought to travel down the microtubules (Stephens and Edds, 1976) propelled in a somewhat jerky fashion by the breakdown of individual molecules of ATP (Cooper and Smith, 1974). The net result is that the synaptic terminals remain well stocked with synaptic vesicles for release except in certain disease states.

One essential factor in the release of transmitter is $Ca^{++}$ ions. Only $Sr^{++}$ can substitute for $Ca^{++}$ ions, and the similar divalent ion $Mg^{++}$ serves to block transmitter release in many instances (Hubbard, 1973). The only apparent exception to this requirement for $Ca^{++}$ ions is the fact that a low level of spontaneous release of transmitter can continue even with no $Ca^{++}$ ions in the bathing medium (Quastel *et al.*, 1971). Some $Ca^{++}$ ions are sequestered internally by mitochondria and other organelles, and the internal $Ca^{++}$ concentration need not fall to zero for a long period of time in a $Ca^{++}$-free medium. Whether this internal $Ca^{++}$ is sufficient to maintain a low level of spontaneous release, or whether other ions can substitute to some extent for $Ca^{++}$ in spontaneous release remains uncertain.

The amount of transmitter released by a presynaptic action potential or other similar potential change increases steeply as the third or fourth power of external $Ca^{++}$ ions (Dodge and Rahamimoff, 1967; Katz and Miledi, 1970). However, the concept of a cooperative action of several $Ca^{++}$ ions in releasing transmitter has recently been questioned by Llinás (1977). He and his colleagues directly measured Ca currents at the squid giant synapse and found a linear relationship between release and these currents (Fig. 7.6B).

The mechanism of action has been studied at the frog neuromuscular junction and at the squid giant synapse in elegant studies by Katz and Miledi (1967, 1969, 1970, 1971). In the squid giant synapse, direct intracellular recording from the presynaptic terminal indicated that a regenerative inward $Ca^{++}$ current persisted after the $Na^{+}$ current was blocked by TTX and the $K^{+}$ current was blocked by TEA. The release of transmitter depended on the inward movement of $Ca^{++}$ ions, because no transmitter was released as long as the membrane potential was maintained above the equilibrium potential for $Ca^{++}$ ions ($E_{Ca}$). Transmitter was released as soon as the voltage was allowed to return to more negative values than $E_{Ca}$. Transmitter could also be released by direct intracellular injection of $Ca^{++}$ ions, or by application of $Ca^{++}$ ions just outside the terminal with a

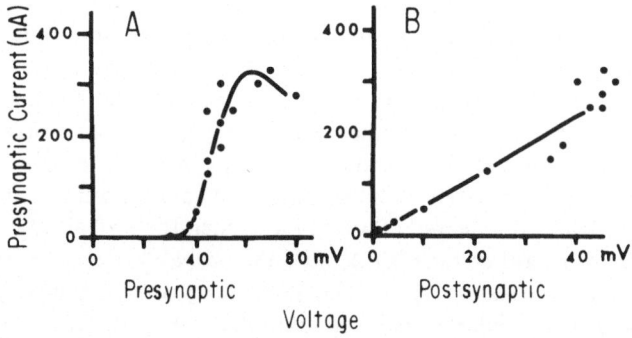

Fig. 7.6. (A) S-shaped curve relating presynaptic $Ca^{++}$ currents to the presynaptic depolarization. (B) Linear relation between the postsynaptic depolarization and the presynaptic current. (From Llinás, 1977.)

microelectrode, as long as the application preceded or was concurrent with the permeability increase due to a presynaptic voltage change. The permeability change has an S-shaped activation curve (Fig. 7.6A) very similar to those shown for $Na^+$ and $K^+$ ions in Chapter 5. Various models have been proposed to describe the reactions of $Ca^{++}$ ions in the release process (Hubbard, 1973; Llinás, 1977), but there is not sufficient evidence to decide unequivocally among them. Nor is it known for certain just how $Ca^{++}$ ions manage to release transmitter. Do $Ca^{++}$ ions simply bind at one side to a vesicle and at the other side to a site on the membrane so that the two membranes are brought close enough so that a spontaneous fusion reaction takes place with consequent release of transmitter? This remains an intriguing question for future study.

## Statistics of Transmitter Release

If one records from a postsynaptic cell intracellularly or with an extracellular microelectrode very close to the synaptic membranes, small potential changes of roughly constant size and time course are observed occurring at "random." Shortly after these *quantal events* were first observed, Del Castillo and Katz (1955) suggested that they represented the release of single vesicles of transmitter. Nearly a quarter of a century later Heuser *et al.* (1979) have confirmed this suggestion using an elegant technique in which they were able to freeze a muscle within milliseconds after stimulating it. Quick freezing was accomplished by contact with a metal plate cooled by liquid helium. They also varied the concentration of a drug, 4-aminopyridine, which greatly increases the amount of transmitter release. In these experiments, Heuser *et al.* (1979) were able to observe a one-to-one correspondence between the estimated number of vesicles opening into the synaptic cleft, as measured with the electron microscope, and the number of quanta released, as measured electrophysiologically. Thus, there is now strong evidence that the basic unit or quantum of synaptic action represents the release of transmitter stored in a single synaptic vesicle.

The hypothesis for the random release of vesicles leads to a number of predictions. Assume that the release of one vesicle does not affect the release of other vesicles. This assumption may not be quite right in certain instances, but is associated with the notion of random release mentioned above. If vesicles are released at a rate $r\ s^{-1}$, then the probability of release in some small instant of time is $r\,\Delta t$, and the probability of a vesicle failing to be released will be $1 - r\,\Delta t$. The probability that the next vesicle is released at a time $t$, where $t$ is some multiple, $m + 1$, of $\Delta t$, is

$$p(t) = (1 - r\,\Delta t)^m r\,\Delta t \tag{7.1}$$

i.e., there must be failures in the next $m$ intervals followed by a success.

The probability density function for the time to the next release of a vesicle of transmitter, $f_1(t)$, is conventionally defined in the limit of $\Delta t \to 0$ (and $m \to \infty$) as

$$f_1(t) = \lim_{\Delta t \to 0} \frac{p(t)}{\Delta t} = r\,e^{-rt} \tag{7.2}$$

Similarly, the probability density for the time to release of the second vesicle, $f_2(t)$, can be found since the release of two vesicles must be accompanied by the release of one vesicle at an intermediate time $u$, with a probability $f_1(u)$, followed by the release of a second vesicle after a time $t - u$, with probability $f_1(t - u)$. Considering all possible values of $u$, we have that

$$f_2(t) = \int_0^t f_1(t - u)f_1(u)\,du \tag{7.3}$$

Finally, the probability density function for the time to release of the $k$th vesicle will be

$$f_k(t) = \int_0^t f_1(t - u)f_{k-1}(u)\,du \tag{7.4}$$

Equations (7.3) and (7.4) are called *convolution integrals* and are easily solved using the method of Laplace transforms introduced in the last chapter. It is a well-known property of Laplace transforms (e.g., Sokolnikoff and Redheffer, 1958) that the Laplace transform of a convolution integral such as Equation (7.3) is given by

$$f_2^*(s) = f_1^*(s)f_1^*(s) = \left[ f_1^*(s) \right]^2 \tag{7.5}$$

where $f_k^*(s) = \int_0^\infty e^{-st}f_k(t)\,dt$ is the Laplace transform of $f_k(t)$. Also from Equation (7.4)

$$f_k^*(s) = f_1^*(s)f_{k-1}^*(s) = \left[ f_1^*(s) \right]^k \tag{7.6}$$

The Laplace transform of the exponential function in Equation (7.2) is $r/(s + r)$, so

$$f_k^*(s) = r^k/(s + r)^k \tag{7.7}$$

Finally, the inverse transform of Equation (7.7) can be found in a table of Laplace transforms (Selby, 1975) in order to solve for the probability density function, $f_k(t)$

$$f_k(t) = \frac{r^k t^{k-1} e^{-rt}}{(k - 1)!} \tag{7.8}$$

where $k!$, read as $k$ factorial, is $k(k - 1)(k - 2) \cdots (1)$. Equation (7.8) is known as the *gamma distribution* and gamma distributions for different values of $k$ are shown in Fig. 7.7. The gamma function with $k = 1$ is an exponential whereas the gamma functions for increasing values of $n$ become increasingly symmetrical and approach a normal or gaussian

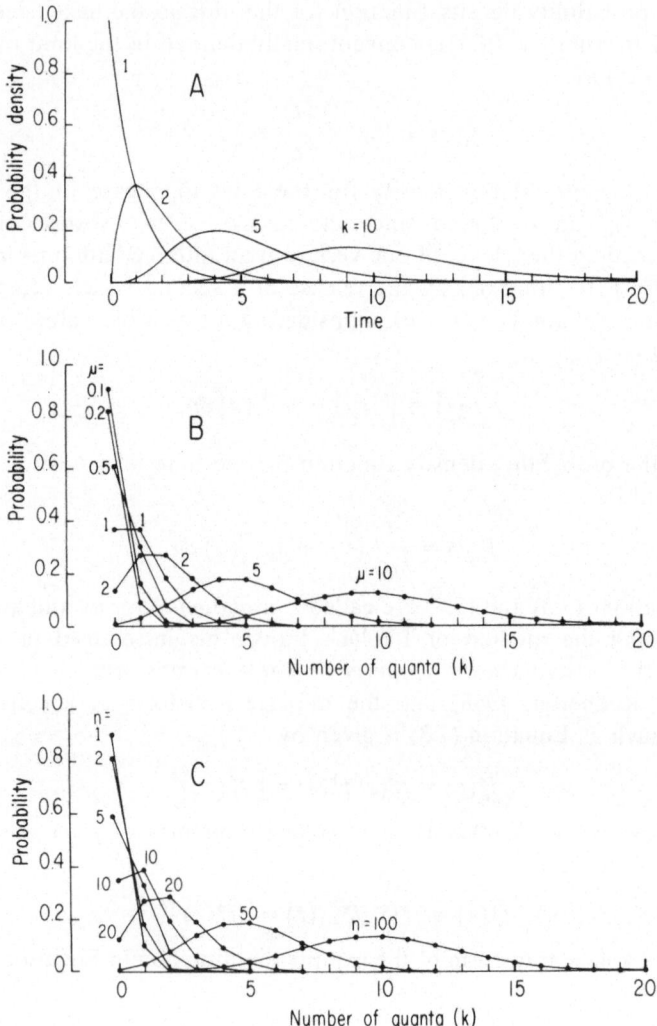

Fig. 7.7. (A) The gamma distribution [Equation (7.8)] gives the probability density that the
$k$th vesicle will be released at a time $t$, given that the mean rate is $r$. The distribution is a
continuous function of time which varies from an exponential shape ($k = 1$) to a normal or
bell-shaped distribution for large $k$. In the examples shown, $r = 1$ and $k$ varies from 1 to 10.
(B) The Poisson distribution (Equation 7.9) gives the probability of $k$ quanta being released
over a period when the mean number is $\mu$. Values of $\mu$ from 0.1 to 10 are shown and lines
have been used to join the points to avoid confusion. (C) The binomial distribution (Equation
7.10) gives the probability that $k$ vesicles will be released, where the probability is $p$ for release
at each of $n$ sites. Note that for large $n$, the binomial distribution approaches the Poisson
distribution (Problem 7.2) and both discrete distributions have the same shape as the
continuous gamma distribution. In this example, $p = 0.1$ and $n$ varies from 1 to 100. The
mean number released is $\mu = pn$.

distribution. Fatt and Katz (1952) confirmed that the times between successive quanta were exponentially distributed. However, Cox and Lewis (1966) noted some deviations from predictions when longer times were considered over which larger numbers of quanta were likely to occur (see also Hubbard, 1973).

One further property of the gamma function may be helpful. The probability that the $k$th vesicle is released in a short time interval $\Delta t$ is $f_k(t)\,\Delta t$ [see Equation (7.2)]. This probability can be thought of as the product of two probabilities, the probability that $k - 1$ quanta had previously been released in time $t$, which will be called $g(k - 1)$, and the probability that one further quantum will be released in the time $\Delta t$, which is given by Equation (7.1). Then, from Equation (7.8), we can find $g(k - 1)$ or $g(k)$, which is

$$g(k) = \frac{\mu^k e^{-\mu}}{k!} \tag{7.9}$$

where $\mu = rt$ is the mean number of vesicles that is released in time $t$. This function is the Poisson distribution, which is also well known in statistics (Fisz, 1963). Boyd and Martin (1956) verified that the number of quanta released in a short period of time following an action potential agreed with the Poisson distribution, although later workers have found some deviations (e.g., Wernig, 1975).

One reason for these deviations is that the number of vesicles available for release, or the number of release sites, is limited. Let $n$ be the number of readily releasable vesicles or the number of release sites (the mathematics does not distinguish between these two possibilities). Also, let $p$ be the probability of a vesicle being released from any one site (and $1 - p$ be the probability that no vesicle is released). Then the probability of $k$ vesicles being released requires that there be $k$ successful releases with probability $p^k$ and $n - k$ failures with probability $(1 - p)^{n-k}$. Various combinations of vesicles could be released to produce the number $k$, and the number of ways of arranging $k$ successes and $n - k$ failures is well known from probability theory (Feller, 1957) to be $n!/[k!(n - k)!]$. Thus,

$$g_n(k) = \frac{n!}{k!(n - k)!}\, p^k (1 - p)^{n-k} \tag{7.10}$$

which is the binomial distribution. The mean of this distribution $\mu$ is the product of the probability $p$ and the number $n$. The binomial distribution for various values of $n$ is shown in Fig. 7.7. Note that for large values of $n$ the binomial distribution approaches the Poisson distribution of Equation (7.9) (Problem 7.2) and the distribution becomes quite insensitive to the value of $n$. Thus, it is difficult experimentally to distinguish changes in the probability $p$ of release from changes in the number of $n$ of vesicles available for release.

One method to distinguish changes in $p$ and $n$ uses the known values for the variances of these distributions. The value of the variance $\sigma^2$ is by definition $\sigma^2 = \sum_{k=0}^{n} (k - \mu)^2 g(k)$, where $\mu$ is the mean. The variance of the Poisson distribution is simply

$$\sigma^2 = \mu \qquad (7.11)$$

while that of the binomial distribution is

$$\sigma^2 = np(1 - p) = \mu(1 - p) \qquad (7.12)$$

Note that the variance of the Poisson distribution is identical to its mean, whereas the variance of the binomial distribution varies slightly from it. In the limit of large $n$ (and small $p$), the difference is small.

## Facilitation and Depression of Transmitter Release

The rapid release of transmitter during a nerve impulse is thought to be due to a large increase in the probability of release (Liley, 1956) as a result of the inflow of $Ca^{++}$ ions. Furthermore, one action potential can facilitate the release of transmitter by a subsequent action potential. At least the early stages of this *facilitation* are due to an increase in the probability of release (Mallart and Martin, 1968). The increased probability could result from the slow decay of intracellular $Ca^{++}$ levels from one impulse, and summation with the $Ca^{++}$ inflow from the second impulse. Direct measurements of intracellular $Ca^{++}$ levels following single impulses have been made by Llinás and Nicholson (1975) using the Ca-sensitive, light-emitting protein, aequorin. Facilitation is, however, a complex process which has been modeled by the sum of several exponentials (Mallart and Martin, 1967) with varying time constants from tens to hundreds of milliseconds (see Fig. 7.8).

If a substantial amount of transmitter is released, further release of transmitter may be depressed. This *depression* appears to be due, at least in part (Betz, 1970), to a decrease in the number of releasable vesicles. This has been correlated in some cases with electron micrographs which show a decrease in the number of vesicles adjacent to the synaptic membrane (Hubbard, 1973). This does not prove that the number $n$ represents the number of vesicles close to the synaptic membrane. Instead, it may represent the number of release sites, which take some time to reload after releasing a vesicle.

Finally, if one stimulates a presynaptic fiber at high frequency for a prolonged period under conditions (e.g., high $Mg^{++}$ concentration) where small amounts of transmitter are released per stimulus, the amount of transmitter released is greatly increased. This increase differs from facilitation in that effects of past stimulation affect not only the magnitude, but

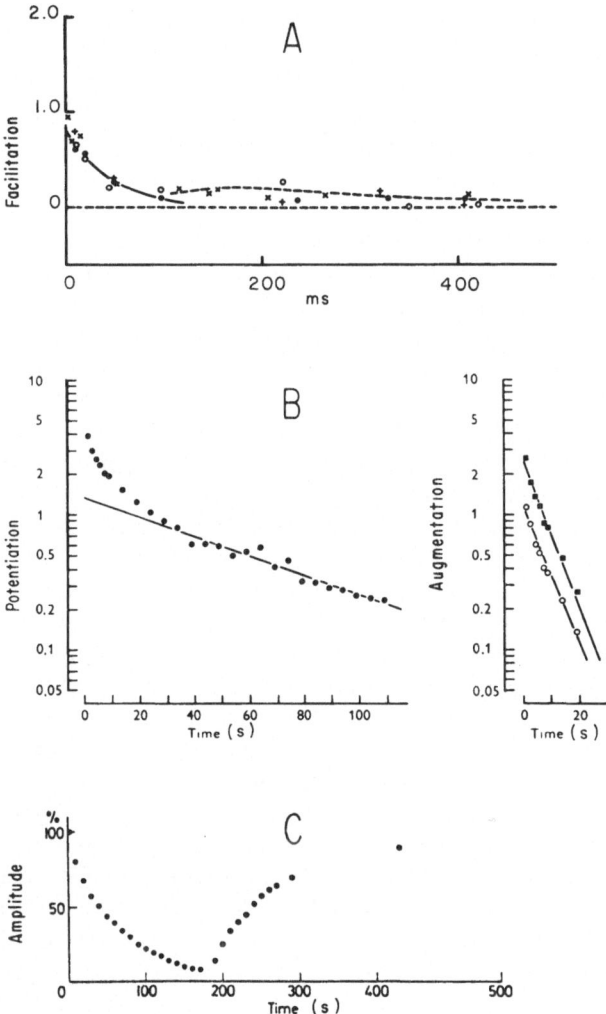

Fig. 7.8. (A) Release of transmitter by a second impulse (relative to that released by the first impulse) is facilitated with an exponential decay of about 45 ms. A later phase of facilitation begins after a delay and decays with a slower time course. (From Mallart and Martin, 1967.) (B) After 300 conditioning impulses at 20/s, release is greatly increased and decays with a much slower time course, which can be subdivided into two exponential processes referred to as augmentation (time constant 7–8 s) and potentiation (time constant about 1 min). The time constant for augmentation does not depend greatly on whether an additive (filled symbols) or multiplicative (open symbols) relation is assumed between augmentation and potentiation. (From Magleby and Zengel, 1976.) (C) Depression of end-plate potentials during and after stimulation at a rate of 20/s for 180 s. The depression and recovery can be fitted with simple exponentials with time constants of the order of a minute. (From Lass *et al.*, 1973.) The experiments in (B) above were carried out at low $Ca^{++}$ and high $Mg^{++}$ concentrations to reduce the depression in transmitter release seen in (C) to very small amounts.

also the time course of the changes (Magleby and Zengel, 1975). The time course of these changes is also much slower in general than for facilitation and has been divided into two phases: *augmentation*, with a time constant of the order of seconds, and *potentiation*, with a time constant of tens of seconds up to minutes depending on the past history of stimulation (Magleby and Zengel, 1976). $Ca^{++}$ ions or some related substance seems to be involved in these slow changes (Rosenthal, 1969).

Associated with these increases in stimulus-evoked release are increases in the rate of spontaneous release of quanta. These increases also have several phases and somewhat different requirements for $Ca^{++}$ ions (Cooke and Quastel, 1973). The long-term potentiation of evoked and spontaneous release during and after a prolonged period of stimulus might represent increases in the probability $p$ and/or the number $n$ of quanta available. Extreme positions have been published claiming either that all the increases are due to increased $p$ (Rosenthal, 1969) or that they all result from increased $n$ (Birks, 1977).

The second possibility (increased $n$) is often referred to as *mobilization* of transmitter. Whereas statistical estimates for the number of vesicles available for release are generally in the range of a few hundred, Heuser *et al.* (1979) have shown that a single stimulus can release at least 5000 quanta in the presence of 4-aminopyridine. Thus, more than one population of vesicles may exist and in the absence of drugs or other methods of mobilizing transmitter, only a relatively small fraction of the vesicles are readily available for release. Hopefully a clearer picture of the mechanisms underlying these complex events will emerge in the next few years. The long duration of potentiation has also suggested that it might serve as a model for the prolonged electrical changes underlying learning and short-term memory (Eccles and McIntyre, 1953). This possibility remains speculative, and will not be considered here. Kandel (1976) has reviewed in some detail the evidence he and others have accumulated on the cellular basis of learning.

## Presynaptic Inhibition

Another way in which transmission is modified at the presynaptic level is by axo–axonic connections. These are not found in the vertebrate neuromuscular system, but are found in invertebrate systems (Dudel and Kuffler, 1961) and are common in the early synaptic relays for sensory information (Eccles, 1964). These connections are useful in regulating the transmission of information to higher centers. A controversial theory of pain (Melzack and Wall, 1965) is based on the concept that different groups of fibers can facilitate or inhibit the transmission of information that higher centers interpret as pain.

Presynaptic inhibition appears to be more common than presynaptic facilitation and may be responsible in part for the ability to attend to one sensory modality while ignoring other sensory information. Thus, one can attend to visual information concerning a pretty girl while ignoring auditory information from the ventilating system in a building or cutaneous information about the weight of one's clothing. The ionic mechanisms of presynaptic inhibition were worked out in the early 1960s (reviewed by Eccles, 1964) and this was the first type of more complex synaptic interaction (i.e., involving more than two cells) to be studied.

Activity in one axon, which will be referred to as the *presynaptic inhibitor*, depolarizes a second presynaptic fiber by increasing its permeability, but to a subthreshold level. No action potential is produced by the presynaptic inhibitor in the second fiber, but conduction of impulses in this fiber will be affected in two ways: (1) $Na^+$ currents near the synaptic junction of the presynaptic fiber will be partially inactivated by the depolarization, and (2) $Na^+$ currents generated by an invading action potential will be shunted by the increased conductance that the presynaptic inhibitor generates. These two effects will reduce the height of the action potential in the presynaptic junction (i.e., it will start from a depolarized level and probably approach the $Na^+$ equilibrium potential less closely) and so reduce the release of transmitter. Thus, the excitation produced in the postsynaptic cell will be reduced and transmission will be inhibited. The amino acid gamma-amino-butyric acid, or GABA, has been implicated as a presynaptic inhibitory transmitter in the mammalian spinal cord (reviewed by Krnjević, 1974). Recently, other types of complex synaptic interactions have been described such as reciprocal synapses, triads, etc. (Shepherd, 1979), but these will not be considered here. Instead, the changes taking place on the other or postsynaptic side of the synaptic cleft will be considered.

## Postsynaptic Mechanisms

Following stimulation of presynaptic fibers, permeability changes begin to take place on the postsynaptic side of the synaptic cleft after a short delay of 0.3–0.5 ms. This delay is characteristic of chemical synapses and is absent in electrical synapses. The delay is highly temperature-dependent and is unlikely to arise solely from the diffusion of transmitter across the short gap (20 nm) separating the two sides of the synapse. Much of the delay probably arises from presynaptic mechanisms such as the permeability changes to $Ca^{++}$, the inflow of $Ca^{++}$, and the binding of $Ca^{++}$ needed to release vesicles (Llinás, 1977).

The postsynaptic membrane contains large numbers of receptor molecules specialized to interact with the synaptic transmitter. In an extreme example such as the electric organ of the torpedo, the ACh receptor molecules may be almost as densely packed as in a gap junction (Fig. 7.1B). The channels for ion passage (Fig. 7.1) also appear similar under the electron microscope except that the receptor molecules seem to have five rather than six subunits, and the channel is about 0.8 nm in diameter, rather than 2 nm for the gap junction (see Stevens, 1979). The receptor has been characterized as a molecular protein with a molecular weight of at least 250,000, and its subunit structure and amino acid composition is being studied (Vandlen *et al.*, 1979).

A toxin has also been isolated from snake venom known as α-*bungarotoxin* which binds very specifically and almost irreversibly with the ACh receptors. From the number of molecules bound, the estimated number of receptors at the neuromuscular junction is greater than $10^4/\mu m^2$. This is at least an order of magnitude larger than the number of Hodgkin–Huxley channels in the membrane (Chapter 5) and the number of sodium pump sites (Chapter 3). Furthermore, this high density of ACh receptors is very localized in normal tissue (but see the section *Modified Interactions between Nerve and Muscle* below). Kuffler and Yoshikami (1975a) applied small amounts of ACh from a micropipette at various distances from a synaptic site which was directly visualized with a high-power microscope and a type of phase contrast attachment known as Nomarski optics. They found that even 10 $\mu m$ from the rim of the synaptic contact, the sensitivity to ACh was diminished to one-hundredth of its value directly under the synapse.

At the skeletal neuromuscular junction, ACh increases the permeability or the conductance of the membrane to small cations such as $Na^+$ and $K^+$. The current $I_s$ that flows across the postsynaptic membrane often obeys a simple linear relation; i.e.,

$$I_s = G_s(V - E_s) \tag{7.13}$$

where $G_s$ is the conductance change produced by the transmitter and $E_s$ is its equilibrium potential. In the case of ACh at the neuromuscular junction, the equilibrium potential is about $-10$ to $-20$ mV (i.e., somewhat closer to $E_{Na}$ than $E_K$) because the permeability to $Na^+$ is somewhat greater than the permeability to $K^+$ (Del Castillo and Katz, 1956). However, the same transmitter ACh is the inhibitory transmitter to the heart. In that tissue, the permeability increase is almost exclusively to $K^+$ ions, so $E_s$ is near $E_K$ (Hutter, 1961). In general, *a transmitter at a postsynaptic membrane is excitatory if it has an equilibrium potential greater than the threshold for generating a nerve impulse. An inhibitory transmitter*, on the other hand, *has an equilibrium potential less than threshold for a nerve impulse.* Note that an inhibitory transmitter can have an equilibrium potential more positive than the resting potential and can actually produce a depolarization. However, it tends to stabilize the membrane at a sub-threshold level and therefore prevents the generation of impulses.

Common postsynaptic inhibitory transmitters include the amino acids glycine and GABA. These substances have been implicated as the inhibitory transmitters utilized, for example, at invertebrate neuromuscular junctions (GABA; Boistel and Fatt, 1958), in the spinal cord of vertebrates (glycine; Werman *et al.*, 1968), and for Purkinje cells of the cerebellum (GABA; Obata *et al*, 1970). In many cases they act largely by increasing the permeability to $Cl^-$ ions whose equilibrium potential is close to the resting membrane potential. Whether excitatory or inhibitory, the effect of postsynaptic transmitters is often to increase the permeability of the membrane and "clamp" the membrane at the net equilibrium potential of the ions involved. Linearly increasing the conductance change produced by a transmitter will produce less and less extra change in voltage, as the synaptic equilibrium potential is approached. However, as shown in Fig. 7.9, these conductance increases will minimize the deviations produced by mechanisms such as action potentials which cause permeability changes to $Na^+$ and $K^+$ ions. The relevant equations are similar in form to the rectangular hyperbola of enzyme kinetics (Chapter 3) and are considered in Problem 7.3.

In principle, an excitatory transmitter could produce its effect by *decreasing* the conductance to ions, such as $K^+$, with equilibrium potentials more negative than threshold, as well as by *increasing* the conductance to ions, such as $Na^+$, with equilibrium potentials more positive than threshold. An excitatory effect produced by a decrease in conductance is the action of *adrenaline* (also called epinephrine) on the heart (Hauswirth *et al.*, 1968). Fig. 7.9 shows the action of adrenaline on the activation curve for a variable $s$ regulating postassium conductance in the pacemaker region of the cardiac action potential. This effect is reminiscent of the action of $Ca^{++}$ ions as a result of binding to the membrane. However, the effect of adrenaline is specific to one type of $K^+$ channel, whereas $Ca^{++}$ appears to affect all channels. Thus, during the pacemaker potential, the $Na^+$ currents are unchanged while the $K^+$ currents at any given voltage are reduced. As a result, the membrane depolarizes more rapidly and the heart rate increases. There are other changes in the plateau potential and in the contractility of the heart due to other actions of adrenaline on the permeability to $Ca^{++}$ ions.

These will not be considered here, but are dealt with in detail elsewhere (Noble, 1975). Tsien *et al* (1972) suggested that these effects on the heart require not only adrenaline as a transmitter or messenger, but also a *second messenger* inside the cell, *cyclic AMP*. This is of considerable interest because cyclic AMP has been shown to be a second messenger in many hormone systems (Robinson *et al.*, 1971). How adrenaline "passes on

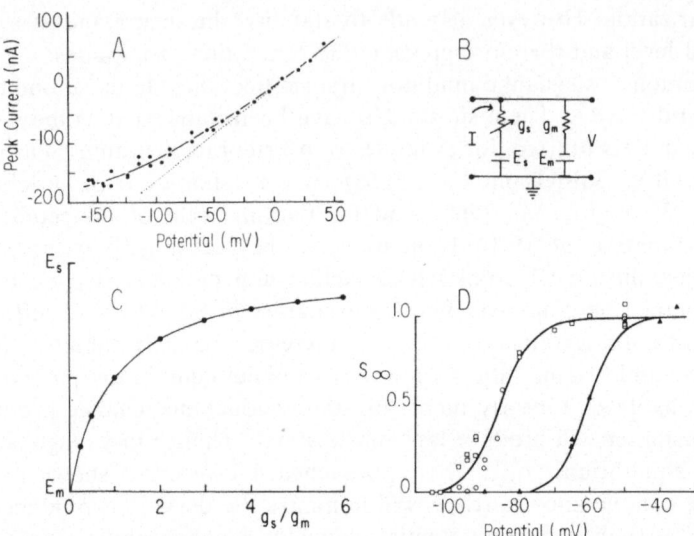

Fig. 7.9. (A) Peak postsynaptic currents generated by nerve stimulation at a voltage-clamped neuromuscular junction. The reversal or equilibrium potential ($E_s$) for the synaptic currents is discussed in the text. (From Magleby and Stevens, 1972b.) (B) Schematic diagram of a patch of synaptic membrane containing a membrane conductance $g_m$ and reversal potential $E_s$. If $g_s$ depends on the concentration of transmitter and not on voltage $V$, a linear current–voltage ($I$–$V$) relationship such as shown by the dashed line in part (A) would be expected. The conductance $g_m$ can, of course, contain $Na^+$ and $K^+$ action potential pores and so depend on voltage, but not on transmitter concentration. (C) As the synaptic conductance is increased, the membrane potential increases linearly at first, but then approaches $E_s$ asymptotically according to the equation given in Problem 7.3. (D) One effect of the transmitter adrenaline on the heart is to shift the steady state curve for the Hodgkin–Huxley variable $s$ regulating $K^+$ conductance to the right (▲) over the pacemaker range of potentials. Thus, there is less potassium current at any voltage over this range and the heart depolarizes more quickly. The action is reversed by the specific $\beta$-blocker of adrenaline action, pronethalol (□). (From Noble, 1975.)

the message" to cyclic AMP, and how this important compound can produce the required permeability changes, is still unknown.

There is a growing list of compounds for which there is strong evidence of their action as synaptic transmitters. In addition to ACh, which was mentioned in detail above in regard to vertebrate neuromuscular transmission, there are whole families of synaptic transmitters. Adrenaline is one of the monoamines, many of which including noradrenaline, dopamine, and 5-hydroxytryptamine (5-HT) function as synaptic transmitters. All but 5-HT are also referred to as catecholamines (Blaschko and Muscholl, 1972). Glycine and GABA, which were mentioned earlier, are amino acids (or their immediate derivatives) which serve as inhibitory transmitters (Davidson, 1976). Other amino acids such as glutamic acid serve as excitatory transmitters. Small polypeptides also serve as transmitters (Gainer, 1977), and among these the role of the endorphins (naturally occurring morphine-like substances) have recently attracted a lot of attention in regard to the regulation of painful sensations. Understanding the role of transmitters in the brain has led to marked advances in treating diseases such as Parkinson's disease (Hornykiewicz, 1966) and schizophrenia (Snyder *et al.*, 1974).

Some pharmacological agents such as LSD, which specifically blocks the action of 5-HT, have featured prominently in the illicit drug field. The improper use of drugs which produce "highs" by knocking out the "lows" (synaptic inhibition) are all too familiar. Whole books have been written on many of these transmitters, and they will not be considered individually here. However, it is worth noting that most nerve cells have many synaptic terminals, but according to Dale's principle (see page 89) each cell releases only one of these transmitters at all of its terminals. Kandel *et al.* (1967) described a cell in the mollusc *Aplysia* that excited one cell and inhibited another while releasing ACh at both. This shows even more clearly than the example with skeletal muscle and cardiac muscle that the transmitter serves like a key opening a door or channel through the membrane. It can open many doors, but does not specify the size and shape of the door (i.e., what ions or other molecules can pass through).

## Synaptic Channels

Despite their varying properties, synaptic channels through the membrane have some common features which distinguish them from the $Na^+$ and $K^+$ channels described by Hodgkin and Huxley. The $Na^+$ and $K^+$ channels, which were discussed in Chapter 5, will be referred to as action potential channels to distinguish them from synaptic channels. One important difference is that the conductance change $G_s$ is not voltage-dependent in the way $G_{Na}$ and $G_K$ are. Even though Equation (7.13) is identical in form to Equation (5.5), the conductance increase is produced by a synaptic transmitter, not a depolarization, and its magnitude does not vary with depolarization. However, by voltage clamping the neuromuscular junction, Magleby and Stevens (1972a) showed that the time course of the conductance change does vary according to an equation of the form

$$G_s = k\, e^{-\alpha t} \tag{7.14}$$

where

$$\alpha = b\, e^{aV} \tag{7.15}$$

and *a*, *b*, and *k* are constants. These equations state that the synaptic currents decay exponentially following the application of ACh and that the decay rate increases exponentially with voltage.

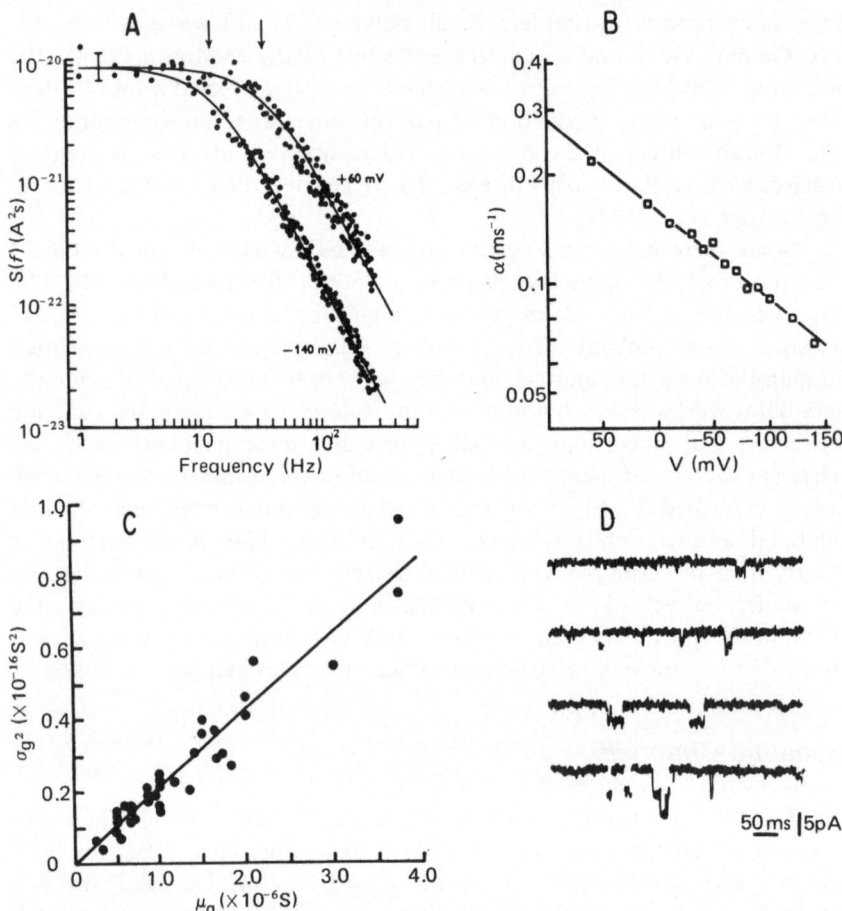

Fig. 7.10. (A) Power spectrum $S(f)$ for the current fluctuations of a frog neuromuscular junction as a function of frequency when the membrane was clamped at two different voltages ($+60$ mV and $-140$ mV). The values for $+60$ mV have been shifted vertically for comparison with those at $-140$ mV. Note the difference in the frequencies (arrows) at which the power falls to half the low frequency values. (B) The half-power points in (A) permit the rate constant $\alpha$ for the closing of ACh channels to be calculated for different membrane potentials. The rate constant increases with voltage according to Equation (7.15). (C) Variance ($\sigma_g^2$) of the conductance fluctuations as a function of the mean conductance ($\mu_g$) produced by applying different levels of ACh. The slope gives the conductance of single ACh channels according to Equation (7.31). Parts (A) to (C) are from Anderson and Stevens (1973). (D) Discrete changes in membrane current recorded with focal extracellular electrodes from denervated frog skeletal muscle in the presence of suberyldicholine (an ACh-like substance) at different membrane potentials ($-60$, $-80$, $-110$, and $-150$ mV). These changes represent the opening and closing of single channels, a large number of which superimpose to produce the membrane noise shown in (A) to (C) above. The amplitude of the current steps varies with voltage, as expected from Equation (7.13). Successive traces show increasing degrees of hyperpolarization. (From Neher and Stevens, 1977.)

The exponential decay of the current does not necessarily imply that the conductance decreases exponentially in each channel. Neher and Sakmann (1976) found evidence in frog muscle fibers only for an open and a closed state with an exponential distribution for the times each channel remained open (Fig. 7.10D). We saw earlier in discussing vesicle release that an exponential distribution could arise if the "lifetime" of an open pore was a random or Poisson process. In fact, these have been well studied by statisticians under the title of *birth and death processes* (see, for example, Cox and Miller, 1965). Thanks to modern medicine, the life span of humans is not a Poisson process. If we get past the first year of life, we have a reasonable chance of living our biblically allotted three score and ten years. However, in more primitive societies where disease can strike at any time, lifetimes are distributed more closely according to a Poisson process.

In order to get more information about the molecular events underlying the behavior of single channels, Katz and Miledi (1972) turned to analyzing the power spectrum of the *noise* generated by application of ACh from a microelectrode at synaptic sites. They reasoned that the opening of pores would occur randomly according to a Poisson process, but that the duration of the open states would serve to "filter" the voltage fluctuations that occurred. Despite the fact that the individual events were too small to detect in their experiments (but see Fig. 7.10D), some properties could be measured indirectly through the overall voltage or current fluctuations. This process is familiar to physicists who have long studied the kinetic properties of gas molecules by measuring macroscopic properties such as pressure and temperature.

The power spectrum of a signal is a function of frequency and can be obtained by measuring over a period of time $T$ the amplitudes of the components of the signal $x(t)$ which are in phase with a sine or cosine wave of that frequency (Bendat and Piersol, 1971). Thus,

$$a(\omega) = \frac{1}{T} \int_0^T \sin(\omega t) x(t)\, dt \tag{7.16}$$

$$b(\omega) = \frac{1}{T} \int_0^T \cos(\omega t) x(t)\, dt \tag{7.17}$$

$a(\omega)$ and $b(\omega)$ are referred to as the Fourier sine and cosine transforms. Both the sine transform $a(\omega)$ and the cosine transform $b(\omega)$ can be obtained simultaneously using an exponential function $e^{-j\omega t}$ since there is a well-known relation between the exponential function and the harmonic functions (Selby, 1975), namely

$$e^{-j\omega t} = \cos(\omega t) - j \sin(\omega t) \tag{7.18}$$

where $j = (-1)^{1/2}$. Thus,

$$\frac{1}{T} \int_0^T e^{-j\omega t} x(t)\, dt = b(\omega) - ja(\omega) \tag{7.19}$$

Note also that there is a close relation between the Laplace transform used previously and the exponential Fourier transform above. Formally,

$$x^*(j\omega) = b(\omega) - ja(\omega) \tag{7.20}$$

i.e., if the Laplace transform is known, the sine and cosine components can be determined by substituting $j\omega$ for the Laplace variable $s$ and determining the real and imaginary parts. The power spectrum $S(\omega)$ is a real function equal to the magnitude of $x^*(j\omega)$ *squared*, that is

$$S(\omega) = a^2(\omega) + b^2(\omega) \tag{7.21}$$

The power spectrum has the units of $V^2$ or may be expressed as a spectral density in $V^2/Hz$, which gives the amount of power per unit of frequency.

To illustrate the application of these methods to the problem of single channels in a postsynaptic membrane, consider the simple exponential function of Equation (7.14). If $x(t) = ke^{-\alpha t}$, then from our previous discussion of Laplace transforms,

$$x^*(s) = \frac{k}{s + \alpha} \tag{7.22}$$

and

$$x^*(j\omega) = \frac{k}{j\omega + \alpha} = \frac{k(\alpha - j\omega)}{\omega^2 + \alpha^2} \tag{7.23}$$

The second expression on the right is obtained by multiplying the top and bottom by a factor of $\alpha - j\omega$. Comparing Equations (7.20) and (7.23),

$$a(\omega) = \frac{k\alpha}{\omega^2 + \alpha^2} \ ; \qquad b(\omega) = \frac{k\omega}{\omega^2 + \alpha^2} \tag{7.24}$$

and

$$S(\omega) = \frac{k^2}{\omega^2 + \alpha^2} \tag{7.25}$$

Figure 7.10A shows a measured spectrum of the current generated at a neuromuscular junction of the frog under voltage clamp conditions at two different voltages when ACh was applied from a microelectrode. Note that the power in the current fluctuations is quite constant at all frequencies up to about 10 Hz, but then falls off sharply with increasing frequency. The fitted curves are of the form given by Equation (7.25). In this equation the power will fall to half at the frequency where $\omega = \alpha$ (points indicated by arrows in Fig. 7.10). From this figure, $\alpha = 75\ \text{s}^{-1}$ at $-140\ \text{mV}$ and $214\ \text{s}^{-1}$ at $+60\ \text{mV}$ ($\alpha$ and $\omega$ are measured in radians/s, which is $2\pi$ times the frequency in Hz). The time constant or lifetime of an open channel is the inverse of this value, 13.3 ms at $-140$ mV and 4.7 ms at $+60$ mV. Anderson and Stevens (1973) conducted the experiment shown at 8°C and the time constant was even briefer (on the order of 1 ms) at room temperature. Thus, the action of a single molecule of ACh at a single postsynaptic site is very brief, but still depends on membrane potential according to Equation (7.15) (see Fig. 7.10B). Anderson and Stevens (1973) interpreted this voltage dependence as indicating that the closing of the gate involves movement of a charged dipole. A positive membrane potential facilitates movement of this gate.

Katz and Miledi (1972) did further studies using pharmacological agents to determine more of the properties of the individual channels. When they applied the drug *curare*, they found that the value of $k$ was reduced but $\alpha$ remained the same. This is consistent with the idea that curare is a *competitive antagonist* of ACh for the postsynaptic sites. By occupying some of the sites, fewer ACh molecules could produce conductance changes, but the lifetime of any that did occur was unchanged. More interesting were the results obtained with *anticholinesterase* drugs such as *prostigmine*. Normally ACh is rapidly hydrolyzed after its action by enzymes known as *cholinesterases*. Anticholinesterases, which block the action of these enzymes, have long been known to prolong the duration of the end-plate potential (i.e., the synaptic potential) at the neuromuscular junction or end-plate (Eccles *et al.*, 1942). However, Katz and Miledi (1972) found that the power spectrum was shifted vertically under these conditions, i.e., $k$ was increased but $\alpha$ was virtually unchanged. This could occur either by prostigmine increasing the conductance of a single channel or by the drug permitting each ACh molecule to act several times.

To distinguish between these two possibilities, further ideas from statistical theory are required to determine the conductance of the single channels. If this conductance is called $\gamma$, and the probability of a channel being open is $p$, then the mean conductance $\mu_1$ of the single channels is simply

$$\mu_1 = p\gamma \tag{7.26}$$

The mean square conductance is similarly

$$\mu_2 = p\gamma^2 \tag{7.27}$$

and the variance of the conductance changes is

$$\sigma^2 = \mu_2 - \mu_1^2 = p\gamma^2(1 - p) \tag{7.28}$$

Neher and Stevens (1977) give a fuller description of these statistical results and references to other literature. If there are $n$ such channels which all behave independently, the total mean conductance $\mu_g$ and its variance $\sigma_g^2$ will simply be

$$\mu_g = np\gamma \tag{7.29}$$

and

$$\sigma_g^2 = np\gamma^2(1 - p) \tag{7.30}$$

The ratio of these two quantities under conditions where $p \ll 1$ (relatively low ACh concentrations) gives the single-channel conductance directly:

$$\sigma_g^2/\mu_g \sim \gamma \tag{7.31}$$

Figure 7.10C shows a plot of $\sigma_g^2$ against $\mu_g$ determined experimentally (Anderson and Stevens, 1973) and the straight line represents a conductance of about $2 \times 10^{-11}$ S = 20 pS, where the siemen (S) is the basic unit of conductance (formerly the mho). Neher and Stevens (1977) compared the conductance of single synaptic channels and action potential channels obtained in various ways. Most estimates for both types of channels were within one order of magnitude of this value (between 2 and 230 pS), which represents a reasonable agreement considering the wide variety of techniques and tissues used.

Knowing the single channel conductance, Anderson and Stevens (1973) were able to estimate that about 1700 ACh molecules from one vesicle bind to postsynaptic sites. This represents between 10 and 20% of the contents of the vesicle and provides a measure of the efficiency of synaptic transmission. This efficiency is normally more than sufficient to generate action potentials postsynaptically in vertebrate twitch muscle fibers. The efficiency is maintained by having a large excess of receptors postsynaptically. Anderson and Stevens (1973) estimated that only between 0.01 and 0.1% of the ACh channels in the membrane were open during the peak of the end-plate current (i.e., $p$ is very small, as assumed). However, if the number of ACh receptors is reduced, as appears to occur due to an immunological reaction in a disease known as *myasthenia gravis*, synaptic transmission becomes tenuous and easily fatigued (Fambrough *et al.*, 1973). Synaptic transmission can be improved by administering anticholinesterase drugs to prolong the activity of ACh with the remaining receptors.

From these studies, a simple picture emerges of the interaction between ACh and single receptors in the postsynaptic membrane. First, the binding of ACh molecules to the receptors appears to be extremely rapid [Equation (7.14) in fact assumes that the conductance increases instantaneously to its maximum value]. Secondly, the factor limiting the rate at which the current decays under normal conditions is a voltage-dependent conformational change in the membrane, which determines the lifetime of an open pore. Thirdly, cholinesterase normally breaks down the molecules of ACh after they have acted a single time at the postsynaptic membrane. However, in the presence of anticholinesterases, the ACh can act at several sites and so prolong the total duration of the end-plate currents.

## Modified Interactions between Nerve and Muscle

The normal pattern of interaction between nerves and muscles can be altered in various ways. The most studied effect results when the motor nerve fibers innervating a muscle are cut (*axotomy*). The changes in the cell body (*chromatolysis*) were described histologically in the last century. A reduction in synaptic connections onto the cell body also occurs (Mendell *et al.*, 1976). If severe enough, the chromatolytic changes can lead to cell death (reviewed by Sunderland, 1972). Other changes, such as the thinning of the axons and the associated reduction in conduction velocity (the relation between fiber diameter and conduction velocity was considered in Chapter 6), are only reversed if the nerve fiber can reinnervate a muscle (Cragg and Thomas, 1961). A signal must be transmitted back across one synapse (the neuromuscular junction) that contact has been made, back along the axon to the cell body, and even back across a severed junction between a presynaptic fiber and a motoneuron. The nature of these signals remains unknown, with the possible exception of transport back along the axon to the cell body which has been extensively studied in recent years (Ochs, 1974).

The loss of synapses from the cell body is so marked that the motoneurons are much more difficult to activate during normal movements (Gordon *et al*, 1980). These changes are not restricted only to motoneurons, and similar changes take place in other nerve cells when their axons are cut. In the autonomic nervous system, cutting the postganglionic fibers leads to such a loss of synapses that the postganglionic cells can often not be activated by a maximal preganglionic stimulus (Purves, 1975), although they do generate action potentials with direct stimulation. In contrast to the neuromuscular system, one of the substances responsible has been clearly identified as the well-studied protein, nerve growth factor (Levi-Montalcini and Angeletti, 1968). Chronic application of nerve growth factor to the cut nerves prevents the synaptic depression (Purves and Njå, 1976).

Interesting changes also take place in the cell deprived of innervation. If its motor axon is cut, a muscle fiber goes through a remarkable series of changes. The high sensitivity to ACh, which was limited to a few $\mu$m from the end-plate, spreads within a few days to cover the entire fiber, and normally remains high until reinnervation takes place. Reinnervation still takes place preferentially at the old end-plate for several months after denervating a muscle, but junctions can be formed at other locations (Bennett and Pettigrew, 1976). The high sensitivity to ACh is obviously related to the signaling system used by the muscle to indicate its receptivity to new innervation. However, in the last few years, a number of ways have been discovered to "fool the system." For example, the high sensitivity to

ACh can occur even if there is physical continuity of the axons and the synapse. This has been produced by (1) placing a cuff around a nerve which contains a local anesthetic to block action potentials (Lømo and Rosenthal, 1972), (2) using colchicine to block axonal transport (Cangiano, 1973; but see also Lømo and Westgaard, 1976), or (3) adding curare in sufficient doses that action potentials are not generated postsynaptically (Berg and Hall, 1975). All these results suggest that the release of ACh in sufficient quantities to produce postsynaptic action potentials is necessary to prevent an increased sensitivity to ACh. This rate of release is only achieved normally by presynaptic action potentials and the supply of ACh needs to be replenished by axonal transport. The small rate of release of single vesicles which occurs spontaneously is not sufficient (Lømo and Rosenthal, 1972).

The development of high chemosensitivity in a muscle fiber requires a massive synthetic effort if the density of receptors at the end-plate region is $10^4/\mu m$, and the size of a muscle fiber can be 100 $\mu m$ in diameter and several cm in length (see Problem 7.4). The onset of chemosensitivity can be delayed by drugs which block protein synthesis (Fambrough, 1970). Interestingly, stimulating the muscle to produce action potentials at a regular or an intermittent rate can reverse the high chemosensitivity so that it disappears except at the old end-plate region (Lømo and Rosenthal, 1972). Very low mean rates of stimulation will suffice down to one burst every $5\frac{1}{2}$ hours (Lømo and Westgaard, 1975) if the stimuli occur in high frequency bursts. The occasional action potentials, known as *fibrillation potentials*, which are characteristic of denervated muscle, are apparently not sufficient to reduce chemosensitivity (Purves and Sakmann, 1974). Thus, the activity of the muscle itself plays a large role in regulating its chemosensitivity and its receptivity to reinnervation, but there must be other factors at the former end-plate. What these factors are remains to be determined, but various possibilities have recently been reviewed in detail (Vrbová et al., 1978). These so-called "trophic interactions" are one of many areas in neurobiology which are difficult to include in textbooks because developments are so rapid that most accounts are out of date even before publication. However, this is one of the challenges of the field in general, and of synaptic interactions in particular.

## Problems

7.1. Show that 6000 molecules in a sphere 50 nm in diameter would produce a solution with a molarity of about 0.15 M or a tonicity of about 0.3 osmoles. [Hints: The volume of a sphere is $4/3\ \pi r^3$ and Avogadro's number is $6 \times 10^{23}$ molecules/mole.]

7.2. Prove that for large values of $n$, but $\mu = np$ constant, the binomial distribution approaches the Poisson distribution. [Hint: The limiting definition of an exponential used in obtaining Equation (7.2) may be helpful.] Verify the values for the variances of the Poisson and binomial distributions given in Equations (7.11) and (7.12).

7.3. Prove from the model of Fig. 7.9B that the change in voltage, $\Delta V = V - V_m$, produced by a synaptic conductance $g_s$, is given by

$$\Delta V = \frac{g_s(E_s - E_m)}{g_s + g_m}$$

(Hint: In the absence of applied currents, the net current flow across the membrane must be zero.) This equation is analogous to the Michaelis–Menten equation of enzyme kinetics [see also Equation (3.5)]. What corresponds to the Michaelis constant and what is its significance in this example?

7.4. If there are about $10^9$ ACh receptors at a frog neuromuscular junction (Miledi and Potter, 1971) and the density of receptors is about $10^4/\mu m^2$, what is the order of magnitude of the surface area covered by the neuromuscular junction? What fraction of the total surface area does this represent in a fiber 75 $\mu m$ in diameter and 4 cm in length? If in a mammalian muscle cell of similar dimensions the density of extrajunctional receptors reaches $10^3/\mu m^2$ in less than a week (Kuffler and Nicholls, 1976), what is the rate of synthesis of new receptors? Is it possible that muscle cells could synthesize this number of receptors *de novo* (see Devreotes and Fambrough, 1976)?

# 8

# Muscular Contraction

In the last chapter, interactions between nerve and muscle were considered in which neuronal action potentials release a chemical transmitter. For many muscles of lower animals and a few specialized muscles of mammals, graded depolarizations of the muscle fiber membrane are produced which lead to graded contractions. In most types of skeletal muscle in mammalian systems, the transmitter produces enough ionic current through the muscle membrane to generate action potentials. As will be discussed in this chapter, the action potential leads to a characteristic twitch contraction. The process whereby the electrical activity of the muscle membrane activates the contractile proteins contained within the muscle cells is referred to as *excitation–contraction coupling*.

Skeletal or *striated* muscle, which will be the main concern of this chapter, represents only one of three major types of muscles. The term striated refers to the pattern of striations or bands observed under the microscope. This pattern results from the regular organization of contractile proteins into two types of filaments. Although other muscles lining the gut, blood vessels, etc., have similar types of contractile proteins, these proteins are not organized in such regular fashion and the muscles appear more uniform under the microscope. These types of muscles are referred to as *smooth* muscles. The muscles of the heart have sufficiently different properties from both the striated muscles of the body skeleton and the smooth muscles of the internal organs that *cardiac* muscle is generally classed as a third, separate type of muscle. The regular organization of skeletal muscle has assisted in understanding its function, which is one reason for concentrating on this type of muscle. Those readers interested in comparisons with smooth muscle and cardiac muscle are referred to detailed reviews of their characteristics (Katz, 1970; Ross and Sobel, 1972; Prosser, 1974; Bülbring and Shuba, 1976; Fozzard, 1977; Gibbs, 1978).

Muscles function at a variety of levels and these levels and the terminology associated with each level should be distinguished before

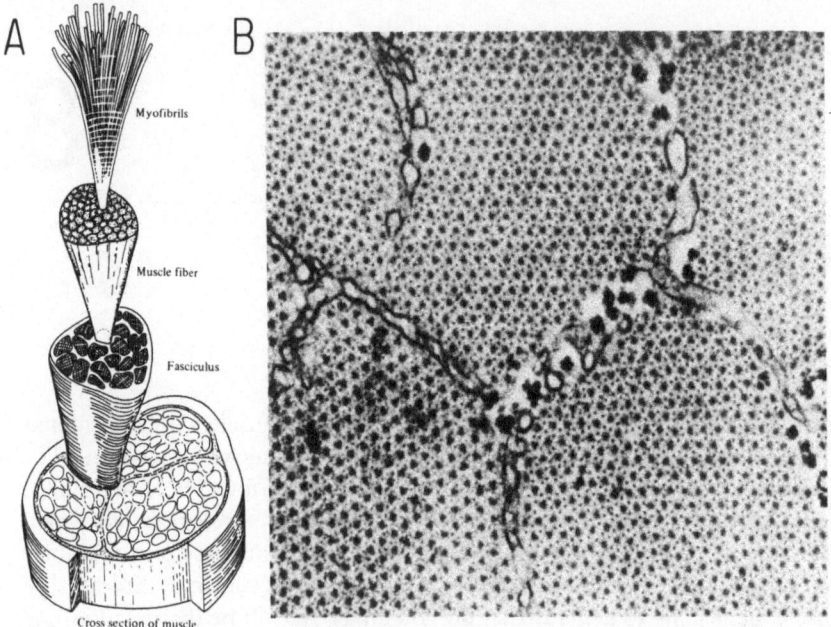

Fig. 8.1 (A) A diagram of a muscle cross section showing its gross organization and terminology. (B) An electron micrograph of several myofibrils each surrounded by tubules of the sarcoplasmic reticulum. Within each myofibril a hexagonal array of thick and thin filaments is seen. (From Dowben, 1969.)

describing the details of muscular contraction. Within a muscle, there are groups of muscle cells or *fibers* which collectively are referred to as a fasciculus or *fascicle* (see Fig. 8.1). Fascicles are surrounded by blood vessels and connective tissue, and they insert into portions of the specialized connective tissue referred to as *tendons* at each end of a muscle. In different muscles, fascicles and fibers are arranged in various patterns depending on the function of the muscle (Basmajian, 1976).

Within each muscle fiber are a large number of *myofibrils* whose cross section is shown in the electron micrograph of Fig. 8.1B. A hexagonal close-packed array of thick and thin *filaments* is seen which is surrounded by a series of tubules known as the *sarcoplasmic reticulum*. Sarco- and myo- are prefixes which refer to muscle, so the sarcoplasmic reticulum is simply a reticulum or network of tubules within the muscle cytoplasm or sarcoplasm. The sarcoplasmic reticulum is important in storing and releasing $Ca^{++}$ ions needed for muscular contraction, as will be described later. The thick and thin filaments are macromolecular arrays of the contractile proteins, and the regular organization of these proteins in skeletal muscle gives it a striated appearance.

Figure 8.2 shows a longitudinal section through a muscle with its characteristic light and dark bands, together with the common representation of these bands below. The broad dark regions are referred to as *A-bands*, because they appeared *a*nisotropic to the early light microscopists, in contrast to the light *I-bands* between them which appeared *i*sotropic. Isotropic means having the same shape and appearance from whatever angle observed, and the isotropy of the I-bands is associated with the presence of only one kind of filament, the thin filament. The denser part of the A-bands contains both thick and thin filaments and will not appear similar from any angle (i.e., they will be anisotropic). The cross

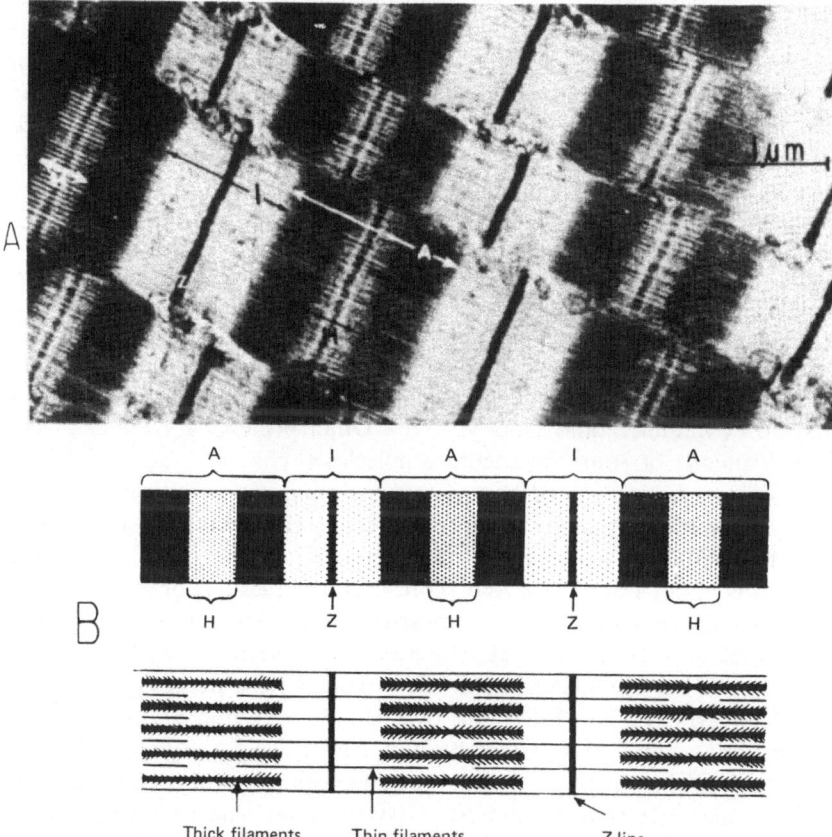

Fig. 8.2 (A) An electron micrograph of a longitudinal section through a skeletal muscle in the rabbit. Within each myofibril a regular pattern of light and dark bands is observed, although the pattern shifts somewhat from fibril to fibril. Letters have been used to label these bands, as described in the text. (From Woodbury *et al.*, 1965.) (B) A diagram of the thick and thin filaments within a myofibril (bottom) to explain the observed bands (top). The basic repeating structure is called a sarcomere, which is the distance from one Z-line to the next. (From Guyton, 1971.)

section shown in Fig. 8.1A was taken in this dense region and a hexagonal array of thick and thin filaments is clearly seen. Generally, a somewhat light region occurs in the center of the A-band, referred to as the *H-zone*, where only thick filaments are found. There is also a narrow dense region in the center of the I-band which is referred to as the *Z-line*. The Z-line contains proteins such as α-actinin (reviewed by Mannherz and Goody, 1976) and other material which extend in a plane perpendicular to the long axis of the myofibril and provide points of attachment for the thin filaments. Therefore, the Z-line is responsible for maintaining the orderly arrays observed in striated muscle. It also serves as a useful marker in defining the basic repeating structure of the muscle, the *sarcomere*, which is the distance from one Z-line to the next within a myofibril.

With this general organization and terminology in mind, we can turn to a more detailed treatment of muscular contraction. Following the general plan of the book, we will begin at the molecular level and work systematically up to the level of muscle fibers and eventually whole muscles in vertebrates, including man.

## Muscle Proteins

The thick filaments in vertebrate muscles contain a major protein, *myosin*, plus another protein (referred to as C-protein), whose function is unknown (Mannherz and Goody, 1976). Other proteins are found in the thick filaments of some invertebrate muscles (Lehman *et al.*, 1972), but these will not be considered here. The description of the filaments follows closely the outline orginally proposed by H. E. Huxley (1971). The reader should consult this source for further details and references. Each thick filament is about 1.6μm long and represents aggregations of myosin molecules. A myosin molecule is composed of two identical subunits (Fig. 8.3A), each about one-tenth the length of the whole filament. The myosin molecule and its subunits can be divided chemically by the enzymes *trypsin* or *papain* (Lowey *et al.*, 1969) into two parts, a long thin rod-like portion referred to as *light meromyosin,* with a molecular weight of about 150,000, and a second shorter, more globular fragment known as *heavy meromyosin* (with a molecular weight of about 340,000). Heavy meromyosin can be further subdivided enzymatically into subfragments S1 and S2. The light meromyosin is often referred to as the *tail* of the molecule and it is the tails which aggregate to form the main structure of the filament. The heavy meromyosin can project out from the filament (Fig. 8.4) and the S1 subfragments, also known as the *heads* of the molecule, can make bonds or crossbridges with actin molecules on the thin filaments.

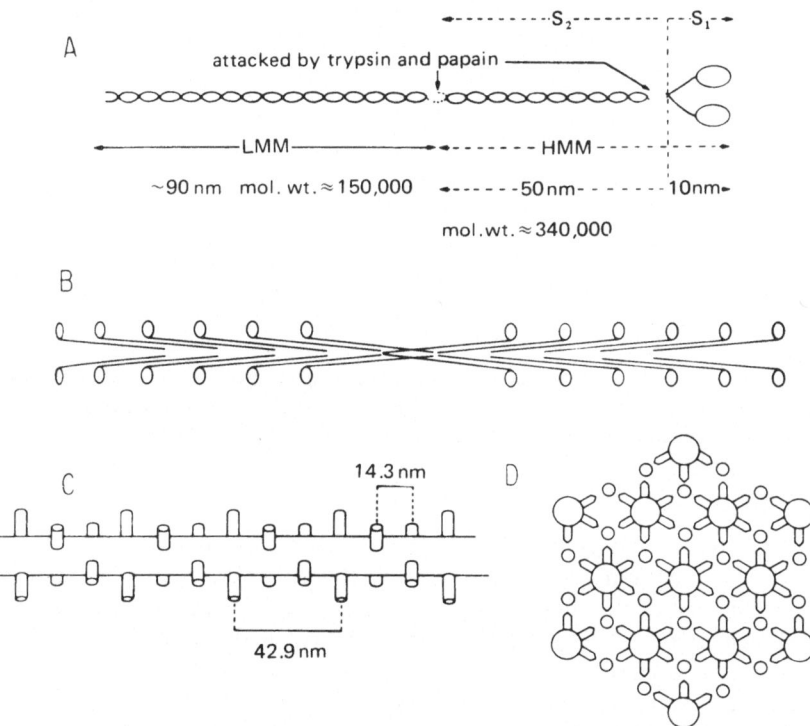

Fig. 8.3 (A) A diagram of the myosin molecule with its two identical subunits. The enzyme trypsin preferentially cleaves the molecule in two between the parts labeled as heavy meromyosin (HMM) and light meromyosin (LMM) because of their relative molecular weights. Further enzymatic digestion of HMM produces subfragments S1 and S2. (B) A diagram showing the arrangement of myosin molecules in a filament. The heads of the molecules are oriented towards the two ends, while the tails of the molecules are oriented towards the center. (From Huxley, 1971.) (C) X-ray diffraction patterns indicate that the filament in part (B) is twisted into a helix with a spacing of about 43 nm. The heads of the myosin molecules, here indicated by pegs, would then be spaced just above 14 nm apart and rotated 120° from each other. (From Huxley, 1971.) (D) Cross-sectional diagram showing that over a distance of 43 nm crossbridges from each thick filament will project toward all six neighboring thin filaments in the hexagonal array. (From Huxley and Brown, 1967.)

The projections are shown diagramatically in Fig. 8.2B at regular intervals except in the central region of the thick filaments. The reason for the orientation of the projections can be understood as follows. When myosin molecules are precipitated from solution, they tend to aggregate so that the tails of two molecules are next to one another while the heads point in opposite directions. Further molecules add on in such a way that the heads are all oriented towards the end of the filament (Fig. 8.3B). This

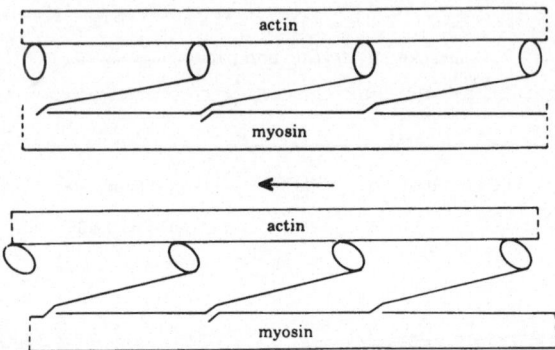

Fig. 8.4 Contraction is thought to occur by rotation of the head of the myosin molecule relative to the rest of the molecule. This rotation would produce a movement of the actin-containing thin filament with respect to the myosin-containing thick filament. (From Huxley, 1971.)

would produce a region approximately 0.2 μm long in the center of the fiber which contains only tails and is therefore devoid of projections. Along the rest of the filament the projections occur with a regular spacing, as shown in Fig. 8.3B. Not adequately shown in this figure is the three-dimensional organization. To show this organization, the thick filaments would have to be twisted to form a helix (Fig. 8.3C), and a distance of 43 nm represents the basic repeating length of this helix. Each thick filament has two projections in opposite directions every third of this distance (14.3 nm) and the direction of succeeding projections is rotated 120° around the filament axis. Thus, over a distance of 43 nm there are six projections, or one for each of the six thin filaments that form a hexagonal array around the thick filament (see Fig. 8.3D and 8.1B). How the tail ends of the myosin molecules are attracted to each other to begin assembly of a thick filament, and how the thick filament is terminated when a length of 1.6 μm is reached, remains unclear (Huxley, 1971).

The thin filaments, which are about 1 μm in length, contain three proteins: *actin, tropomyosin*, and *troponin*. The protein present in the largest quantity is actin, which is a nearly spherical or globular molecule with a molecular weight near 50,000. These molecules are also arranged in the form of a double helix, which gives the appearance (see Fig. 8.5) of a twisted double string of beads. Actin molecules are placed every 5.5 nm on each string or chain, so the repeating distance of the myosin crossbridges (14.3 nm) corresponds to a distance of about 2.6 actin molecules. There are two S1 subfragments per projection and two projections on opposite sides of each thick filament every 14.3 nm. However, the thin filament is also a double chain of actin molecules and there are twice as many thin filaments as thick (in Fig. 8.3D each thick filament is surrounded by six thin filaments, but a thin filament is surrounded by only three thick filaments). Thus the ratio of actin/S1 = 2.6, which is not an integer. When a crossbridge of one thick filament is opposite an actin molecule, the next crossbridge does not line up exactly (this point was ignored in the diagrams of Fig. 8.4). However, if one crossbridge attaches to an actin molecule and bends or rotates with respect to the long axis of the filament, the next crossbridge can come into line and make an attachment. If this process of attachment and bending of successive crossbridges takes place, there will be a sliding of one set of filaments with respect to the other. This *sliding filament* theory was proposed simultaneously by A. F. Huxley, H. E. Huxley, and their collaborators as the basis for muscular contraction (Huxley and Niedergerke, 1954; Huxley and Hanson, 1954). It has received wide support in the intervening quarter century, and will be discussed in more detail later.

If S1 fragments are added to a solution of pure actin filaments, they will spontaneously and stably bind in a ratio of one S1 fragment per actin molecule. Dissociation requires the high energy phosphate compound ATP, which also binds stably to the S1 fragment of the myosin molecule in a 1:1 ratio (Chock *et al.*, 1976). After death, the level of ATP drops and actin and myosin spontaneously bind together in a state of *rigor mortis*. The tight coupling of the actin and myosin molecules in that state accounts

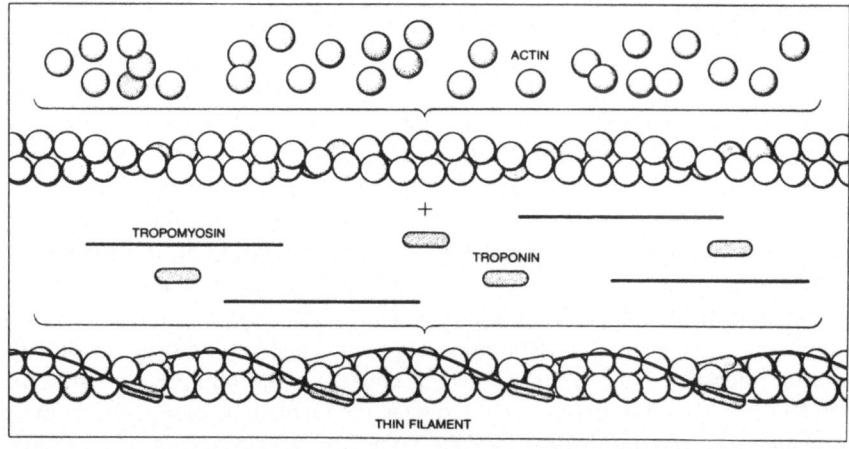

Fig. 8.5. A thin filament contains actin, tropomyosin and troponin molecules, assembled as shown schematically here. The spherical actin molecules are arranged like a double string of beads twisted to form a helix. A tropomyosin molecule extends along seven actin molecules, and there is one troponin molecule near the end of each tropomyosin (From Murray and Weber, 1974.)

for its rigidity. The association of the S1 fragments of myosin to actin in the absence of ATP is sometimes referred to as a *rigor complex* (Weber and Murray, 1973).

In addition to ATP, some divalent ion such as $Mg^{++}$ or $Ca^{++}$ is required for the dissociation of pure actin and myosin. However, the interaction of actin and myosin is normally regulated by the other two proteins (tropomyosin and troponin) which comprise the thin filaments. Regulation by other proteins on the thick filaments also occurs in some invertebrates (Lehman *et al.*, 1972), but this has not been described in vertebrates.

Tropomyosin is a long, very thin molecule with a molecular weight near 50,000. Tropomyosin associates with a single strand of actin molecules, as shown in Fig. 8.5. In this figure, taken from Murray and Weber (1974), one tropomyosin molecule is shown for each seven actin molecules, and the tropomyosin molecules are arranged end-to-end. Near the end of each tropomyosin is found one troponin molecule. Thus, the spacing of troponin molecules is at $7 \times 5.5 = 38.5$ nm intervals along the thin filaments. With these three proteins assembled into a normal thin filament, the thick and thin filaments remain detached in the absence of $Ca^{++}$ ions, but attach in the presence of $Ca^{++}$. Note that this is opposite to the situation described above for pure actin filaments. The association of thick and thin filaments is quite specific for $Ca^{++}$ ions [only $Sr^{++}$ ions can replace $Ca^{++}$ and their affinity is only $1/30$ that of $Ca^{++}$ ions (Edwards *et al.*, 1966)].

The action of $Ca^{++}$ involves the troponin molecules (Ebashi *et al.*, 1969), which have more than one site for binding $Ca^{++}$ per molecule (Hartshorne and Pyun, 1971; Ashley and Moisescu, 1972). In fact, each troponin molecule can be split into three fragments which are referred to as TN–C, TN–I, and TN–T (Greaser and Gergely, 1971; Weber and Murray, 1973), the final letter indicating the major function of the fragment. TN–T binds strongly to the tropomyosin molecule. TN–I, together with tropomyosin, can inhibit the association of a string of approximately seven actin molecules to S1 fragments of myosin. TN–C binds $Ca^{++}$ ions strongly and regulates the inhibition. The inhibition of binding between actin and myosin by the troponin–tropomyosin complex may be due to steric hindrance; i.e., the tropomyosin covers up the binding site on actin molecules for myosin. A simple suggestion, which is consistent with a considerable amount of data (see the review by Weber and Murray, 1973), is that addition of $Ca^{++}$ ions shifts the tropomyosin toward the groove between the two chains of actin molecules and so exposes the binding site for myosin. Very low levels of intracellular $Ca^{++}$ ions are effective and even $10^{-7}$ M $Ca^{++}$ can produce some contraction (Fig. 8.10B). If the whole chain shifts, the binding sites on all seven actin molecules would then be exposed and available to form either *active complexes* with myosin and ATP present or *rigor complexes* as mentioned above with myosin in the absence of ATP. The rigor complexes do not have a specific requirement for $Ca^{++}$, but Weber and Murray (1973) have described a cooperative phenomenon between the two types of complexes. Formation of rigor complexes potentiates formation of active complexes at adjacent sites and vice versa. Movement of the tropomyosin–troponin complex which is overlying seven actin molecules could account for this cooperative phenomenon.

## Contraction Cycle and Sliding Filaments

In a normal resting muscle, the supply of ATP is adequate and the $Ca^{++}$ concentration is maintained at a very low level (about $10^{-7}$ M; Hasselbach, 1964). These two factors favor the dissociation of actin and myosin, but ATP will be bound to virtually all the myosin crossbridges. The depolarization from an action potential releases $Ca^{++}$ ions into the sarcoplasm, as will be described later (in the section *Excitation–Contraction Coupling*). The binding of $Ca^{++}$ to troponin starts a cyclic process which can be divided into several stages, as shown in Fig. 8.6. This figure represents a simplification of the considerable number of possible reactions which have been described by Lymn and Taylor (1971) and others. One could start at any point of a cyclic process, but perhaps it is easiest to start with the sequence which follows the attachment of $Ca^{++}$ to the troponin molecule. This attachment exposes the active binding site, and leads to the following sequence of reactions:

(1) *Association* of actin with the myosin, which will normally have an ATP molecule bound to it. Association produces the active complex referred to previously.

(2) *Rotation* of the crossbridge. At this stage the inorganic phosphate and ADP are removed, and the energy from the ATP molecule is converted into mechanical work. A low-energy or rigor complex results, which is the form found in the absence of high-energy phosphate, for example after death.

(3) *Binding* of another molecule of ATP permits the

(4) *dissociation* of actin and myosin. Once dissociated, the myosin head spontaneously

(5) *returns* to its original perpendicular position. This last reaction is actually composed of more than one step (see, for example, Eisenberg and Hill, 1978) and is related to the formation of a *charged intermediate* $M \cdot ADP \cdot P^*$. This notation is used to indicate that the high-energy ATP bond is already partially cleaved. The return of the myosin head to the

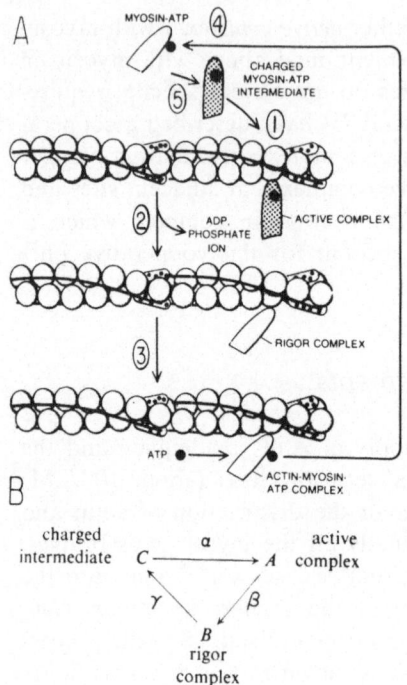

Fig. 8.6. (A) Interactions between the head of a myosin molecule and a thin filament during one contraction cycle. The parts of the cycle are numbered and described in more detail in the text. (Modified from Murray and Weber, 1974.) (B) For analysis of the overall kinetics of the cycle, it can be simplified, perhaps to as few as three rate-limiting steps (from Oğuztöreli and Stein, 1977.)

perpendicular position may then simply result from electrostatic repulsion between the charged myosin head and the thick filament. This step has been likened to cocking a gun, which requires some energy, although the final release of the "bullet" of phosphate doesn't occur until the "trigger" is pulled in Step 2 (rotation of the crossbridges).

As mentioned in the last section, the net effect of this cycle is to slide the thin filament a short distance past the thick filament. Other myosin heads on the thick filament can then line up with other actin molecules on the thin filaments. The pattern of projections from the thick filaments has also been likened to the oars in a multi-oared rowboat. However, the oars (myosin heads) are oriented in opposite directions in the two halves of the thick filament (since the tails of the myosin butt against one another in the center). A rowing movement won't necessarily move the thick filament, but the thin filaments at the two ends of the sarcomere will be pulled toward the center. Thus, the net effect of the sliding filaments is a shortening of the sarcomere.

If contraction represents a sliding of filaments past one another, rather than a folding or shortening of the filaments themselves (Huxley, 1974), certain changes in the pattern of bands can be predicted. For example, as the muscle shortens, the amount of overlap between the thick and thin filaments should increase. The I-band, which contains only thin filaments, should become narrower, as should the H-zone, which contains only thick filaments. The A-band, which represents the length of the thick filament, should not change. These and other predictions were verified by Huxley and Niedergerke (1954) in proposing the sliding filament theory initially.

A more exacting prediction of the sliding filament hypothesis is that the amount of force generated should be proportional to the amount of overlap between thick and thin filaments, or more precisely, the number of crossbridges that can be formed between the two filaments. Gordon *et al.* (1966) verfied this prediction as shown in Fig. 8.7. At lengths greater than 3.6 $\mu$m, no overlap between thick and thin filaments was observed, and no force was produced. The absence of overlap agrees with a measured length of 1.6 $\mu$m for the thick filament and 1 $\mu$m for each of the thin filaments at the two ends of the sarcomere (Fig. 8.7A). As the muscle was shortened in successive steps, the force increased linearly until a length of 2.2 $\mu$m was reached. At that length the maximum number of crossbridges could be made (Fig. 8.7B) and the tension reached its maximum value. Note that this is slightly longer than the length of two thin filaments since myosin molecules join tails at the center in forming a thick filament and produce a short region in the middle without any heads to form crossbridges. As the muscle was shortened further, no further change was observed (Fig. 8.7C) until a length of less than 2 $\mu$m was reached. At that length, a thin filament from one half-sarcomere extended sufficiently into the opposite half-

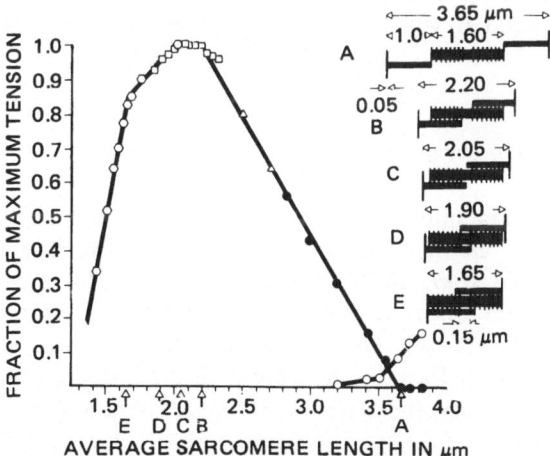

Fig. 8.7. Length–tension curve for sarcomeres in single frog muscle fibers. The letters at the bottom of the figure refer to the diagrams on the right which are explained in more detail in the text. The open circles give the tension generated just by stretching the muscle to the length indicated in the absence of stimulation (passive tension). During maximal stimulation at the lengths indicated (isometric conditions), additional tension was produced, as indicated by the other symbols (active tension). (From Woodbury *et al.*, 1965.)

sarcomere to interfere with formation of some crossbridges (Fig. 8.7D). At still shorter lengths, the sarcomere spacing becomes less than or equal to the length of the thick filament (Fig. 8.7E). Then, the thick filament must either extend into the next sarcomere (with further reduction in tension) or become wavy, if it cannot penetrate the Z-line but must remain within a single sarcomere (González-Serratos, 1971).

These experiments provide strong confirmation of the basic ideas underlying the sliding filament hypothesis. This hypothesis can also be reconciled to a large extent with the hypotheses based on electrostatic repulsion of filaments (e.g., Yu *et al.*, 1970), if details of the contraction cycle are included. The return of the crossbridges to a nearly perpendicular position is associated with the formation of a charged intermediate, and could be due simply to electrostatic repulsion of the myosin heads by the rest of the thick filament. When the ATP is split, and the charge is removed, the myosin head would be free to rotate to form an acute angle. This model could operate under a variety of muscle lengths and filament spacings so long as the myosin heads are able to swing out far enough from the main portion of the thick filament to reach neighboring thin filaments (Huxley, 1971). Muscle length ($l$) and interfilament spacing are normally related since the muscle cell retains a constant volume ($V$) as it is lengthened. Thus,

$$V = \pi r^2 l \tag{8.1}$$

where $r$ is the radius of the fiber and is directly related to the spacing between filaments (Elliott *et al.*, 1963, 1970). The constant volume of muscles is associated with the common observation that muscles bulge (i.e., the radius increases) as they shorten in length.

## Contraction Kinetics

In recent years, much work has gone into determining the kinetics of the contraction cycle *in vitro* (e.g., Lymn and Taylor, 1971; Eisenberg *et al.*, 1972; Bagshaw *et al.*, 1974; Chock *et al.*, 1976). This work will only be summarized briefly following the reviews of Weber and Murray (1973) and Eisenberg and Hill (1978). The cycle is also referred to as the *actomyosin cycle* to distinguish it from the hydrolysis of ATP in the absence of actin (*myosin cycle*). The myosin cycle, which proceeds at a slower rate, will be ignored here.

In the presence of sufficient $Ca^{++}$ and ATP (Step 1 in the previous section), the association of actin and myosin is very rapid. Step 2, the rotation of the head with the release of inorganic phosphate, is less rapid with a rate constant of about 100 s$^{-1}$. Step 3, the binding of ATP, is again rapid with the normal concentration of ATP of about 5 mM. The rate constant is about $10^6$ s$^{-1}$M$^{-1}$. Once bound, the dissociation of actin and myosin is very rapid, with a rate constant greater than $10^3$ s$^{-1}$. The final step in the sequence, the formation of the charged intermediate, and the return of the myosin head to its original position, proceeds at a slow rate which is rate-limiting, with high actin concentrations *in vitro*.

The rate-limiting step *in vivo* remains uncertain, but several steps probably contribute significantly to the overall kinetics of the cycle. The kinetics were originally described using a scheme proposed by Huxley (1957), and since updated by Huxley and Simmons (1971, 1972) and Huxley (1974). Several variations of this scheme have been proposed based on more recent physiological and biochemical information (Lymn and Taylor, 1971; Podolsky and Nolan, 1972; White and Thorson, 1974; Julian *et al.*, 1974; Eisenberg and Hill, 1978). The approach given here is based on the fairly general formulation of Oğuztöreli and Stein (1977). The notation (Fig. 8.6B) assumes that three of the five steps in the cycle contribute significantly to the overall kinetics. Huxley's original work (1957) only included rate constants for the formation and breaking of crossbridges between actin and myosin, although more recently Huxley and Simmons (1971) included a third step in which the crossbridge is rotated. The rate constants in Huxley's original formulation, $f$ and $g$, correspond more or less to the rate constants $\alpha$ and $\gamma$ in Fig. 8.6B.

From this figure, two differential equations for the change in the fraction of active complexes $A$ and of rigor complexes $B$ can be written:

$$dA/dt = \alpha(1 - A - B) - \beta A \qquad (8.2)$$

$$dB/dt = \beta A - \gamma B \qquad (8.3)$$

The quantity in parentheses in Equation (8.2) is simply the fraction of charged intermediates $C$ since

$$A + B + C = 1 \qquad (8.4)$$

Thus, Equation (8.2) simply states that active complexes are formed from charged complexes with a rate constant $\alpha$ and broken down with a rate constant $\beta$.

If the muscle is held at a constant length (referred to as *isometric conditions*) and is producing a steady force, the fractions of $A$ and $B$ will be constant (i.e., $dA/dt = dB/dt = 0$). Then, Equations (8.2) and (8.3) can be easily solved to give the steady state fraction of active complexes and rigor complexes:

$$A = \frac{\alpha\gamma}{\alpha\beta + (\alpha + \beta)\gamma} \qquad (8.5)$$

$$B = \frac{\alpha\beta}{\alpha\beta + (\alpha + \beta)\gamma} \qquad (8.6)$$

If $\alpha = \beta = \gamma$, it is clear that the fraction of active complexes, rigor complexes, and charged intermediates would each be $1/3$. Both X-ray and metabolic data indicate that even during a maximal contraction less than half the crossbridges are in the form of active complexes (Huxley and Brown, 1967; Curtin *et al.*, 1974), although the exact values of $\alpha$, $\beta$, and $\gamma$ under physiological conditions are unknown.

Another commonly studied situation is a contraction proceeding against a constant load or tension (*isotonic conditions*). After some initial transients (Podolsky *et al.*, 1974), the rate of contraction usually settles down to a constant velocity $v$. The Equations (8.2) and (8.3) become

$$v \, dA/dx = \alpha(1 - A - B) - \beta A \tag{8.7}$$

$$v \, dB/dx = \beta A - \gamma B \tag{8.8}$$

since $x = vt$ and $dx = v \, dt$. However, these equations are much more difficult to solve than Equations (8.2) and (8.3) because $A, B, \alpha, \beta$, and $\gamma$ all depend on $x$, and some form for the dependence on $x$ must be assumed. For example, the rate of formation of active complexes $\alpha$ has never been measured as a function of $x$, but is probably maximal when the crossbridge is perpendicular to the long axis of the filament opposite an actin site ($x = 0$ in Fig. 8.8A; see also Eisenberg and Hill, 1978). Huxley and Simmons (1971) measured the rate of transition $\beta$ from the active complex to the rigor complex as a function of position; $\beta$ is a monotonically decreasing function and a simple exponential form has been assumed in Fig. 8.8A. As indicated earlier, $\gamma$ is a condensation of several steps, some of which depend on $x$ (e.g., Step 4) while others will be independent of $x$ (Step 5). A weak dependence on $x$ has been indicated for the overall reaction.

Once assumptions are made about $\alpha$ and $\beta$ as a function of $x$, Equations (8.7) and (8.8) can be solved. Analytic solutions for several choices of $\alpha$ and $\beta$ have been given by Oğuztöreli and Stein (1977) and numerical solutions can be obtained if analytical solutions are not available. Other important quantities can also be calculated. For example, the total number of bonds $n$ made at any time will be

$$n \propto \int (A + B) \, dx \tag{8.9}$$

where $A$ and $B$ are the fractions of bonds in attached states, and the integration will be over a suitable length of the muscle. The number of bonds made determines the stiffness of the muscle when it is stretched a short distance (Huxley, 1974). The rate that inorganic phosphate is released, and hence the rate of ATP splitting and energy expenditure ($dE/dt$), will be

$$dE/dt \propto \int \beta A \, dx \tag{8.10}$$

Fig. 8.8. A plausible, but hypothetical, model for the kinetics of the contraction cycle as a function of (A) the displacement $x$ from the equilibrium position of the myosin head, or (B) its angle with respect to the long axis of the fiber. (A) Crossbridges are formed with a maximum rate constant $\alpha$ near $x = 0$ (near an angle of 90°; Eisenberg and Hill, 1978). The rate constant $\beta$ for transition from an active complex $A$ to a rigor complex $B$ is a monotonically decreasing function of the displacement $x$ from equilibrium (Huxley and Simmons, 1971). The rate constant $\gamma$ for dissociation of the filaments and the formation of charged intermediates $C$ occurs at a low rate relatively independent of position $x$. (B) Free energy changes associated with a contraction cycle (heavy line). A myosin head in state $C$ will bind to actin when the angle approaches 90° and follow the curve for the active complex $A$. At a smaller angle a transition will take place to state $B$ and the

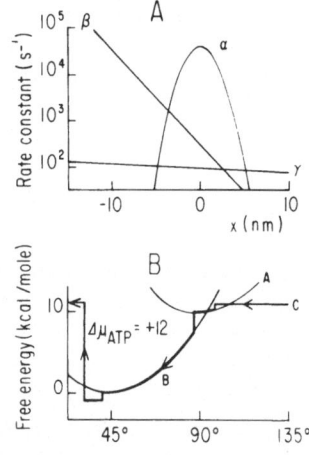

curve for the rigor complex will be followed. Downward movement over some distance $x$ represents generation of force over distance (i.e., work). Shortly after the minimum for this state at 45°, another ATP molecule will be bound and the head will detach. The energy of the ATP (12 kcal/mole; Curtin *et al.*, 1974) will then be transferred to the myosin head to form the charged intermediate $C$ for the next cycle. (Adapted from Eisenberg and Hill, 1978.)

since release of ATP depends on the transition (with rate constant $\beta$) from states $A$ to $B$.

The output of the muscle will be the force $F$ generated. To calculate $F$, some relationship between force and position must be assumed. In the work done to date, a linear relation according to Hooke's law has been assumed, although the two types of complexes may have different proportionality constants ($k_1$ and $k_2$) and equilibrium positions ($x_1$ and $x_2$),

$$F = \int \left[ k_1 (x - x_1) A + k_2 (x - x_2) B \right] dx \qquad (8.11)$$

where $k_1$, $k_2$, $x_1$, and $x_2$ are constants. The potential energy stored in a spring is proportional to $k(x - x_0)^2$, where $x_0$ is the equilibrium position. Thus, in a diagram of the free energy (Fig. 8.8B), states $A$ and $B$ will show parabolic curves. The minimum or equilibrium positions in this figure have been chosen so that $x_1 = 0$ (an angle of 90°) and $x_2 = -10$ nm (an angle of 45°), in line with the experimental data; $k_1 = k_2$ was assumed for simplicity.

Movement of a crossbridge (indicated by arrows) downward along one of these curves represents force generation, while the vertical jumps represent transitions from one state to another. A muscle will operate efficiently when most of the energy of the ATP goes into work (i.e., force generation by a crossbridge over a distance $x$) rather than in vertical transitions from state to state. Any molecular model of muscle contraction must be tested to see if it gives reasonable values for the energetics of muscular contraction. T. L. Hill (1974, 1975) has analyzed in detail the constraints imposed on models of muscular contraction by the thermodynamics and energetics of the processes involved.

## Energetics of Contraction

The immediate source of energy for muscular contraction, as for other processes in the body, is ATP (Infante and Davies, 1962). However, for many years, reduced ATP levels after muscular contraction were difficult to demonstrate, since the ATP required for contraction can be rapidly regenerated in muscle through various processes. For example, muscles contain another high-energy source, creatine phosphate (CP), and an enzyme, creatine phosphokinase, to rapidly transfer the high-energy phosphate from CP to ADP, according to the Lohmann reaction (Lohmann, 1934),

$$CP + ADP \xrightarrow[\text{phosphokinase}]{\text{creatine}} \text{creatine} + ATP \qquad (8.12)$$

In addition, muscles contain varying amounts of glycogen which can be broken down to pyruvic acid with further regeneration of ATP, a process known as *glycolysis*. Pyruvic acid can be reversibly converted to lactic acid, which is released from muscles during exercise. This release contributes to the metabolic acidosis following exercise, and lactic acid can be measured in the blood.

Muscular contraction from stores of ATP, CP, and glycogen can proceed even in the absence of respiration (*anaerobic* processes). If oxygen is available through respiration, pyruvic acid can be broken down further to $CO_2$ and water with further production of ATP. The chemical pathways for this *aerobic* process have been well worked out and are referred to as the tricarboxylic acid cycle. Details of these reactions are given in many biochemistry texts (e.g., Lehninger, 1975). The relative proportions of aerobic and anaerobic metabolism used in exercise will vary with the severity of the exercise. Extreme exercise may far overrun the capacity for aerobic metabolism, and an *oxygen debt* is built up through the use of anaerobic metabolism which must be "paid back" after completion of the exercise.

The energy utilized is also consumed in several ways which were studied for many years by A. V. Hill (1965). Some energy in muscle cells is consumed in active transport of $Na^+$ and $K^+$ ions which was described in Chapter 3. To maintain a muscular contraction requires further expenditure of energy, even under isometric conditions. This energy is given off in the form of heat and was called *maintenance heat* by Hill. If the muscle is allowed to shorten against a load, still more heat is given off which he called *shortening heat*. Shortening heat was empirically found to be proportional to the distance shortened. Finally, work will be done when a muscle shortens while actively generating a constant force $F$ against a load (*isotonic contraction*), so that the total extra energy $E$ produced in shortening a distance $x$ will be

$$E = (a + F)x \tag{8.13}$$

where $a$ is a constant. The rate of extra energy production $dE/dt$ during shortening at velocity $v$ will be,

$$dE/dt = (a + F)v \tag{8.14}$$

If a muscle is shortening against a load, it cannot generate the maximum amount of force $F_0$ that will be generated under isometric conditions. Hill also found experimentally that the rate of extra energy production was proportional to the difference between the maximum force $F_0$ and the actual force a muscle could produce when shortening at a velocity $v$. Thus,

$$dE/dt = b(F_0 - F) \tag{8.15}$$

where $b$ and $F_0$ are constants. From Equations (8.14) and (8.15), a relationship between force and velocity can be obtained,

$$(F + a)(v + b) = (F_0 + a)b = c \tag{8.16}$$

where $c$ is a constant defined by Equation (8.16). The equation is usually rearranged in a normalized form (see Problem 8.3):

$$\frac{F}{F_0} = \left[ 1 - \frac{v}{v_{max}} \right]\left[ 1 + \left( \frac{F_0}{a} \right)\left( \frac{v}{v_{max}} \right) \right] \tag{8.17}$$

In this form, only a single constant $a$ must be determined since the isometric force $F_0$ and maximum velocity $v_{max}$ are easily measured. Equation (8.17) is often called Hill's characteristic equation or simply the *force-velocity curve*, since it has been found to apply to such a wide range of muscles (Fig. 8.9). Note that the hyperbolic relationship between force and velocity predicted by Equations (8.16) or (8.17) was derived completely from energy measurements on muscle. Thus, the good fit to experimental force-velocity curves represents a strong confirmation of Hill's basic ideas. However, the data points in Fig. 8.9A were not taken from experimental measurements, but from Huxley's (1957) calculations based on the sliding filament theory and from earlier work (Fenn and Marsh, 1935).

Furthermore, using the same constants in his calculations, Huxley (1957) was able to show that the predicted rates of energy consumption

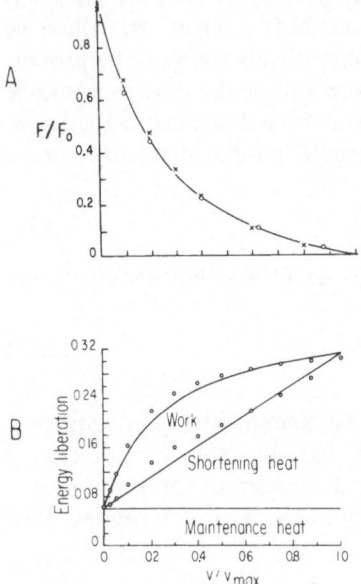

Fig. 8.9. (A) Force–velocity curves from Hill's equation (8.16) (solid line), the sliding filament theory of Huxley (1957)(circles), and an earlier relation of Fenn and Marsh (1935) (×'s), which is given by $F = W_0 e^{-av} - kv$ (see Problem 8.3). (From Huxley, 1974.) (B) Total rate of energy liberation is divided into the rate of producing maintenance heat and shortening heat (the sum equals the rate of heat production) and the rate of doing useful mechanical work at different velocities of shortening $v$, relative to the maximum velocity $v_{max}$. (The scale for $v/v_{max}$ applies to both parts of this figure.) Solid lines are again Hill's (1938) predictions (but see Hill, 1964) and the data points are from Huxley's (1957) sliding filament model. (From Huxley, 1957.)

and heat production were consistent with Hill's (1938) equations. Figure 8.9B shows the three components separated by solid lines: a maintenance heat independent of shortening, a shortening heat whose rate varies directly with the speed of shortening, and a rate of work production which also depends on the speed of shortening. The rate of work production (i.e., mechanical power) depends on the product of force and velocity. This product will be zero at the two extremes of the curve, when either velocity is zero (under isometric conditions) or force is zero (at the maximal rate of shortening possible with no load). The efficiency (work done/total energy consumed) is maximal at intermediate speeds of shortening. Although experimental estimates vary considerably (Curtin *et al.*, 1974), the optimal efficiency of frog muscle is about 25%, which compares well with artificial motors (see, for example, the introduction to Chapter 27 of Lehninger, 1975).

Some deviations are observed in the fit of Huxley's model to Hill's equations in Fig. 8.9, but the form of Hill's equations is partially derived empirically. More recently, Hill (1964) himself observed deviations experimentally from his earlier predictions at high rates of shortening. Deviations are also found for negative velocities of shortening (i.e., when a muscle is stretched while it is attempting to contract; Katz, 1939; Joyce *et al.*, 1969), and in considering brief contractions such as a twitch (Jewell and Wilkie, 1958). Hopefully, in the future the experimental observations can be predicted from the sliding filament model without requiring curve fitting and unmeasured functions. To fit a brief contraction such as a twitch, the process of *excitation–contraction coupling* by which a nerve normally activates a muscle must also be understood in detail.

## Excitation–Contraction Coupling

The trigger for contraction in a *twitch* muscle fiber is a conducted action potential. The muscle action potential is normally initiated by a nerve impulse via the processes of synaptic transmission at the neuromuscular junction. Twitch muscle fibers also have a single, localized end-plate region, since the action potential can spread from there over the rest of the fiber. Other muscle fibers, which are referred to as *nontwitch* or *slow* muscle fibers, do not normally generate action potentials. Nontwitch muscle fibers are found in only a few specialized skeletal muscles in mammals (Hess, 1961; Fernand and Hess, 1969), but have long been known and are much more common in lower vertebrates and invertebrates (Kuffler and Vaughan-Williams, 1953; Wiersma, 1953). These fibers have neuromuscular junctions distributed over the surface of the muscle fiber. Stimulation of the nerve produces a graded depolarization and graded contraction.

If a twitch muscle fiber is depolarized to various levels, for example by changing external $K^+$ ions, the contractile force is also continuously graded with depolarization (e.g., Hodgkin and Horowicz, 1960). Thus, the all-or-none nature of the muscle action potential is responsible for the characteristic twitch contraction in these fibers. The relation between tension and depolarization follows an S-shaped curve (Fig. 8.10A) much like that described for the action potential channels in Chapter 5. A similar curve is found if myofibrils are bathed in different concentrations of $Ca^{++}$ ions (Fig. 8.10B). Thus, one might assume that the effect of the depolarization is to produce an increased permeability to $Ca^{++}$ ions. Using the $Ca^{++}$-sensitive protein aequorin mentioned in the last chapter, the time course of the changes in $Ca^{++}$ concentration have been measured directly (Fig. 8.10C; Blinks *et al.*, 1978), which supports the idea of $Ca^{++}$ serving as an intermediary between excitation and contraction of muscle.

A. V. Hill (1949) pointed out that $Ca^{++}$ ions did not have time to diffuse from the surface to central myofibrils in a large, fast twitch muscle fiber. This problem was solved with the discovery of the extensive system of tubules in twitch muscle fibers. Two types of tubules have been described: the *transverse tubules* are open to the extracellular space and continuous with the surface membrane (Fig. 8.11A). These tubules could form a pathway for conducting the electrical activity of the surface

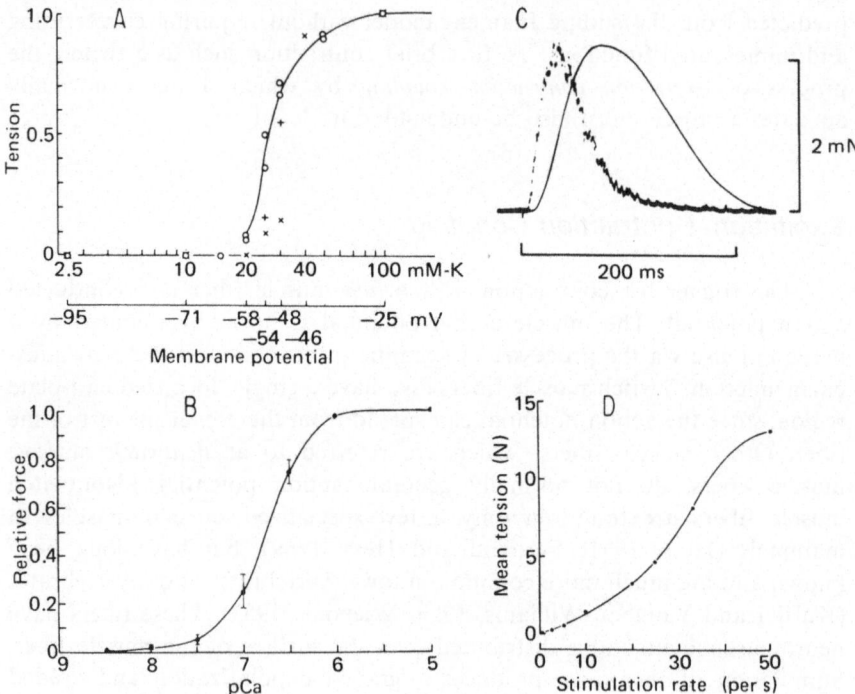

Fig. 8.10. (A) Peak tension produced by bathing single frog muscle fibers in different concentrations of K⁺ ions which depolarized the fibers to the levels indicated. The scale for concentration is logarithmic while that for potential is approximately linear. Tension was measured relative to the maximum value for large depolarizations in several fibers (different symbols). (From Hodgkin and Horowicz, 1960.) (B) Relation between force and the negative logarithm of Ca⁺⁺ concentration (pCa) in skinned muscle fibers of the frog. (From Hellam and Podolsky, 1969.) (C) Force and luminescence from frog muscle fibers filled with the Ca⁺⁺-sensitive protein, aequorin. Average of seven twitches. Note the rapid decline of the luminescence before peak tension is reached. (From Blinks *et al.*, 1978.) (D) Mean force as a function of stimulus rate in a cat muscle. A sigmoid relation is observed, as in parts (A) and (B), although the mechanism could be quite different. (From Mannard and Stein, 1973.)

membrane deep into the muscle fiber, although the electrical properties of the transverse tubules appear to be distinct from those of the surface membrane (Adrian *et al.*, 1970). Conduction inward along these fine tubules is relatively slow (González-Serratos, 1971; Sugi, 1974) and perhaps decremental (see also Problem 8.2).

The second type of tubule, the *longitudinal tubule,* runs parallel to the long axis of the muscle fiber. Two longitudinal tubules are often seen in close apposition to a single transverse tubule to form what is referred to as a *triad* (see Fig. 8.11). The longitudinal tubules are known to actively take up

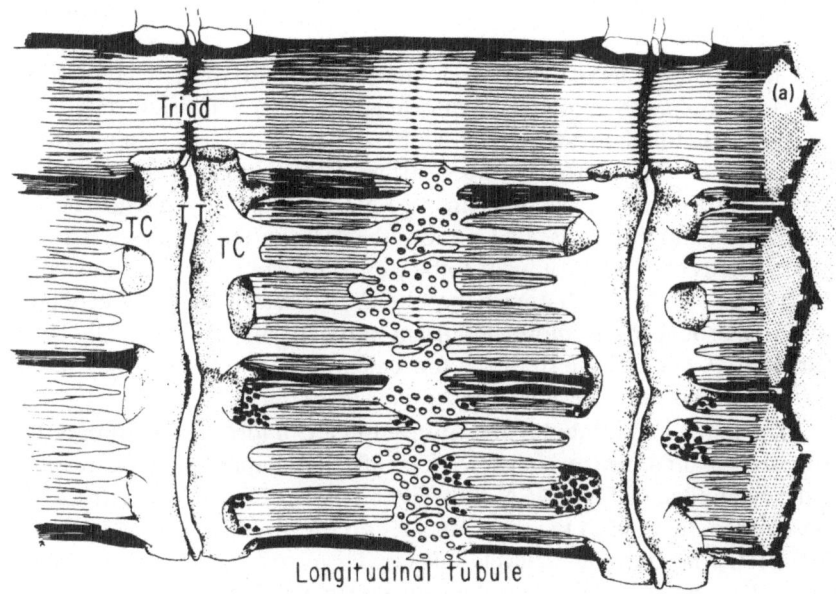

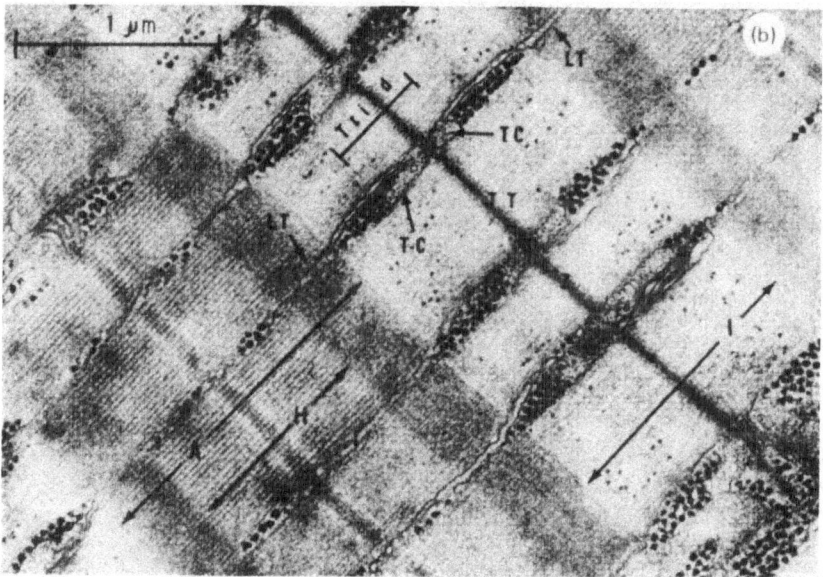

Fig. 8.11. (A) Diagram illustrating the structure of the transverse tubules and the longitudinal tubules which extend along the long axis of the muscle fiber. (B) Longitudinal section through muscle showing the longitudinal tubules (LT) expanding to form terminal cisternae (TC). The longitudinal tubules plus terminal cisternae are also known as the sarcoplasmic reticulum. Two terminal cisternae end in close apposition to the transverse tubules (TT) to form a triad. The band pattern is also shown and in this muscle (frog sartorius) the triads are at the level of the Z-line. (From Peachey, 1965.)

$Ca^{++}$ ions (Hasselbach, 1964), and the triads could form a connection between the depolarization of the muscle action potential (conducted down the transverse tubule) and the release of $Ca^{++}$ ions. Several mechanisms for the release of $Ca^{++}$ ions have been suggested (reviewed by Endo, 1977; Caputo, 1978), but the best documented are the release by membrane de-polarization and the release by $Ca^{++}$ ions themselves. The release of $Ca^{++}$ by drugs such as caffeine is thought to be closely related to the $Ca^{++}$-mediated $Ca^{++}$ release. Endo (1977) argues that the $Ca^{++}$ release induced by depolari-zation is probably the more important mechanism physiologically, and Adrian *et al.* (1976) have studied a movement of charge, which they feel is related to this process (reviewed by Adrian, 1978).

The longitudinal tubules are arranged in a cylindrical ring around the groups of myofilaments which form a myofibril (compare the cross section at the right edge of Fig. 8.11A with Fig. 8.1B), so that the released $Ca^{++}$ ions can rapidly bind to the troponin sites on the thin filaments. As seen in Fig. 8.10C, the free $Ca^{++}$ levels fall rapidly to near control levels, even before the twitch tension reaches its maximum. The rapid decline is presumably due to the active reuptake into the longitudinal tubules of $Ca^{++}$ that does not find binding sites on the myofilaments. Since the binding of $Ca^{++}$ to the myofilaments is reversible, $Ca^{++}$ ions that come off the myofilaments will also be pumped back into the longitudinal tubules, leading to a gradual relaxation of the muscle fiber. Viewed in this way, relaxation also requires energy, since the active transport of $Ca^{++}$ into the longitudinal tubules requires ATP, as do other transport systems (reviewed by Tada *et al.*, 1978). By assuming reasonable kinetics for $Ca^{++}$ movements and for the sliding filament model, Julian (1969) could match the time course of a twitch contraction. However, values of several un-measured constants were chosen to obtain a good fit.

Because of the slow time course of the twitch (tens to hundreds of ms in different muscles), the motor axons and the muscle fibers are able to generate action potentials even before the observed tension begins to increase. The excitation–contraction delay is of the order of 10 ms, during which the tension may even decline somewhat (Sandow, 1944). A second impulse releases further $Ca^{++}$ ions and the contractile events can sum with one another. More impulses can be generated until enough $Ca^{++}$ is released to saturate the troponin molecules. At this point, a maximum tension output or *tetanus* is reached. Thus, the rate of nerve impulses determines the rate of tension output in a muscle according to an S-shaped curve (Fig. 8.10D) similar to that obtained by varying $Ca^{++}$ concentra-tions or varying depolarization.

The central nervous system can set the desired level of tension by specifying the rate of firing of motoneurons, a phenomenon referred to as *rate coding* (Stein, 1974). However, a number of factors determine the actual tension produced, including muscle length (Fig. 8.7) and muscle load or velocity (see the force–velocity curve of Fig. 8.9A). Even if the rate of stimulation is maintained constant, the tension can increase or decrease with time. An increase in tension is usually referred to as *potentiation*, in analogy with the potentiation or facilitation of synaptic potentials. Potentiation of force output has not been studied as extensively as that associated with neuromuscular transmission (Cooper and Eccles, 1930; Burke *et al.*, 1976; Stein and Parmiggiani, 1979), although modulation of force is functionally more important. Increasing a synaptic potential above that required to generate a muscle action potential will not affect the output, whereas force output can be continuously graded.

A decrease in muscular force over time with maintained stimulation is often referred to as *fatigue*. If the depolarization of a muscle fiber is maintained by a voltage clamp, muscular contraction does not continue indefinitely, but gradually decays. This decay could be due to a depletion of the releasable $Ca^{++}$ from the ends of the longitudinal tubules (i.e., the *terminal cisternae*) which are the regions in contact with the transverse tubule. This depletion may be responsible for the common phenomenon of muscular fatigue in intact organisms, although availability of ATP, blood supply, and other factors may play a role (Stephens and Taylor, 1972). Once the tension declines, a muscle fiber requires time to regain the ability to contract, a phenomenon known as *repriming* (Adrian *et al.*, 1976). Similarly, $Ca^{++}$ ions are taken up all along the longitudinal tubule and need some time to diffuse or be transported to the terminal cisternae (Weingrad, 1968).

In summary, the action potential conducted down a motor axon produces a complex series of steps leading to muscular contraction and relaxation. These include: (1) neuromuscular transmission with the pre- and postsynaptic mechanisms discussed in the last chapter, (2) initiation of an action potential in twitch muscle fibers or a graded depolarization in nontwitch muscle fibers, (3) spread of the depolarization along the muscle fiber membrane and down into the transverse tubules, (4) release of $Ca^{++}$ ions from the longitudinal tubules, mediated in some way via the triads connecting the two types of tubules, (5) binding of $Ca^{++}$ ions to troponin molecules on the thin filaments, which initiates the sliding of the thin and thick filaments past each other, (6) reuptake of $Ca^{++}$ ions into the longitudinal tubules, and (7) relaxation of the muscular contraction.

## Motor Units

Except in certain pathological conditions in which single fibers contract in isolation (*fibrillations*; Lenman and Ritchie, 1970), groups of muscle fibers normally contract in groups, referred to as *muscle units* (Burke, 1967). A mammalian muscle unit is composed of fibers which may be widely dispersed throughout a muscle (Burke and Tsairis, 1973), but are innervated by a single motor axon (Fig. 8.12B). Each twitch muscle fiber is normally innervated by one and only one motor axon, although this is not true in early life (Redfearn, 1970), or shortly following reinnervation of a muscle by nerve axons (Brown *et al.*, 1976). The muscle unit, together with its associated motoneuron, is known as a *motor unit*. All movements are composed of the contractions of muscle units, as signaled by the impulses in the attached motoneurons. These motoneurons represent the "final common pathway" for controlling movements, as first elaborated by Sherrington (1906).

Motor units vary widely in their properties, even within one muscle. However, the muscle fibers within a single muscle unit are quite homogeneous, and the properties of a motoneuron are normally well matched to the properties of its muscle unit. For example, large motor axons tend to produce large twitch tensions when stimulated, because they innervate large muscle fibers (Burke and Tsairis, 1973). Whether large axons also innervate more muscle fibers is less certain. For the medial gastrocnemius muscle of the cat, large and small axons tend to innervate about the same number of muscle fibers (Burke and Tsairis, 1973). However, the predominantly large motor axons in medial gastrocnemius innervate more muscle fibers than the axons of the predominantly slow soleus muscle (Burke *et al.*, 1974), which is a synergist (i.e., the two muscles tend to be activated together during movements).

Large muscle fibers also tend to have a well-developed system of transverse and longitudinal tubules to permit the inward spread of the action potential and the rapid activation of myofibrils deep within the muscle (Schiaffino *et al.*, 1970). In fact, the large motor units which contain large muscle fibers tend to contract and relax more rapidly than small motor units (McPhedran *et al.*, 1965; Wuerker *et al.*, 1965). The briefer twitch will be less effective in generating externally measurable tension, so the ratio of tetanic tension to twitch tension tends to be greater for fast motor units or muscles (Close, 1972). The precise rules for laying down transverse and longitudinal tubules are not known, but large muscle fibers seem to have not only a greater density of sarcoplasmic reticulum, but also a higher specific activity for accumulation of $Ca^{++}$ (Fiehn and Peter, 1971). Thus, more $Ca^{++}$ can be released, but it can be pumped back into the sarcoplasmic reticulum more quickly to produce the characteristic large, brief twitches (Close, 1972).

Close (1972) also reviewed the evidence that large muscle fibers are specialized in many other ways, in that they have a faster myosin ATPase and the enzymes for glycolytic metabolism. Small muscle fibers have a slower myosin ATPase and are richly endowed with the enzymes for

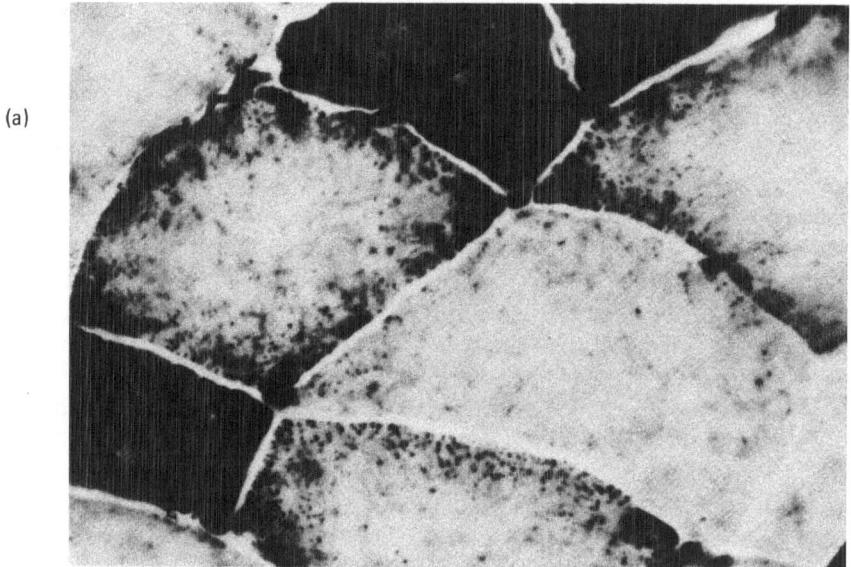

(a)

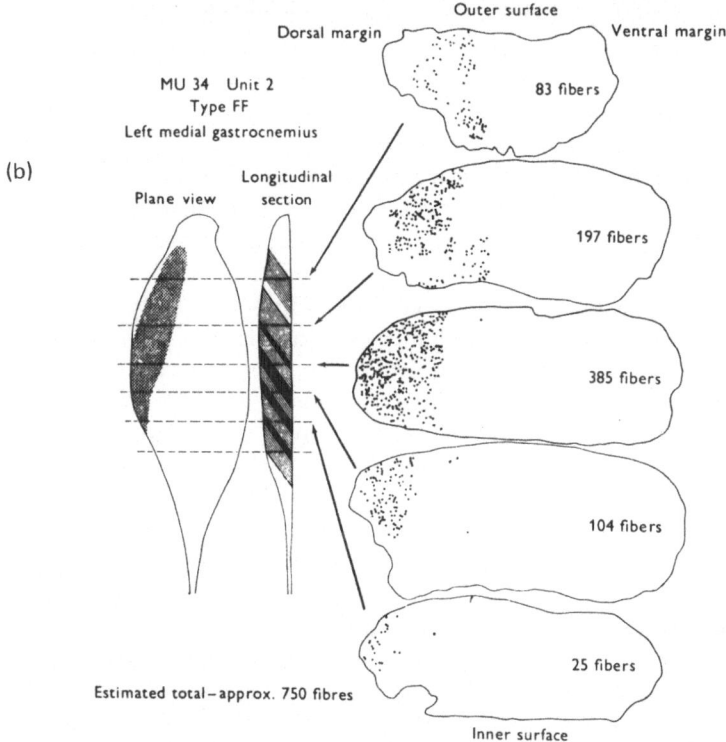

(b)

Outer surface

Dorsal margin — Ventral margin

MU 34   Unit 2
Type FF
Left medial gastrocnemius

83 fibers

Longitudinal
section

Plane view

197 fibers

385 fibers

104 fibers

Estimated total – approx. 750 fibres

25 fibers

Inner surface

Fig. 8.12. (A) Cross section of cat medial gastrocnemius (MG) muscle stained for the presence of mitochondrial ATPase activity (an oxidative enzyme). The large, pale fibers have very little of this enzyme (type FF), the small dark fibers contain large amounts (type S) and intermediate staining is also seen (type FR). (From Hennenman and Olson, 1965.) (B) A whole MG muscle showing distribution of approximately 750 fibers belonging to one type FF motor unit. A single motor axon had been stimulated repetitively to deplete the muscle unit of glycogen, and the muscle was then stained for glycogen. (From Burke and Tsairis, 1973.)

oxidative metabolism (Fig. 8.12A). These enzymes permit them to contract for long periods of time and small muscle fibers are therefore resistant to fatigue. Still other muscle fibers have a mixture of both systems, which makes them relatively fast and relatively resistant to fatigue. Based on their histochemistry and electrophysiology, muscle fibers have been generally classified into three groups which are given various names. For example, Burke *et al.* (1971) refer to them as FF (*f*ast contracting, *f*atigable), FR (*f*ast contracting, fatigue *r*esistant) and S (*s*low contracting).

What determines the matching of motor axons to their muscle units? Much of the differentiation of muscle fibers takes place after innervation by motoneurons, and is lost after denervation (Vrbová *et al.*, 1978), which suggests an important role for the motoneuron. Large motoneurons might supply different amounts or a different kind of substance to maintain their muscle fibers (referred to as a *trophic* substance). The correlation between the size of a motor axon and the size of its twitch tension is poorer following denervation and reinnervation (Bagust and Lewis, 1974), and no significant correlation could be demonstrated in adult human subjects following nerve injury and nerve repair (Milner-Brown *et al.*, 1974). Other explanations of these findings are possible, since adult muscle fibers might be less sensitive to the hypothetical trophic substance. Alternatively, extraneous factors such as the growth of connective tissue might prevent some motor axons from finding their way back to a suitable number of muscle fibers in adult life.

The gross properties of adult mammalian muscles can be changed by suturing a nerve from a predominantly slow muscle into a predominantly fast muscle, and vice versa (*cross-innervation*; Buller *et al.*, 1960). Even changes in the detailed biochemistry of muscle proteins are produced (Amplett *et al.*, 1975; Weeds *et al.*, 1974). Many of these changes have also been produced by chronically stimulating nerves, *without* changing their synaptic connections (Salmons and Vrbová, 1969; Lewis *et al.*, 1977). Thus, the pattern of muscle activity is important in determining muscle properties. Certainly, isometric exercise can increase the bulk of a muscle and the number of myofilaments contained in each muscle fiber. Other exercise patterns such as running predominantly increase the oxidative capacity of muscle, including its blood supply (Burke and Edgerton, 1975). However, the experiments to date cannot really specify which modifications are due to the activity of the muscle and which are due to the action of the transmitter (ACh) or some other trophic substance which is released during activity (see also Chapter 7).

## Viscoelastic Properties

Up to this point, the contractile properties of muscle fibers have been considered, but a contractile force must be transmitted via a tendon to a

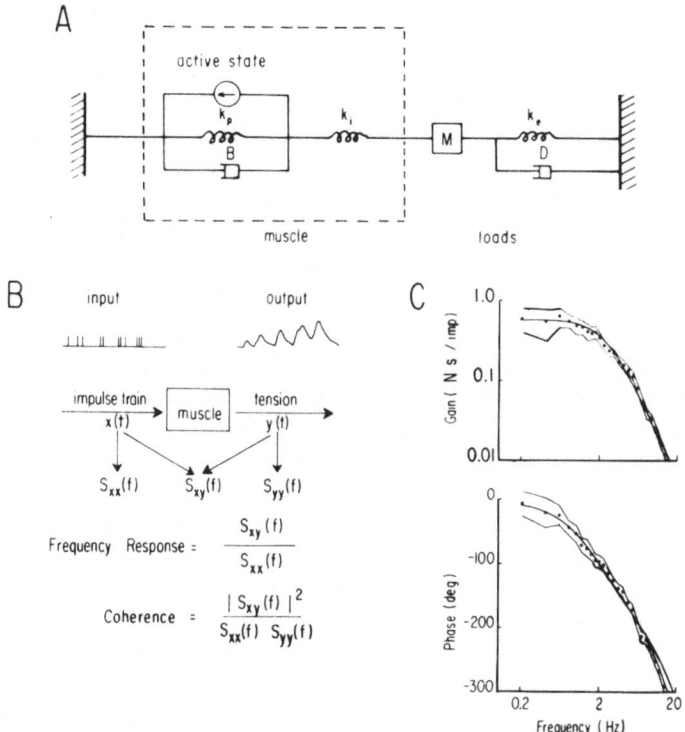

Fig. 8.13. (A) A simple model of muscle containing internal series ($k_i$) and parallel ($k_p$) elastic elements (springs), a viscous element ($B$) or dashpot, and an active state element, which determines the muscle's ability to contract. In Hill's (1938) model, the viscous element obeyed his nonlinear force–velocity equation (8.16) rather than being a linear dashpot. The active state was assumed to turn on rapidly and then decay exponentially. Also shown in this diagram are various loads consisting of springs ($k_e$), dashpots ($D$), and masses ($M$) against which a muscle may work. (From Oğuztöreli and Stein, 1976.) (B) Random stimulation of a muscle nerve as a means of determining the frequency response function or input–output properties of a muscle. The method involves computation of the input spectrum $S_{xx}(f)$, the output spectrum $S_{yy}(f)$ and the cross-spectrum $S_{xy}(f)$ over a range of frequencies. [See Bendat and Piersol (1971) for further details.] (C) The gain [upper curve, measured as the newtons (N) of force produced per impulse/s modulation at each frequency] and phase (lower curve in degrees) of the frequency–response function for human soleus muscle. The solid curve is the response predicted if there were two limiting rate constants or time constants (see text and Problem 8.5). The outer brackets are 95% confidence limits, which indicate that the curve fits the data well. (From Bawa and Stein, 1976.)

recording apparatus before it can be observed *in vivo*. The recorded signal will inevitably be filtered to some extent by the viscous and elastic properties of the muscle and tendon. In an early model, Hill (1938) considered that muscle consisted of an active state element in parallel with viscous and elastic elements (Fig. 8.13A). A second elastic element was in series with the other elements, which he considered to be undamped (no measurable viscosity). The active state element represented the ability of a muscle to contract and was assumed to rise quickly to a peak and decay exponentially. In more modern terms, the active state would be related to the $Ca^{++}$ ions bound to the myofilaments. $Ca^{++}$ ions will bind rapidly and be released more slowly for reuptake into the sarcoplasmic reticulum.

The force generated in a viscous fluid will be proportional to the velocity of movement through the fluid (Stokes' law). However, the relation between force and velocity in muscle is nonlinear, so Hill included a nonlinear viscous element which obeyed his characteristic equation (8.17). The parallel elasticity was much more compliant (less stiff) than the series elasticity, which Hill assumed resided within the tendon or other structures functionally in series with the contractile elements.

More recently, D. K. Hill (1968) showed that the crossbridges themselves are an important source of elasticity. Even in resting muscle, some bonds are made between thick and thin filaments. Stretching these bonds a short distance produces substantial force changes. Beyond their elastic limit, the bonds rupture and the force declines, even though the muscle is stretched further. This *short-range elasticity* is the major source of the incremental stiffness in a contracting muscle, since the stiffness has been shown to depend on length in exactly the same way as muscle force (i.e., both depend on the number of crossbridges that can be formed; Flitney and Hirst, 1978). This elasticity is also important in determining the properties of muscle receptors, which are discussed in the next chapter. If a muscle is stretched at constant velocity while it is trying to contract, new bonds will be formed and ruptured by the stretch, so the short-range elastic forces will also contribute a viscous drag for larger stretches.

With the enormous complexity of whole muscles, which processes are most important in determining the kinetics of contraction and relaxation following nervous stimulation? To answer this question, muscles can be stimulated randomly at intermediate rates between those which give almost isolated twitches and those which give an almost fused tetanus. A random stimulation pattern is preferable (Fig. 8.13B) since the force fluctuations associated with a wide range of input patterns can then be analyzed by spectral analysis. This technique was introduced in Chapter 7 and provides a measure of the power contained in a signal over a range of sinusoidal frequencies in Hertz or $\omega = 2\pi f$ in radians/s. In the present example there are two signals, an input $x$ (the pattern of nervous stimulation) and an output $y$ (the pattern of force fluctuations). One can compute an input spectrum $S_{xx}(f)$, an output spectrum $S_{yy}(f)$, and a cross-spectrum $S_{xy}(f)$, which depends on the product of the frequency components in the input and output (Bendat and Piersol, 1971).

From these spectra (Fig. 8.13B), the *frequency–response* function of the muscle can be estimated. This function and various methods of obtaining it will be considered further in the next chapter (see also

Problem 8.5). Whether derived using random or other inputs, this function measures the ability of the muscle to follow different frequencies. The frequency response functions for muscles in the frog, cat (Mannard and Stein, 1973), and human (Bawa and Stein, 1976) were all described by curves having two limiting rate constants or time constants. Although the evidence was inevitably indirect, it suggested that the two limiting processes were the decay of the $Ca^{++}$ bound to the myofilaments (i.e., the decay of the active state in Hill's terminology) and a viscoelastic time constant (Mannard and Stein, 1973). Computations based on the sliding filament model with added kinetics for the activation process (Julian, 1969) were consistent with these ideas (Stein and Wong, 1974). Thus, of the many processes going on in muscle, only the decay of the active state and the muscle's viscoelastic properties may suffice to determine the overall kinetics of contraction and relaxation.

## Problems

8.1. Assume that bonds can be formed only over a very narrow range of positions (Podolsky and Nolan, 1973), or in the limit $\alpha = \alpha_0 \delta(x)$, where $\delta(x)$ is the Dirac $\delta$ function. Assume also that $\beta$ and $\gamma$ do not change appreciably over the range of positions of interest (i.e., $\beta$ and $\gamma$ are constants) and $v \neq 0$.

(a) Show that

$$A = (1 - e^{-\alpha_0/v})e^{-\beta x/v}$$

$$B = \frac{\beta(1 - e^{-\alpha_0/v})(e^{-\gamma x/v} - e^{-\beta x/v})}{(\beta - \gamma)}$$

$$n \propto v(1 - e^{-\alpha_0/v})\left(\frac{\gamma + \beta}{\beta\gamma}\right)$$

(b) At what velocity $v$ does $n$ reach a maximum?
(c) What is this maximum number?

8.2. Consider a cable of length $2K$ as shown below and which receives symmetric-current inputs of magnitude $I$ at the two ends.

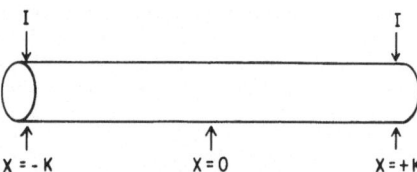

(a) Using the notation given in Chapter 6 (e.g., $X = x/\lambda$, where $\lambda$ is the length constant), show that the transfer function for this cable is

$$Z^*(s) = \frac{r_a\lambda \cosh[X(s + 1)^{1/2}]}{(s + 1)^{1/2}\sinh[K(s + 1)^{1/2}]}$$

(b) What is the effective capacitance of the cable for constant currents at short times? (Hint: Use the initial value theorem of Laplace transforms.)

(c) What is the significance of these results for the function of the transverse tubules in skeletal muscle (see, for example, Constantin, 1975)?

8.3. (a) Show that Equation (8.17) follows from Equation (8.16). (Hint: Eliminate the constant $b$ by considering the situation when $v = v_{max}$.)

(b) Expand both Hill's equation (8.16) and Fenn and Marsh's equation (see Fig. 8.9) in power series (Selby, 1975). Up to what power can the two equations be made to agree exactly by suitable choice of constants?

8.4. Huxley's (1957) model has two states which can be denoted $A$ (attached) and $D$ (detached) with rate constants $\alpha$ and $\beta$:

Huxley tried various forms for $\alpha$ and $\beta$, including the following:

$$\alpha = \begin{cases} \alpha_1 x/h, & x \leqslant h \\ 0, & \text{elsewhere} \end{cases}$$

$$\beta = \begin{cases} \beta_1 x/h, & x \geqslant 0 \\ \beta_2, & x < 0 \end{cases}$$

(a) Write the equation for the fraction $n$ of bonds made during a steady contraction with velocity $v$.

(b) Solve this equation for $n$ [cf. Huxley (1957), Equations (7) and (8)].

(c) If the force $F$ generated is $F \propto \int nx\,dx$, derive the force–velocity curve for this model [cf. Huxley (1957), Equation (11)]. It is this function which is plotted in Fig. 8.9A.

8.5. The general form for the transfer function of a second-order system is

$$Z^*(s) = \frac{Z_0}{1 + 2\zeta s/s_0 + (s/s_0)^2}$$

where $Z_0$ is the gain at low values of $s$ (i.e., $s \ll s_0$), $s_0$ is the cutoff frequency, and $\zeta$ is the damping ratio. The term *second-order system* refers to the fact that the response declines as the second power of $s$ for large $s$.

(a) What is the approximate value of $Z^*(s)$ for $s \gg s_0$?

(b) Show that the approximations for large and small $s$ cross at $s = s_0$ [$s_0$ is the frequency $s$ at which $Z^*(s)$ begins to decline or "cut off" rapidly].

(c) Show that the system has two real time constants if and only if $\zeta > 1$. [Hint: Consider the Laplace transform of a function $Z(t) = e^{-t/\tau_1} - e^{-t/\tau_2}$.]

(d) What are the time constants $\tau_1$ and $\tau_2$ in terms of $s_0$ and $\zeta$?

# 9

# *Sensory Receptors*

This chapter will consider mechanisms for the initiation of action potentials in sensory receptors, and will emphasize their significance in the transmission of information to the central nervous system. In keeping with the theme of the book, the properties of muscle receptors will be stressed, but general methods will also be described for analyzing sensory receptors. The treatment of muscle receptors can be considered as an example of how many other types of sensory receptors can be studied quantitatively.

Sensory receptors have a variety of complex geometries specialized to the particular types of stimuli they receive. Nonetheless, certain general features of sensory receptors can often be described and divided into three stages, as illustrated in Fig. 9.1A. The first stage consists of a *transformation* of the stimulus without changing the form of the energy. The light energy impinging on the visual system must be focused on the receptors for maximal effect. Sound waves impinging on the ear must be converted to pressure waves in a fluid before ultimately deflecting the hair cells in the inner ear. Even with signals which are already internal to the body, such as the length of individual muscles, these signals are often transformed by specialized muscle fibers designed to produce particular patterns of stress on the sensory nerve fibers innervating them. This transformation stage serves to match the incoming stimulus energy to the properties of the sensory receptor, whether by focusing, amplifying, filtering, or modifying the signal in some other way. The transformed stimulus is often referred to as an *adequate stimulus*, because it is then adequate to activate the sensory receptor maximally and quite selectively.

The second stage consists of a *transduction* process by which the stimulus energy is converted into an electric current across a cell membrane. This is the stage which is most poorly understood. The chemistry of the breakdown of rhodopsin molecules by light may be fairly well known, but how does this breakdown produce the currents and potentials recorded

A

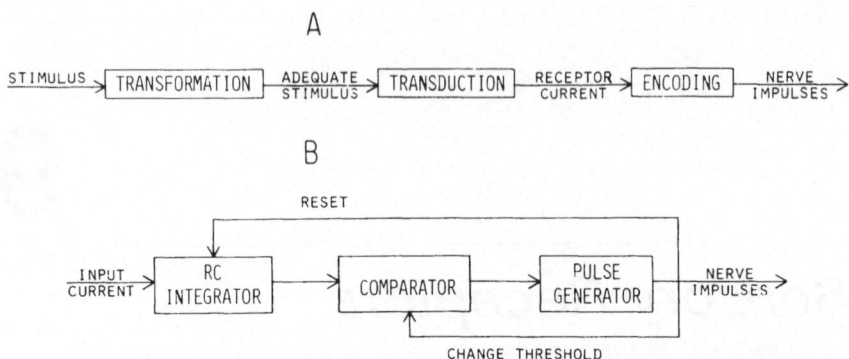

Fig. 9.1. (A) Block diagram of a sensory receptor. (B) The impulse generation or encoding process can itself be expanded in a block diagram. (Adapted from French and Stein, 1970.) The different processes and terminology are discussed further in the text.

in visual receptors? How does the bending of a hair cell in the inner ear produce a potential change in these cells? How exactly does the tension in a specialized muscle fiber cause current to flow in the sensory fibers innervating them? These questions are largely unanswered, although quantitative descriptions of the effects of these transduction processes are rapidly becoming available.

The final stage, which is common to most sensory systems, is the generation of nerve impulses by the membrane currents. This stage is often referred to as the *encoder* (Poppele, 1973), since the membrane currents are coded as a series of nerve impulses for transmission to the central nervous system. All three stages may take place in the dendrites (input processes) of a single sensory neuron, or they may involve a chain of cells, some of which may not even originate from nervous tissue. For example, the hair cells in the skin and in the inner ear, as well as the taste buds in vertebrates, are modified epithelial (skin) cells (Tamar, 1977). In more complex systems, such as the vertebrate retina, much processing in several different types of cells may be involved before nerve impulses are generated (Werblin and Dowling, 1969).

Nerve impulses are required in cells which are long compared to their length constant λ. Otherwise the signals would be greatly attenuated in spreading from the receptor to the central end of the nerve fiber. Since the transfer function of a cable [Equation (6.19)] decreases very rapidly with frequency, rapidly varying signals would be attenuated even more than steady inputs. A mammalian nerve fiber may be a meter or more in length, and have a length constant which requires boosting the signal every few millimeters at the nodes of Ranvier. Rushton (1961) pointed out that the amplifiers would have to be of extremely high tolerance if the signals at the central end were to bear any relation to the receptor potentials. By generating all-or-none action potentials, rather than having analog amplifiers, this problem can be overcome. *Thus, information transfer, particularly about rapidly varying signals over considerable distances, must take place by means of nerve impulses* (Stein, 1970). Where distances are shorter, synaptic transmission can be regulated directly by the receptor potential. Bush and Roberts (1968) described reflex effects of a crustacean stretch receptor which are produced without the receptor generating any nerve impulses.

A fourth stage might also have been added to the block diagram since information is often *recoded* several times during synaptic transmission from cell to cell before reaching higher centers of the brain. This topic is outside the scope of this chapter, although some aspects of it will be dealt with in regard to the control of movement in Chapter 10.


## Crustacean Stretch Receptors

The slowly adapting stretch receptors of crayfish and lobsters are the muscle receptors which have probably been studied most intensively in relation to the stages of sensory processing. Fast-adapting receptors also exist (Fig. 9.2A) that end on different, faster muscle bundles. Both muscle bundles are innervated by excitatory and inhibitory motor axons. The inhibitory axons (not shown in Fig. 9.2A) also end on the dendrites of both receptor cells (Jansen *et al.*, 1971). Only the excitatory mechanisms in the slowly adapting receptor will be treated here.

Stretching the muscle bundle containing the dendrites of the receptor produces a tension change which rapidly reaches a peak and then declines with time (Fig. 9.2B). The reason for this response is easily understood in terms of the block diagram of a muscle fiber (Fig. 8.12A; see also Nakajima and Onodera, 1969b). In this diagram there is a series elastic element which can be rapidly stretched and a second elastic component which is in parallel with a viscous element and the active state element which produces the contractile force. Muscle receptors are often on a region of the muscle which contains less contractile material. This region can be rapidly stretched with a resultant large deformation of the dendrites. The viscous elements associated with the contractile apparatus can only be stretched slowly. As this happens, the deformation is spread more evenly along the fiber and the stress or tension on the dendrites is reduced somewhat.

High-speed cinemicrographs have been taken of vertebrate muscle receptors (Smith, 1966; Boyd and Ward, 1969; Poppele, 1973) to study this process quantitatively, and mechanical models have been considered (see, for example, Rudjord, 1972). The transformation from length to tension in

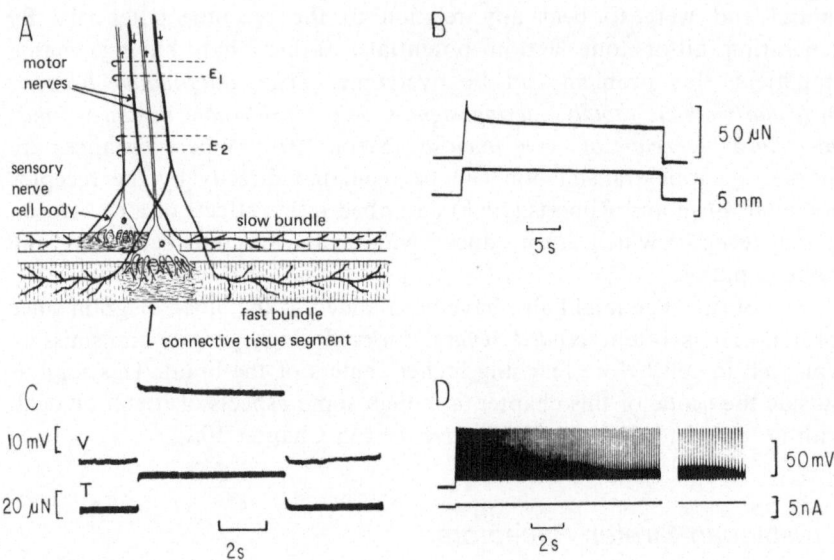

Fig. 9.2. (A) Abdominal muscles in the crustacea have both slowly adapting and fast adapting receptor cells whose dendrites end on separate muscle bundles. The motor innervation and muscle properties are also different. Not shown in the figure are inhibitory axons which end on the dendrites of the receptors as well as muscle fibers. (From Kuffler, 1954.) (B) A step length change applied to the muscle fiber produces a tension change which declines with time. (From Nakajima and Onodera, 1969b.) (C) If the tension is maintained constant, the receptor potential recorded intracellularly in the presence of TTX (to block nerve impulses) still adapts somewhat. (From Nakajima and Onodera, 1969b.) (D) If a constant current is applied intracellularly without TTX, adaptation is observed in the rate of impulses until a steady state is reached which can be maintained for long periods. The gap between the two halves of this record is 40 s. (From Nakajima and Onodera, 1969a.)

these receptor muscles produces an adequate stimulus which is large for transient stimuli and adapts to maintained stretches. This process of *adaptation* is complex, with changes occurring at each of the stages in Fig. 9.1, as will be seen.

To study the transduction process more carefully, tension can be measured and the length can be automatically adjusted so as to maintain a constant tension by using a servo-amplifier (Wendler, 1963). This method is similar to the way in which the voltage of a cell can be clamped by automatically adjusting the current passed across its membrane. If the tension is changed abruptly, the voltage recorded by an intracellular microelectrode shows an abrupt depolarization (referred to as a *receptor potential*), but this potential declines or adapts over time, even if the tension is maintained constant. The decline is markedly slower and less complete than the inactivation of $Na^+$ channels underlying the action potential, and is like the slow desensitization of synaptic channels (Magazanik and Vyskočil, 1970).

The fact that the crayfish stretch receptor has a cell body large enough to penetrate stably with microelectrodes is a major advantage compared to vertebrate muscle receptors, where this has not been possible. However, the recordings are some distance from the dendritic source of the currents underlying the receptor potential, so the ionic basis of the decline has not been clarified. A slow decline in the permeability changes underlying the receptor potential could be responsible (see below) and/or a permeability to other ions which increases slowly as a result of the depolarization (e.g., a slow delayed rectification of $K^+$ ions; Nakajima and Onodera, 1969a; Husmark and Ottoson, 1971).

What gives rise to the receptor currents in this or other mechanoreceptors? All nerve fibers are sensitive to mechanical deformations and respond with a permeability change and a depolarization (Julian and Goldman, 1962). Perhaps some pores in the membrane are opened or enlarged by the stress, and permit ions to flow through more readily. Presumably the fine dendrites of muscle receptors are specialized in some way to be particularly sensitive to these deformations, but the molecular mechanisms remain obscure.

Nonetheless, the effect of the stimulus is to produce a permeability change and a depolarization. In vertebrate muscle receptors the permeability to $Na^+$ ions, $Ca^{++}$ ions, and perhaps other ions is increased (Ottoson and Shepherd, 1971). The permeability change is much less selective than that of the action potential channels, but resembles the changes produced by synaptic transmitters. In fact, the model of Fig. 7.9 can be applied directly to a sensory receptor if the letter $s$ now represents a sensory process, rather than a synaptic one. There is a reversal potential or equilibrium potential $E_s$ for the combination of ions involved, and the conductance change $g_s$ depends on the strength of the sensory stimulus, but not on voltage. In many sensory systems, such as taste receptors (Beidler, 1961), visual receptors (Rushton, 1972), cutaneous receptors (Loewenstein and Mendelson, 1965), and muscle receptors (Hunt and Ottoson, 1975), the response saturates as shown in Fig. 7.9 according to the Michaelis–Menten equation. However, over a range of muscle lengths, as shown in Fig. 9.3, the receptor potential often increases linearly with muscle length.

The final stage shown in Fig. 9.1 is the encoding of the receptor current into a train of nerve impulses. This stage can again be studied in relative isolation by applying currents intracellularly to the stretch receptor. The rate of nerve impulses increases with increasing current, but even with a constant current, the rate of nerve impulses declines with time. Thus, there is adaptation of the impulse generator as well (Nakajima and

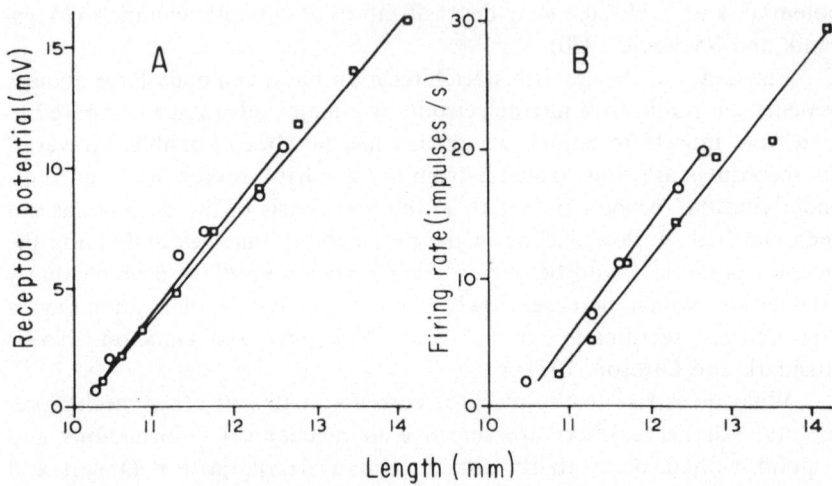

Fig. 9.3. The receptor or generator potential (A) and the firing rate (B) for the slowly adapting stretch receptor are both linear functions of muscle length and are therefore linearly related to one another. The different data points are from different experiments. Lengths are measured relative to the rest length of the muscle. (From Terzuolo and Washizu, 1962.)

Onodera, 1969b). The mechanism of this adaptation is better understood than that which occurs in the transducer or transformation stages. Nakajima and Takahashi (1966) noted that there is an increase in the after-hyperpolarization of nerve impulses with activity. This *post-activation hyperpolarization* is blocked by ouabain or by replacing the $Na^+$ ions in the bathing solution by $Li^+$. $Li^+$ flows through the $Na^+$ channels in the membrane but cannot replace $Na^+$ ions effectively in the $Na^+$ pump. These experiments suggest a mechanism involving the increased internal concentration of $Na^+$ following activity. The increase in substrate concentration increases the rate of the electrogenic $Na^+$ pump (i.e., three $Na^+$ ions are pumped out for every two $K^+$ ions pumped in). Thus, a net outflow of positive charge results from the pump, which leaves the inside of the cell hyperpolarized.

The effects of an electrogenic pump have been verified quantitatively in a snail neuron, which is large enough to insert four intracellular electrodes (Thomas, 1969). Two electrodes were used for current injection and voltage recording in order to voltage clamp the neuron. Current could also be passed *between* the other two electrodes, which were filled for example with sodium and potassium acetate. Thus, depending on the direction of the current, $Na^+$ or $K^+$ ions could be injected into the cell without affecting its membrane potential. A hyperpolarizing current was observed following $Na^+$ injection (but not $K^+$ injection) whose amplitude and time course was just sufficient to remove the extra $Na^+$ ions loading the cell. The $Na^+$ concentration could also be measured directly using $Na^+$-sensitive microelectrodes. With increasing amounts of $Na^+$ injected into the cell, the time course of the current was also prolonged because of the limited capacity of the $Na^+$ pump.

An increased amplitude and prolonged time course has also been observed in the hyperpolarization following activity in the slowly adapting stretch receptor (Nakajima and Takahashi, 1966) and in many vertebrate neurons. It is particularly prominent in small unmyelinated fibers, since the concentration changes will be larger in these fibers (Rang and Ritchie, 1968). (For a cable of unit length, the membrane area and ionic currents will be proportional to the radius $r$ while the volume will be proportional to $r^2$. Thus, the surface to volume ratio, or the concentration change, in terms of ions per unit volume will vary inversely with radius $r$, other things being equal.) The functional effect of this post-activation hyperpolarizing current is clear. The greater the activity, the larger and more prolonged this current will be, and the greater its effect in counteracting the depolarizing current generated by the stimulus. The impulse rate will therefore decline or adapt with time.

Thus, adaptation in the slowly adapting stretch receptor involves each of the stages in Fig. 9.1. The relative percentages of adaptation were approximately 50% (transformation), 20% (transduction), and 30% (encoding) (Nakajima and Onodera, 1969b). Although the fast-adapting stretch receptors show similar receptor potentials, they will stop firing after a few impulses, no matter how strong a current is injected (Nakajima and Onodera, 1969a). With subthreshold depolarizations, slowly adapting receptors show membrane noise associated with well maintained Na currents, but these are absent in the fast-adapting receptor (Sjölin and Grampp, 1975).

Different receptors show different amounts of adaptation, and the difference in adaptation between the two types of stretch receptors seems to arise in the encoding stage. In mammals, there is a continuum from receptors which adapt very little and very slowly to those which adapt very quickly and completely. For example, the Pacinian corpuscle will discharge no more than a few impulses no matter how strongly it is stimulated (Gray and Malcolm, 1950). It is, however, exquisitely sensitive to rapidly changing stimuli and will respond with one impulse per cycle to vibratory stimuli less than 1 $\mu$m in amplitude occurring at several hundred Hertz (Sato, 1961). In general, adaptation permits neurons to respond preferentially to *what's new* or changing in the environment. Different receptors are specialized to different degrees in their ability to respond either to changing or to maintained stimuli.

Human observers are not very good at estimating the absolute magnitude of a maintained stimulus, because of the adaptation going on continuously (hence, the popularity of light meters in photography and sound intensity meters on tape recorders). In a classic paper entitled, "The magical number 7 plus or minus 2," Miller (1956) pointed out that observers could only reliably distinguish about seven different categories for various sensory modalities. This problem will be considered again in relation to the information capacity of sensory receptors at the end of the chapter, but first the mechanisms and representation of the impulse generating process will be examined in more detail.

## Impulse Generation

Schematically, the impulse generator or encoder in Fig. 9.1A can itself be represented by a series of boxes (Fig. 9.1B). To understand the box labeled $RC$ integrator, recall that current can neither be created nor destroyed (Kirchhoff's law). Thus, the current generated by the stimulus must flow back across the membrane as either capacity current or ionic current. If the membrane capacity $C$ and resistance $R$ are assumed constant for the moment (this assumption will be examined later), the relation between the membrane potential $V$ and the membrane (or the sensory) current $I$ is then simply (see Chapter 5)

$$I = C \, dV/dt + V/R \tag{9.1}$$

Taking Laplace transforms of both sides yields

$$I^*R = V^*(1 + sRC) \tag{9.2}$$

where transformed quantities, which are functions of the transform variable $s$, are indicated by an asterisk (*). Equation (9.2) also assumes that the membrane potential is measured relative to its initial or resting value, so that $V = 0$ at $t = 0$. Rearranging and taking the inverse transform gives

$$V = \int_0^t IRe^{-(t-u)/\tau} \, du \tag{9.3}$$

where $\tau = RC$ is the membrane time constant as in previous chapters. Note that the voltage at time $t$ is obtained by integrating the current at earlier times $u$ with an exponential weighting in terms of the time constant $\tau$. This is sometimes referred to as a *leaky integrator* (French and Stein, 1970) because the charge leaks away with a time constant $\tau$. In a *perfect integrator*, all previous times would be given equal weight ($\tau \rightarrow \infty$). For a constant current $I$, the integral can be easily solved, and Equation (9.3) reduces to Equation (5.3):

$$V = IR(1 - e^{-t/\tau})\qquad(9.4)$$

The next stage of the process is referred to as a comparator, in that the voltage is continuously compared with a threshold voltage $V_0$. After rearranging Equation (9.4), the time to reach the threshold voltage is given by

$$t = -\tau \ln(1 - I_0/I)\qquad(9.5)$$

where $I_0$ is the minimum or *rheobasic* current required for the voltage to just reach the threshold level. The inverse of this time, referred to as the firing rate $\nu = 1/t$, is plotted as a function of current as a solid line in Fig. 9.4A. The rate approaches 0 for $I = I_0$, but increases linearly for $I \gg I_0$. For large values of $I$, the membrane depolarizes rapidly to threshold relative to the time constant $\tau$, so the model behaves more like a perfect or linear integrator.

For linear behavior, the output (i.e., the nerve impulses) of the final stage (the pulse generator) must reset the membrane quickly so the process

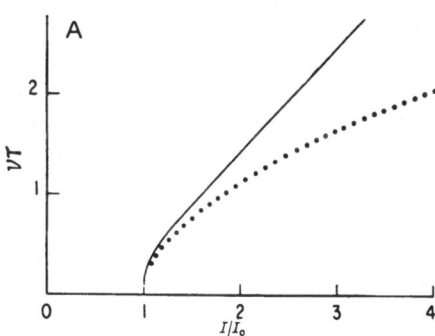

Fig. 9.4. (A) Frequency–current curves for a simplified neural model (Fig. 9.1B) in dimensionless coordinates. These coordinates are obtained by multiplying the firing rate $\nu$ in impulses/s by the time constant in seconds and measuring the current $I$ relative to the minimum or rheobasic current $I_0$ necessary for repetitive activity. The curves shown were computed with (dotted line) and without (solid line) a refractory period. (B) Frequency–current curve for the Hodgkin–Huxley equations at 6.4°C. See further discussion in text. (Modified from R. B. Stein, 1967.)

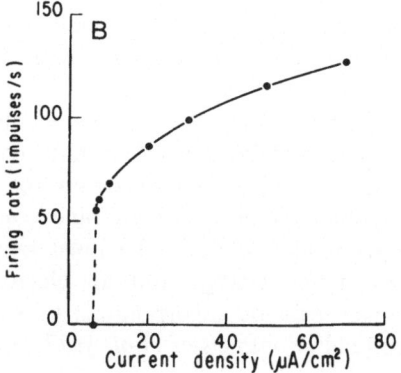

of depolarization to threshold can start almost immediately. However, the nerve impulse has a finite duration, and there is an *absolute refractory period* before the process can start again. If a refractory period $t_0$ is added to the total time in Equation (9.5), the frequency $\nu$ will be limited to a maximum value $1/t_0$, and the curve given by the dotted line in Fig. 9.4A results.

The model of Fig. 9.1B is exceedingly simplified compared to the Hodgkin–Huxley model considered in Chapter 5. For comparison, the rate of nerve impulses for the Hodgkin–Huxley equations was solved as a function of current density across the membrane (R. B. Stein, 1967). The results are shown in Fig. 9.4B and there are three points to note: (1) The model shows no adaptation (i.e., the firing rate remains virtually constant after the first couple of impulses, as long as the current is maintained), (2) only a limited range of firing rates is possible (between about 50 and 130 impulses/s), and (3) the firing rate is a nonlinear function of current density.

The lack of adaptation in the Hodgkin–Huxley equations is not surprising. The longest time constants are only on the order of a few ms, so one interval between impulses already represents several time constants. Hagiwara and Oomura (1958) found that squid axons do adapt to maintained constant currents, and this adaptation may be due to the longer time constant processes described by Chandler and Meves (1970), to the accumulation of $K^+$ ions in the limited extracellular space (Villegas *et al.*, 1962), or to the action of an electrogenic pump. The importance of the pumps is likely to be much smaller in these giant axons than in smaller cells for the reasons discussed above (see also Thomas, 1972), but the slow $Na^+$ inactivation of Chandler and Meves (1970) could well be important. The presence of longer time constants could also be important in extending the range of firing rates.

Connor and Stevens (1971) described a potassium current with a long time constant which is responsible for determining the rate of impulses in a molluscan neuron that shows firing rates over a wide range. This is the only such cell to date in which a full voltage clamp analysis has been performed and computations based on this analysis have been compared with experiments. Fohlmeister *et al.* (1977) have studied two sensory neurons, including the stretch receptor of the crayfish, which fire over a wide range of frequencies. In both neurons, they had to modify the model shown in Fig. 9.1B to account for their data by adding a slow time-dependent change in membrane conductance. Thus, it appears that the simple model of Fig. 9.1B must be modified to include at least a slow conductance change and an electrogenic pump, although these latter changes can be incorporated to some extent as feedback changes in threshold (Fohlmeister *et al*, 1977).

The third point mentioned above, the nonlinear relation between current and frequency, requires more discussion because it is at the heart of understanding the steady state input–output relations of sensory neurons.

## Input–Output Relations

The nonlinearity is not surprising in light of the simplified model discussed previously. However, experimentally many investigators have found a linear relation between firing rate and what is referred to as a generator or *receptor potential*. For example, Terzuolo and Washizu (1962) described such a relation in crayfish stretch receptors (see Fig. 9.3). The receptor potential is the steady depolarization observed near the receptor after nerve impulses have been blocked by passing a hyperpolarizing current or by applying a drug such as TTX. Since TTX has a specific action on the $Na^+$ pores underlying the action potential, its effects can be studied by removing the $Na^+$ permeability system in the Hodgkin–Huxley equations. The current–voltage relation which results (Fig. 9.5A) is still nonlinear in that the delayed rectification of the potassium system remains. The nonlinearity resulting from the delayed rectifier is in the opposite direction to that observed in Fig. 9.4B. If the firing rate that results from a steady current (under normal conditions) is plotted against the voltage that

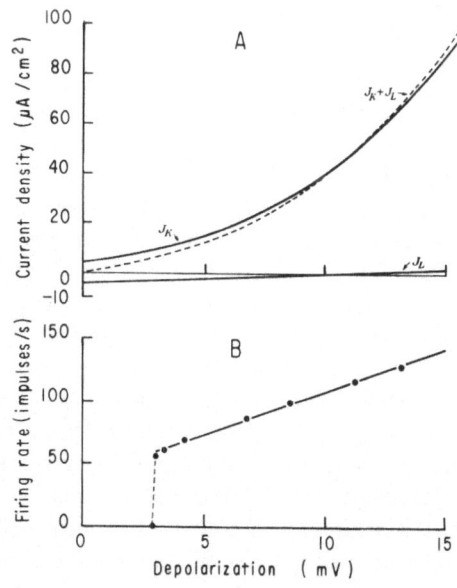

Fig. 9.5. (A) Steady state potassium $(J_K)$ and leakage $(J_L)$ current densities as a function of membrane depolarization for the Hodgkin–Huxley equations. (B) The steady firing rate produced by a constant current density in the presence of $Na^+$ currents (Fig. 9.4B) has been plotted against the steady depolarization this same current would produce after blocking the $Na^+$ currents (from part A using the line marked $J_K + J_L$). The implications of this linear relation are discussed further in the text. (From R. B. Stein, 1967.)

results from the same current in the presence of TTX (the absence of $Na^+$ currents), a linear frequency–voltage relation is found (Fig. 9.5B).

Thus, the linear relation between frequency and receptor potential may not have a simple ionic basis, but may result from cancellation of nonlinearities. Other so-called *psychophysical laws* may have a similar, indirect basis. Psychophysics refers to psychological judgments being made of physical quantities, and so the observations are often much less direct. The *Weber–Fechner law* (Fechner, 1966) predicts a logarithmic relation between stimulus magnitude and the perceived response. A logarithmic relation would have a negative curvature, as observed in Fig. 9.4B, and Agin (1964) found that the relation between firing rate and current for the Hodgkin–Huxley equations could be fitted by an equation of the form

$$v = k \log(I + I_0) \tag{9.6}$$

where $k$ is a constant, $I$ is the current in $\mu A/cm^2$, and $I_0$ is a constant. Agin's relationship does not take into account the properties of the transformation or transduction stages of sensory processing.

For several sensory modalities, Stevens (1961) found that a power function gave a better fit to the relation between stimulus and perceived intensity, a result which has become known as *Stevens' law*. By comparing the responses of single sensory neurons to different stimulus intensities and by making careful psychophysical measurements on monkeys to the same stimuli, Mountcastle and his colleagues (reviewed in Mountcastle, 1974) have tried to provide a physiological foundation for these laws. They found that the response of neurons could be well fitted by a power function of the form

$$v = k(S - S_0)^n + c \tag{9.7}$$

The constant $c$ was included because many neurons have a constant, spontaneous discharge of impulses in the absence of sensory stimulation. However, the exponent in the power function is an extra constant, and the improved fit of Equation (9.7) over (9.6) may simply represent the improvement expected when adding an extra undetermined constant. No convincing rationale for either Equation (9.7) or (9.6) has been given.

The only equation to have a clear physiological basis is the hyperbolic equation related to the Michaelis–Menten equation (3.5), as discussed in Problem 7.3, and in the section on crustacean stretch receptors. Both the Weber–Fechner law and Stevens' law may represent no more than an exercise in curve fitting. Nonetheless, if the same equation applies to single sensory receptors and judgments by whole organisms, further processing of the signals must be reasonably linear to avoid distorting the form of the relationship (but see Knibestöl and Vallbo, 1980). Before describing the functional properties of mammalian muscle receptors in more detail, the classification of fibers will be reviewed briefly.

## Classification of Nerve Fibers

Mammalian sensory and motor neurons are classified, at least in part, by the conduction velocity of their axons. The nomenclature is derived from the peaks of different latencies observed for compound action potentials in the 1930's (Erlanger and Gasser, 1937). This classification often leads to confusion, so it is best to review the notation (summarized in Table 9.1) before proceeding. First, Latin letters are used to refer to the major types of nerve fibers: *A* refers to the myelinated fibers in peripheral nerves, *C* to the small unmyelinated fibers in both peripheral and autonomic nerves, and *B* to the intermediate, often weakly myelinated fibers contained in the autonomic nerves which control many of the internal organs.

Within the *A* category, which will be of most interest here, a subclassification based on Greek letters is used, with *A*α being the largest axons and *A*δ being the smallest axons. These differences in size are correlated with differences in function: the α- and γ-subcategories contain motoneurons, which will be referred to as α- and γ-motoneurons, while the β- and δ-categories largely contain sensory fibers. There are also differences in function associated with fiber size. For example, the β-fibers innervate the large, specialized sensory endings of the skin which respond to light pressure, touch, vibration, etc., whereas the δ-fibers respond to temperature and grosser mechanical stimuli which may produce the sensation of pain.

To further confuse the issue, muscle sensory fibers are classified according to a separate scheme using Roman numerals. In general, group I

*Table 9.1. Classification of Nerve Fibers*

Nerve fibers are (1) classified using Latin letters according to the type of nerve (peripheral or autonomic) and the presence or absence of myelin, (2) subdivided further (type *A* fibers only) according to the size of their axons (using Greek letters), or (3) classified separately using Roman numerals (muscle sensory fibers only).

| Latin letter | Greek letter | Type of fibers | Roman numerals | Muscle afferents |
|---|---|---|---|---|
| *A* | | Myelinated, peripheral nerve | | |
| | α | Motoneurons | I | a. Primary muscle spindle |
| | | | | b. Golgi tendon organs |
| | β | Cutaneous, joint afferents ⎫ | | |
| | | | II | Secondary muscle spindle |
| | γ | Motoneurons ⎭ | | |
| | δ | Temperature, pain afferents | III | Muscle temperature, pain |
| *B* | | Myelinated, autonomic | | |
| *C* | | Unmyelinated | IV | Unmyelinated |

fibers correspond in size to the $A\alpha$-fibers in the other notation, group II fibers to $A\beta$- and $A\gamma$-fibers, group III to the $A\delta$-, and group IV to $C$-fibers. Group III and IV muscle sensory fibers do not differ greatly from the fibers of corresponding size in other tissues. They sense, for example, the temperature of the muscle and signal potential sources of damage to the muscle (such as lack of oxygen) by means of pain sensations. Some group II sensory fibers are also similar to specialized sense organs elsewhere. For example, Pacinian corpuscles, which sense vibration, are found in muscles as well as subcutaneous tissue.

Three types of sensory fibers are unique to muscle: The Golgi tendon organs and the primary and secondary muscle spindle sensory fibers. The first two of these are in the group I range and overlap considerably in diameter, although there is some difference on average in size. To indicate this difference, primary muscle spindle sensory fibers are often referred to as group Ia afferents and Golgi tendon organs as group Ib afferents. The

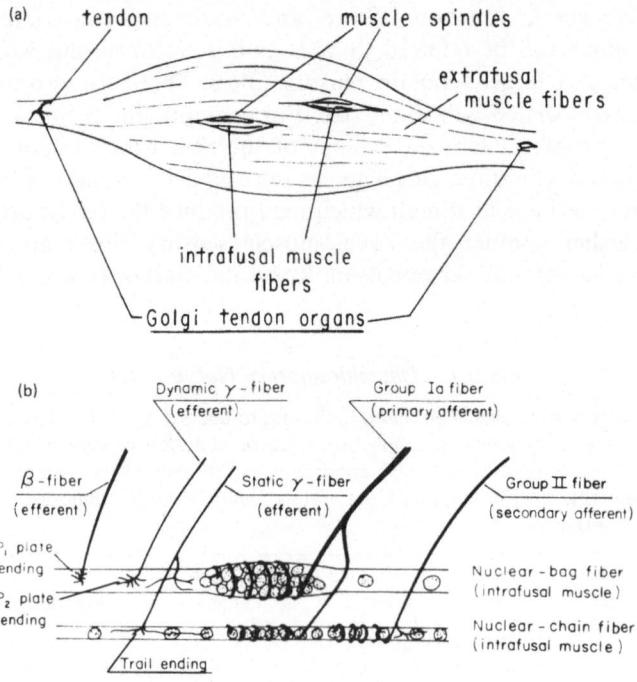

Fig. 9.6. (A) Schematic drawing (not to scale) showing the arrangement of tendon organs *in series* with the large extrafusal muscle fibers, and of muscle spindles which contain intrafusal muscle fibers *in parallel* with the extrafusal fibers. Contraction of the extrafusal fibers will stretch the tendons, but unload the muscle spindles. (B) Internal structure of a mammalian muscle spindle showing the organization of intrafusal muscle fibers, as well as sensory (afferent) and motor (efferent) nerve fibers. (From Stein, 1974.)

term *afferent* is a synonym for sensory fiber and comes from the Latin for "carrying to," since information is carried to the spinal cord. Similarly, motor fibers are often termed *efferents*, since they carry information away from the spinal cord.

Golgi tendon organs are found in small portions of tendons, and stretching the whole muscle sufficiently will produce nerve impulses. However, Golgi tendon organs are much more sensitive to the stretch imposed by stimulation of the few motor units which have muscle fibers inserting into the particular piece of tendon where they are situated (Jansen and Rudjord, 1964; Houk and Henneman, 1967). Golgi tendon organs could be useful in feeding back detailed information on the way in which a contraction directed by the central nervous system is proceeding (see Chapter 10). By responding to the internal contractions of the muscle, the Golgi tendon organs are also referred to as being *in series* with the muscle fibers. This distinguishes them from muscle spindle afferents, which are *in parallel* with the main contractile elements of the muscle (see Fig. 9.6A).

## Muscle Spindles

Figure 9.6B shows the complex internal structure of a mammalian muscle spindle. As well as the two types of afferent or sensory fibers mentioned above, the muscle spindle contains several types of specialized muscle fibers which are referred to as *intrafusal* muscle fibers because they lie within the spindle or fusiform shape of the muscle spindle. By contrast, the normal muscle fibers lying outside the muscle spindle are referred to as *extrafusal* muscle fibers. In the mammalian muscle spindle, but not that of lower vertebrates (Gray, 1957; Matthews and Westbury, 1965), the motor supply to the intrafusal and extrafusal muscle fibers is largely separate. The larger or $\alpha$-motoneurons innervate the extrafusal muscle fibers to form the motor units which were discussed in the last chapter, while the smaller $\gamma$-motoneurons innervate the intrafusal muscle fibers. Even in mammals there are some intermediate-sized motoneurons, often referred to as $\beta$-motoneurons, which innervate both intrafusal and extrafusal muscle fibers.

The functional differences between primary and secondary muscle spindle afferents were first analyzed using ramp changes in muscle length (reviewed by Matthews, 1964). A ramp waveform (see Fig. 9.7) contains a constant velocity portion followed by a constant length portion, and is therefore a good choice for determining the relative sensitivity of the two types of afferents to length and velocity. The sensitivity to length changes of several millimeters is similar in the two types of afferents (Fig. 9.7B), but the primary afferents are markedly more sensitive to the velocity of stretch.

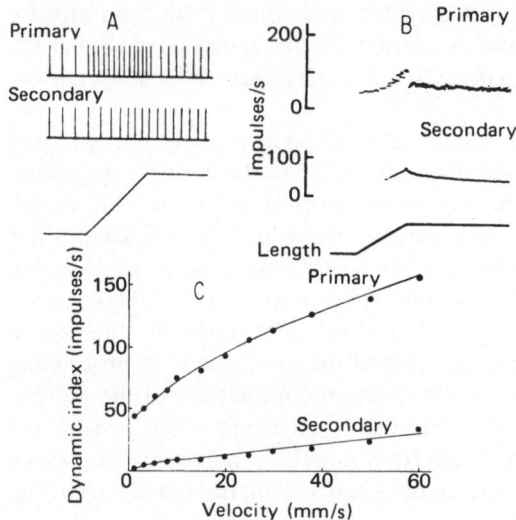

Fig. 9.7. (A) Schematic representation of the response of primary and secondary afferents to a ramp stretch. Note the higher firing rate of the primary ending during the constant velocity of the stretch. (Modified from Matthews, 1964.) (B) Data from primary and secondary muscle spindle afferent endings to stretches of 3 mm length at 5 mm/s. Each dot represents the inverse of the time interval since the last impulse, which gives the "instantaneous" firing rate in impulses/s. (From Brown et al., 1965.) (C) The difference in firing rate at the end of the stretch, compared to its value 0.5 s later (dynamic index), is much larger for the primary ending than for the secondary ending at all velocities of stretch. (From Matthews, 1972.)

One simple explanation for the velocity sensitivity of the primary afferents follows from the viscoelastic properties of the intrafusal muscle fibers, as suggested by Matthews (1933). The primary afferent has endings on both nuclear bag and nuclear chain intrafusal muscle fibers (Fig. 9.6B), whereas the secondary afferent ends almost exclusively on the nuclear chain fibers. The two types of intrafusal muscle fibers are named according to the placement of their nuclei. Intrafusal muscle fibers, like the extrafusal fibers, are formed by the fusion of a number of embryonic muscle cells or *myoblasts* and contain a number of nuclei. The nuclei are arranged in a prominent baglike swelling in the *nuclear bag* intrafusal fibers, or more uniformly in a row or chain along the length of the *nuclear chain* intrafusal fibers.

The swollen bag region in the center of the nuclear bag fiber is almost devoid of myofibrils, and is stretched more easily by the constant velocity portion of the ramp stretch. At a constant length, the more viscous ends of the fiber containing the myofibrils would slowly stretch out and relieve the tension from the central region. Because of the more uniform structure of the nuclear chain fibers, differential viscoelastic effects should be less prominent.

Poppele (1973) verified quantitatively by high-speed cinemicroscopy that the viscoelastic properties of intrafusal muscle fibers can account for the slower changes underlying the high velocity sensitivity of primary afferents. Thus, some of the adaptation at a constant length arises in the transformation stage from muscle length to tension or stress on the afferent ending.

The two types of afferents also differ in their responses to activity in γ-motoneurons. One type of γ-motoneuron, the *dynamic* γ-motoneuron, only affects primary afferents since it innervates nuclear bag intrafusal muscle fibers on which only primary afferents end (Fig. 9.6B). The term "dynamic γ-motoneuron" arises from the greatly increased sensitivity this motoneuron produces in the afferent's response to the velocity component of large ramp stretches (Fig. 9.8). The anatomy again gives a clue to the mode of action of this fiber. By causing contraction at the ends of the muscle fiber, the dynamic γ-motoneuron will stretch the central region where the muscle spindle endings lie. Furthermore, by making the ends stiffer and more viscous, the transient effects of the constant velocity stretch will be even greater in the central regions.

The other type of γ-motoneuron is the *static* γ-motoneuron, which increases the response of primary and secondary muscle spindle afferents

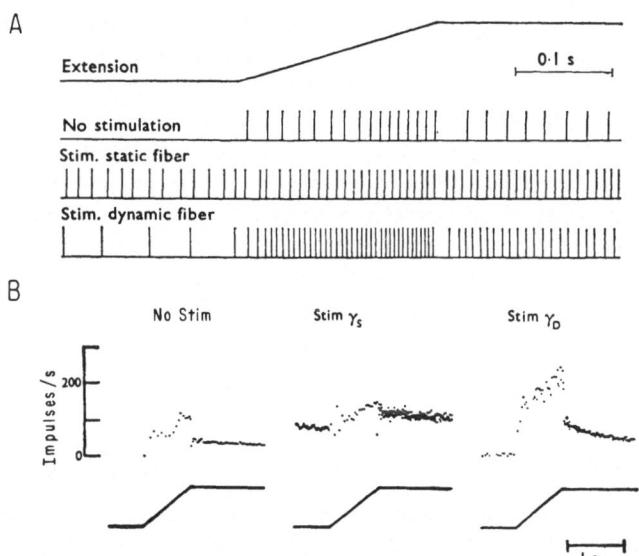

Fig. 9.8. (A) Different effects of stimulating single static and dynamic γ-motoneurons on the response of a primary ending to a ramp stretch. The stimulation was continuous at 70 impulses/s and produced a maintained increase throughout the records shown. Note, however, the extra effect of the dynamic γ-motoneuron during the constant velocity phase of stretch. (B) Plot of results similar to those shown in (A) in terms of the instantaneous firing rate (see Fig. 9.7). The difference in rate between the beginning and ending of the stretch is relatively constant (static response) in all three situations. However, the response during the constant-velocity phase of the stretch (dynamic response) is markedly increased by stimulating the dynamic γ-fiber ($γ_d$), but not the static γ-fiber ($γ_s$). (From Matthews, 1972.)

at a constant length. The response to the constant velocity portion of large ramp stretches (Fig. 9.8) may be unchanged, or even depressed, which is understandable in terms of the more uniform structure of the nuclear chain muscle fibers on which the static $\gamma$-motoneurons end. Why these fibers also end on nuclear bag fibers is less clear, although evidence has been presented in recent years that there are really two types of nuclear bag fiber (termed $bag_1$ and $bag_2$; Barker *et al.*, 1976). Although some degree of overlap may exist, the dynamic and static $\gamma$-motoneurons mainly end on different bag fibers, which themselves have dynamic and static properties respectively (Boyd *et al.*, 1977). The anatomical connections remain somewhat controversial, but the functional differences between the two types of $\gamma$-fibers are clear (Emonet-Dénand *et al.*, 1977). The intermediate or $\beta$-motoneurons innervate small, slow extrafusal muscle fibers and have a predominantly dynamic effect on muscle spindle afferents via innervation of nuclear $bag_1$ intrafusal fibers (Barker *et al.*, 1977).

Table 9.2 summarizes the pattern of excitatory connections, which involve at least three types of fibers in each column. The end result is that three sensory variables: length, velocity, and tension, can be signaled to the central nervous system. The $\gamma$-motoneurons regulate the extent to which two of these, length and velocity, are signaled. The rationale for this complexity of organization will become clearer in the next chapter, "Control of Movement."

Although the ramp stretches discussed in the previous section were important in defining the types of muscle receptors and $\gamma$-motoneurons in mammalian muscles, they represent a somewhat unnatural stimulus. More commonly, muscles go through cyclic contractions and relaxations, so the response to sinusoidal length changes of varying frequency and amplitude is of interest. Figure 9.9 shows the sinusoidal analysis that can be carried out on muscle or other sensory receptors. In response to a small length change (0.1 mm at 1 Hz), the rate of impulses from a primary muscle

### Table 9.2. Pattern of Excitation for Muscle Receptors

Four distinct types of motoneurons are involved, as indicated by the arrows, in innervating four types of muscle fibers and exciting three types of sense organs which can signal three different sensory variables. The connection indicated by the dashed arrow is controversial (Barker *et al.*, 1976; Boyd *et al.*, 1977).

| Motoneuron | Muscle fiber innervated | Sensory fiber excited | Adequate stimulus |
|---|---|---|---|
| $\alpha$ ⟶ | Extrafusal ⟶ | Golgi tendon organ | Tension |
| $\beta$ ⟶ | | | |
| Dynamic $\gamma$ ⟶ | Nuclear $bag_1$ intrafusal ⟶ | Primary muscle spindle | Velocity + length |
| Static $\gamma$ ⇢ | Nuclear $bag_2$ and chain intrafusal ⟶ | Secondary muscle spindle | Length |

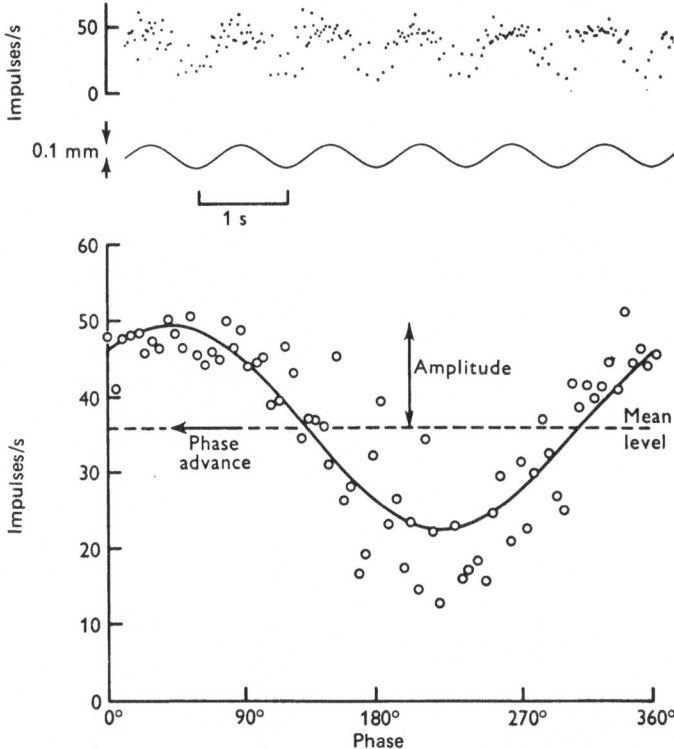

Fig. 9.9. Response of a primary ending to sinusoidal length changes of 0.1 mm amplitude at 1 Hz. The instantaneous firing rate for several cycles is shown above and the average rate for each 5° of the cycle is plotted below for ten cycles. The best fitting sinusoid is also shown (solid line) from which the amplitude and phase of the response can be measured. (From Matthews and Stein, 1969a.)

spindle afferent in the cat is clearly modulated. Each point in the upper part of this figure represents the reciprocal of the time interval since the last nerve impulse. For example, if the time since the last spike was 20 ms, this corresponds to an *instantaneous firing rate* of 50 impulses/s. Averaging the data over ten cycles gives a clearer picture of the modulation in firing rate and the best sinusoid can be fitted to the data. This sinusoid gives the amplitude of the modulation in impulses/s and the phase change of the response with respect to the sinusoidal input, which will have its peak at 90°.

For different amplitudes of stretch at one frequency, a striking difference between the primary and secondary endings was observed, which is illustrated in Fig. 9.10A. The primary afferents were more responsive to very small stretches than were the secondaries. The *sensitivity* of the

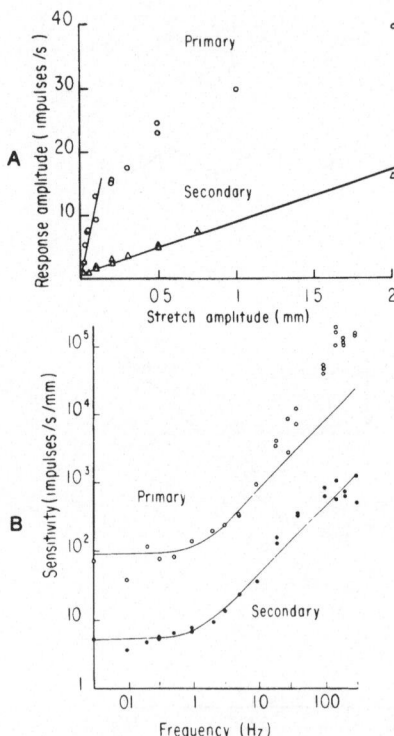

Fig. 9.10. (A) The responses of the primary and secondary endings to different amplitudes of stretching at 1 Hz. Note the marked difference in the responses to small stretches. The fitted straight lines measure the sensitivity of the endings in terms of their responses in impulses/s per mm of stretch. (B) Comparison of the sensitivity to sinusoidal stretching within the linear range of a primary and a secondary ending at various frequencies. The continuous lines represent the vector sums of a length component (horizontal segment) and a velocity component (diagonal segment). As a first approximation, the same curve transposed vertically fits both endings, indicating that they differ mainly in their absolute sensitivities. (From Matthews and Stein, 1969a.)

endings can be measured in terms of the modulation amplitude in impulses/s per mm of stretch, which is the slope of the straight lines shown in this figure. Above 0.1 mm the primary afferent no longer responds linearly, and the slope of the curve declines toward that of the secondary. This explains why a significant difference in sensitivity was not observed using stretches of several millimeters (Fig. 9.7).

In order to compare the responses at different frequencies, the amplitudes of the stimuli were kept small enough that the afferents responded linearly. The sensitivity varies dramatically with frequency as shown in Fig. 9.10B. The fitted curve assumes that the response $y(t)$ is given by the sum of the length $x(t)$ and the velocity $\dot{x}(t)$ of the stretches,

$$y(t) = a[x(t) + b\dot{x}(t)] \qquad (9.8)$$

where $a$ and $b$ are constants. Velocity is the first derivative of length and a derivative is often indicated by a dot as in Equation (9.8). Since $y(t)$ will have the dimensions of impulses/s and $x(t)$ the units of mm, $a$ has the units of sensitivity, $\text{impulses s}^{-1}\,\text{mm}^{-1}$. Similarly, since $\dot{x}(t)$ has the units of $\text{mm s}^{-1}$, the constant $b$ has the units of s. Taking Laplace transforms of

Equation (9.8) gives

$$y^*(s) = ax^*(s)(1 + sb) \qquad (9.9)$$

At lower frequencies ($sb \ll 1$), the response will be independent of frequency and the muscle receptor will have the sensitivity given by the constant $a$. At high frequencies ($sb \gg 1$), the response will increase linearly with frequency. (Note that the velocity of a sinusoidal length change increases linearly with frequency, since the same distance must be covered in a shorter time.) The velocity of a sinusoid also leads the length change by 90° since the peak velocity will occur when the length change is at zero, not its peak. Thus, the response leads the sinusoidal input considerably in phase at higher frequencies. The frequency $s = 1/b$ represents the transition or *break frequency* between the low- and high-frequency regions. For primary afferents $b = 0.08$ s, so $b^{-1} = 12$ radians/s, or 2 Hz.

The curve fitted to the secondary has an identical break frequency, but the absolute sensitivity ($a$) is an order of magnitude lower. Thus, the difference between primaries and secondaries in the small amplitude response is a difference in overall sensitivity, not a difference in velocity sensitivity alone, as was observed with large amplitude ramp stretches. The sensitivity of the primary afferent increases substantially faster than predicted at higher frequencies, which suggests that the ending may be sensitive to acceleration as well as velocity and length. The phase advance at these frequencies may also be greater than 90°. Using ramp stretches, Schäfer (1967) proposed an acceleration response for primary afferents, although Brown *et al.* (1969) provided an alternative explanation of the data, based on the properties of intrafusal muscle. Poppele and Bowman (1970) developed a more complex expression than Equation (9.9) that accounts for most of the curve except where the curve levels out above 100 Hz. [See Goodwin *et al.* (1975) for further details of very high frequencies.]

Unexpected results were also obtained with stimulation of γ-motoneurons (Fig. 9.11). Stimulation of dynamic γ-motoneurons had relatively little effect on the modulation of the afferent discharge for small sinusoidal length changes, although the sensitivity was a little less at low frequencies and a little higher at very high frequencies. In contrast, as mentioned previously, dynamic γ-stimulation markedly increases the velocity sensitivity during large ramp stretches.

The characteristics of the intrafusal muscle fiber may explain these results. Nuclear bag muscle fibers, on which the primary afferents end, are slow muscle fibers which show little or no twitch to single impulses in the γ-motoneurons (Smith, 1966; Bessou *et al.*, 1968). Although the small intrafusal fibers are difficult to study directly, the short-range elasticity (Chapter 8) is much more prominent in slow muscle fibers of the frog than in the faster twitch fibers (Brown, 1971). Because of the stiffness of this

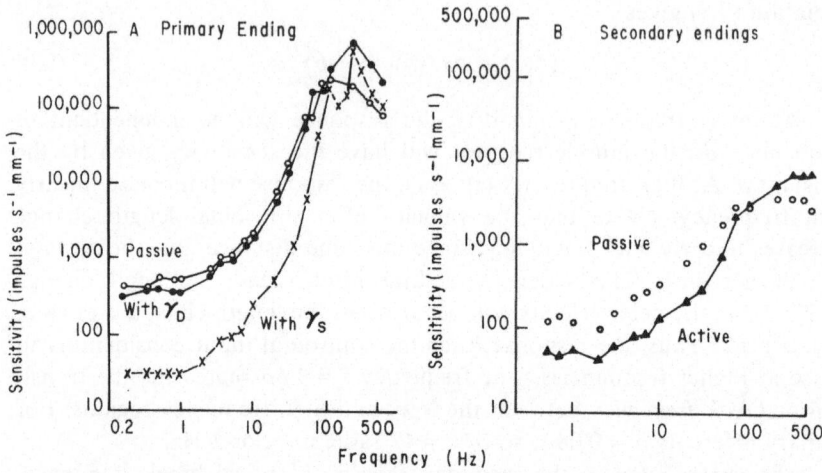

Fig. 9.11. Effect of the sensitivity of (A) primary and (B) secondary muscle spindle afferents to sinusoidal stretches when stimulating single γ-motoneurons. Both dynamic ($\gamma_D$) and static ($\gamma_S$) motoneurons affect primary afferents, while only activity in static γ-motoneurons affects secondaries. [From (A) Goodwin *et al.*, 1975; and (B) Cussons *et al.*, 1977.]

elastic element, substantial tension will be exerted over a short distance, which may account for the increased sensitivity of the primary afferent to small stretches (Fig. 9.10). With larger stretches, the bonds underlying the short-range elasticity will break and reform, thus providing a viscous drag and an enhanced velocity sensitivity in the primary afferents (Fig. 9.7). Huxley and his colleagues (Chapter 8) showed that the bonds can rapidly be converted from one state to another (e.g., from an active complex to a rigor complex) with resultant changes in tension having a time constant near 1 ms. Thus, only with the highest frequencies of stretch (a frequency of 100 Hz corresponds to a rate constant of 200$\pi$ radians/s or a time constant of 1.6 ms) will the full effect of these bonds be evident in the sensitivity of the muscle receptors.

The lowered sensitivity of the muscle receptors to small stretches during static fusimotor activity may represent a switching of impulse generation from a site on a nuclear bag to one on a nuclear chain intrafusal muscle fiber. Stimulating combinations of static and dynamic γ-motoneurons (Lennerstrand, 1968; Hulliger *et al.*, 1977) produces a complex mixture of summation and occlusion. *Occlusion* of two responses can occur if there are separate impulse generators on different intrafusal muscle fibers. The most active generator serves as the pacemaker for the sensory fiber and switching may even occur during a single cycle of a sinusoidal stretch. For example, dynamic effects (transmitted via a nuclear bag fiber) may dominate during a stretch of the muscle, and static effects (transmitted by a nuclear chain fiber) may dominate during the release phase of the sinusoidal length change (Matthews, 1972). Whether the muscle properties can account fully and quantitatively for the exact shape of these curves remains to be tested.

Whatever the mechanisms are, the primary muscle spindle afferent is exceedingly sensitive to small length changes. A sensitivity of $10^5$ impulses $s^{-1}\,mm^{-1}$ means that a stretch with an amplitude of $0.1\ \mu m = 10^{-4}$ mm to a muscle several cm in length will produce a modulation in the discharge of the order of 10 impulses/s. With a vibratory length change of a few $\mu m$ at 100 Hz or more, the afferent may fire one impulse per cycle at a particular phase of the cycle (Brown *et al.*, 1967), a phenomenon known as *phase-locking*. This phenomenon and its significance will be discussed in more detail later. With smaller amplitudes the afferent may fire only one impulse every few cycles and the impulse may occur at most phases of the cycle. The modulation then merely represents a different density of impulses at different phases when averaged over a large number of cycles. Rather than pursuing the details of particular sensory fibers further, the final sections of this chapter discuss general methods which can be applied to analyze a wide range of sensory systems.

## Frequency–Response Curves

Curves such as shown in Figs. 9.10B and 9.11 are referred to as *frequency–response curves*. A frequency–response curve was also given for a mammalian muscle in Fig. 8.13C. A complete *frequency–response function* includes the phase changes as well as the gain (see Fig. 8.13C) and is conventionally displayed in a *Bode plot*. The amplitude and frequency are plotted on logarithmic scales in a Bode plot and the phase on a linear scale. The detailed characteristics of Bode plots and frequency response functions are given in many introductory texts for *control theory* (e.g., D'Azzo and Houpis, 1966; Milsum, 1966). Only a few particularly important points will be discussed here in relation to muscle receptors. The logarithmic scales for amplitude and frequency are useful in condensing a wide range of data. Another useful property of a log–log plot is the fact that a power function is displayed as a straight line. Thus, the response of muscle falls off as the square of frequency since the slope in Fig. 8.13C is $-2$ at high frequencies, whereas the response of muscle receptors increases as the first (velocity) or second (acceleration) power of frequency over certain ranges (Fig. 9.11).

Any input waveform can be constructed from a sufficient number of sinusoids of various frequencies according to the Fourier theorem. Thus, the frequency response function for a linear system can be used to predict

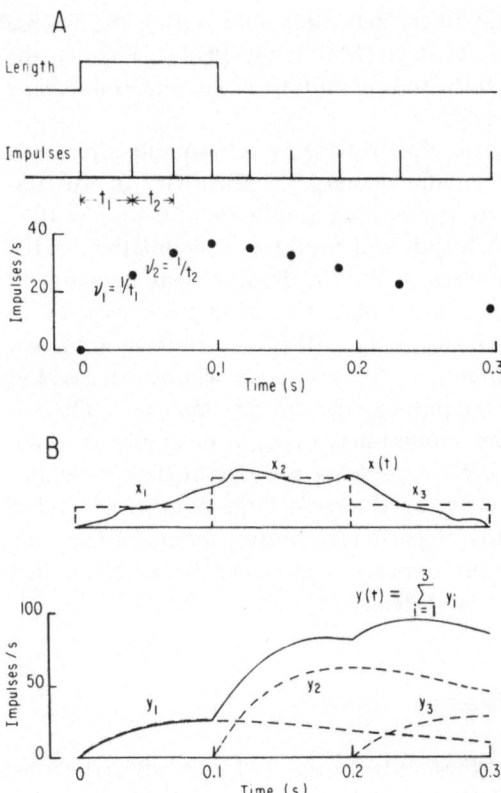

Fig. 9.12. The response to one input (A) can be used to predict the response to a second input of any shape (B) in a linear system. The input in (A) is a brief square wave; the response is a series of nerve impulses. The instantaneous firing rate in impulses/s is plotted after taking the reciprocal of each interval. Other inputs $x(t)$ in (B) can be approximated using a series of these square waves with various amplitudes $(x_1-x_3)$. In a linear system, the responses to a series of inputs will sum linearly so the output $y(t)$ would be produced. By making the square waves in (A) briefer and taller (so that in the limit they approach Dirac $\delta$ functions), the approximation to the input can be improved and the sum approaches the convolution integral given as Equation (9.10) in the text.

the response to any input waveform. The converse is also true in that any input waveform can be used to predict the frequency–response function. To see how this might be done, consider a situation in which only the response of a muscle receptor to a brief perturbation, such as a tendon tap, is known. This perturbation is idealized as a square wave in Fig. 9.12 and the response is referred to as $z(t)$. Then, as shown in the lower part of this figure, the response to any other arbitrary waveform could be constructed by linearly summing or superimposing the responses to successive square waves of varying amplitude and timing. The response $y(t)$ at time $t$ will be obtained by summing or integrating the responses which would be produced by the inputs occurring at all earlier times $u$ according to the convolution integral (9.10):

$$y(t) = \int_0^t x(u)z(t - u)\,du \qquad (9.10)$$

The Laplace transform of a convolution integral in this form is a simple product

$$y^*(s) = x^*(s)z^*(s) \qquad (9.11)$$

Note that, as in Chapters 6 and 8, $z^*(s)$ denotes the *transfer function*, since it determines according to Equation (9.11) how any input $x$ is transferred or transformed into an output $y$. Note also that the form of the input or output in Equations (9.10) and (9.11) has not been specified. $z^*(s)$ might be determined from one input, namely $x_1(t)$, according to the formula

$$z^*(s) = y_1^*(s)/x_1^*(s) \tag{9.12}$$

and then used to predict the response to a second input $x_2(t)$. From Equation (9.11), $y_2^*(s) = x_2^*(s)z^*(s)$, and $y_2(t)$ can be obtained by taking the inverse Laplace transform.

The transfer function is also closely related to the frequency–response function because of the connection between Laplace transforms and Fourier transforms (Chapter 7). Formally, the frequency–response function can be written as $z^*(j\omega)$ where the frequency $\omega = 2\pi f$ in radians/s has been substituted for the Laplace variable [$f$ is a sinusoidal frequency in Hertz and $j = (-1)^{1/2}$]. To simplify the notation somewhat, the letter $j$ will be dropped, but it is important to remember that $z^*(\omega)$ is a complex number as a result of the substitution

$$z^*(\omega) = z_r + jz_i \tag{9.13}$$

where $z_r$ and $z_i$ are real functions of frequency. The amplitude, or *gain*, of the frequency–response function (indicated by vertical lines) is then given by

$$|z^*(\omega)| = \left(z_r^2 + z_i^2\right)^{1/2} \tag{9.14}$$

and the phase angle is given by

$$\angle z^*(\omega) = \tan^{-1}(z_i/z_r) \tag{9.15}$$

The reason for using the term gain should be clear from Equations (9.12) and (9.14) since the magnitude of the ratio of output to input is measured. Whereas the input–output relations considered in a previous section dealt mainly with the steady state situation (near 0 Hz), the frequency–response function, as the name implies, gives the gain at all frequencies. As an example of the use of these formulas, the transfer and frequency response functions for Equation (9.9) are considered in Problem 9.1, and the derivation of another transfer function from data is considered in Problem 9.2. The transfer and frequency response functions of the crustacean stretch receptor and the muscle spindle secondary afferent are discussed further in Problems 9.3 and 9.4.

Up to this point, we have considered three distinct methods of deriving a frequency–response function. First, it can be derived from any simple and convenient waveform. This might be a brief perturbation to a muscle or a flash of light to a photoreceptor. If sufficiently brief, such responses are treated mathematically as Dirac $\delta$ functions. Alternatively, a step or a ramp change in muscle length or light intensity can be applied.

The Laplace transforms of all these input waveforms are well known and tabulated (Selby, 1975). The output must then be Laplace or Fourier transformed, either numerically or after first fitting the data with a simple function whose transform is known. The advantage of using these simple waveforms is speed, but they offer no direct test of linearity.

However, Equation (9.11) can be used to test predictions based on the assumption of linearity. For example, Brown and Stein (1966) derived a transfer function for the slowly adapting stretch receptor of the crayfish from the step response and used it to predict the response to ramps and sinusoids (see Problem 9.3). Another problem is that each waveform has a particular and limited range of frequencies. A step function has a transfer function $x^*(s) = 1/s$, so that high frequencies are under-represented. On the other hand, a Dirac $\delta$ function is so brief that the response may be a single impulse. This response does not adequately represent the frequency response of the neuron because it could not respond to another Dirac $\delta$ function until after the refractory period. Nor could it necessarily respond to a Dirac $\delta$ function of twice the size with two impulses (i.e., the response will not be linearly related to the input).

Some of these difficulties can be overcome by using the second method outlined earlier, direct determination of the frequency response by using sinusoids over a wide range of frequencies. There is then a partial check on linearity since the output of a linear system should always be a sinusoid of similar frequency, only changed in amplitude and phase as indicated by the Bode plot. Nonlinearity becomes apparent as a deviation from a single sinusoid and is referred to as harmonic distortion. With increasing amplitudes, the modulation will increase until there is a silent period in each cycle, since the firing rate can never go negative (the minimum firing rate is zero). Similarly, the maximum firing rate will be limited by refractory periods. These limits produce what are known as saturation nonlinearities. To check comprehensively for linearity, a range of amplitudes must be used at each frequency, as well as some measure of harmonic distortion. This becomes exceedingly tedious, and may be unsatisfactory for many biological systems where recording time is limited and the response varies with time. For example, muscles fatigue, so the response to the initial input will no longer be the same after measuring the response to a number of frequencies.

For this reason, a third method based on *spectral analysis* is gaining increasing popularity (Marmarelis and Marmarelis, 1978). This method was introduced briefly in the last chapter to analyze the response of a muscle to random trains of impulses (see Fig. 8.13B). Similarly, it can be used with receptors to analyze the train of impulses produced by a random stimulus. The input waveform is often band-limited white noise (i.e., a signal which contains equal components of all frequencies within a certain

## Table 9.3. Definition of Terms Used in Spectral Analysis

Method for computing the frequency response function and related quantities from input and output functions and their spectra. Further details are given in the text.

| Term | Input $x(t)$ | Output $y(t)$ |
|---|---|---|
| Fourier transform | $x^*(\omega) = \int e^{-j\omega t} x(t)\, dt$ | $y^*(\omega) = \int e^{-j\omega t} y(t)\, dt$ |
| Fourier coefficients | $= x_r(\omega) - jx_i(\omega)$ | $= y_r(\omega) - jy_i(\omega)$ |
| Complex conjugate | $= x_r(\omega) + jx_i(\omega)$ | $= y_r(\omega) + jy_i(\omega)$ |
| Input spectrum | $S_{xx}(\omega) = [x_r(\omega) - jx_i(\omega)][x_r(\omega) + jx_i(\omega)]$ | |
| | $= x_r^2(\omega) + x_i^2(\omega) = \|x^*(\omega)\|^2$ | |
| Output spectrum | $S_{yy}(\omega) = [y_r(\omega) - jy_i(\omega)][y_r(\omega) + jy_i(\omega)]$ | |
| | $= y_r^2(\omega) + y_i^2(\omega) = \|y^*(\omega)\|^2$ | |
| Cross spectrum | $S_{xy}(\omega) = [x_r(\omega) - jx_i(\omega)][y_r(\omega) + jy_i(\omega)]$ | |
| | $= x_r(\omega)y_r(\omega) + x_i(\omega)y_i(\omega)$ (real part) | |
| | $+ j[x_r(\omega)y_i(\omega) - x_i(\omega)y_r(\omega)]$ (imaginary part) | |
| Frequency–response function | $z^*(\omega) = S_{xy}(\omega)/S_{xx}(\omega)$ | |
| Coherence function | $\gamma^2(\omega) = \dfrac{\|S_{xy}(\omega)\|^2}{S_{xx}(\omega)S_{yy}(\omega)}$ | |
| Cross correlation function | $C(t) = \int e^{j\omega t} S_{xy}(\omega)\, d\omega$ | |

band of frequencies). From the input and the output, the input spectrum, the output spectrum, and the cross spectrum (see Table 9.3) can be calculated through well developed methods (Bendat and Piersol, 1971; French and Holden, 1971a, b).

If the input is band-limited white noise, the input spectrum will be approximately a rectangular function, having a constant value over a range of frequencies and very little power outside this range. The output will be a series of pulses, and the form of its spectrum will be discussed further in the next section of this chapter. The output spectrum will clearly depend on the sensory receptor chosen. The cross spectrum is of particular interest, since it depends both on the input and the output. In fact, the cross spectrum is the Fourier transform of the cross-correlation function, and represents the correlated power between input and output as a function of frequency. The frequency response function can be calculated by dividing the cross spectrum by the input spectrum (see Table 9.3 or Bendat and Piersol, 1971). The input and output spectra are real functions, but the cross spectrum has a real and an imaginary part. Thus, the frequency response function derived in this way contains both gain and phase information (see also Problem 9.5).

Furthermore, a quantity known as the *coherence function* can be computed, which is a normalized measure of the extent to which the linear frequency response function completely accounts for the data. The coherence function will be less than its maximum value of one at any frequency if (1) there are significant nonlinearities or (2) the response shows inherent variability (i.e., it is statistical or *stochastic* in nature). As seen in Fig. 9.9, neuronal responses invariably show some variability, and averaging was used in the sinusoidal analysis to eliminate this variability. The spectral analysis does not eliminate this variability, but it gives some measure of its magnitude through the coherence function. The implications of this variability, and other methods used for describing it, will be discussed in the next sections of this chapter.

Finally, to the extent that there are nonlinearities, the linear analysis can be extended in a logical fashion using the series of Wiener kernels to describe the nature of the nonlinearities (Wiener, 1958). This topic is outside the scope of this volume, but has recently been presented in considerable detail for biological systems (Marmarelis and Marmarelis, 1978). Since all the frequencies are mixed together in the random input waveform, time-dependent changes such as fatigue are minimized and the frequency response can be determined in minutes rather than hours using the sinusoidal analysis. This is still longer than the seconds that are required for analysis with step or ramp inputs, but spectral analysis has several advantages. The linearity of the system can be tested, confidence limits can be determined (see Fig. 8.13 and Bendat and Piersol, 1971), and where necessary, the analysis can be extended to consider nonlinearities.

## Statistical Properties of Impulse Trains

In considering the input–output properties and frequency–response characteristics of nerve cells, the inherent biological variability has been ignored or concealed by averaging. Even under conditions where the mean rate of impulses is virtually constant, all nerve cells show some variability in the intervals between successive impulses (*interspike intervals*). Different nerve cells show very different levels of variability so that systematic methods for analyzing and displaying the variability of nerve cells have been developed (Moore *et al.*, 1966; Gerstein and Perkel, 1972). In this section, some of the methods will be reviewed which describe the statistical properties of those points in time at which impulses occur. This represents one aspect of a general field known as *stochastic point processes* (Cox and Miller, 1965). The implications of neuronal variability for the transmission of information by nerve cells will be discussed in the final section.

One of the commonest methods of displaying data is in the form of an *interval histogram*, which shows the number of intervals from a given sample which have various durations (see Fig. 9.13). The number will clearly depend on the total number of impulses in the sample and the

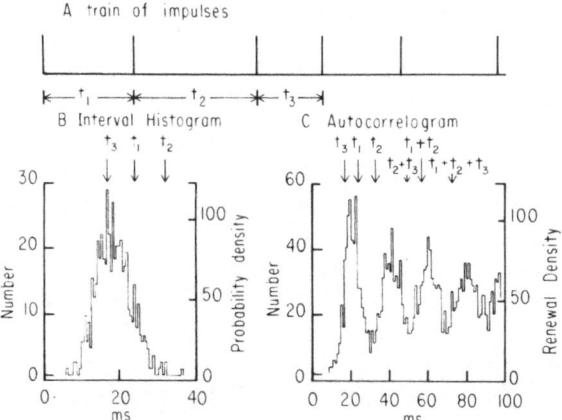

Fig. 9.13. The intervals ($t_1$, $t_2$, etc.) between successive impulses in a train (A) can be used to construct an interval histogram (B). Each interval is measured to a given accuracy [the bin width in (B) is 0.5 ms] and the number of intervals with each duration is plotted. The number can be converted to the form of a probability density measured in impulses/s by dividing by the bin width and the total number (500) of intervals measured. In the interval histogram of (B) the mean interval ($\mu$) is about 20 ms and the standard deviation ($\sigma$) about 4 ms. If the intervals between any two spikes are measured irrespective of whether other spikes have intervened, an autocorrelogram (C) is generated. After adjusting for bin width (1 ms) and total number (500) of spikes used as the starting point for interval measurements, a renewal density in impulses/s is obtained analogous to the probability density in (B). The dispersion around successive peaks increases ($t_1 + t_2$ actually falls within the third peak, not the second) and the peak heights decline until a steady density ($\nu = 1/\mu$) of about 50 impulses/s is reached.

accuracy with which the measurements are made (the width of the bins in ms in Fig. 9.13). This dependence can be eliminated by converting the histogram into a *probability density function* $f_1(t)$ which is formally defined as [cf. Equation (7.2)]

$$f_1(t) = \lim_{\Delta t \to 0} \frac{\text{(probability of an interval between } t \text{ and } t + \Delta t)}{\Delta t} \tag{9.16}$$

Experimentally, the number in each bin is divided by the total number to give a probability (the sum of all the probabilities must equal one). Then, the probability is divided by the bin width ($\Delta t$) to obtain the probability density which has the units of impulses/s. If the sample is sufficiently long, a very small value of $\Delta t$ can be used to improve the resolution.

Various parameters can be calculated; for example, the mean interval $\mu$ can be determined either from the original data or the probability density function

$$\mu = \frac{1}{N} \sum_{i=1}^{N} t_i = \int_0^\infty t f_1(t)\, dt \tag{9.17}$$

Similarly, the standard deviation $\sigma$ can be calculated in two ways:

$$\sigma = \frac{1}{N} \sum_{i=1}^{N} (t_i - \mu)^2 = \int_0^{\infty} (t_i - \mu)^2 f_1(t)\, dt \qquad (9.18)$$

The mean interval is just the inverse of the mean rate of firing ($\mu = 1/\nu$), while the standard deviation gives the dispersion about this mean. A useful quantity is the *coefficient of variation* ($\sigma/\mu$) which gives a relative measure of the variability.

Figure 9.14 shows four different examples ranging from a completely regular train of impulses in which $\sigma = 0$ and hence the coefficient of variation is zero, to an exponential distribution with a coefficient of variation of one. An exponential distribution arises, for example, from the release of transmitter quanta [Equation (7.2)], if the release of one quantum is independent of another (*Poisson process*). A Poisson process is also found in the capture of light quanta by visual receptors, and in many other

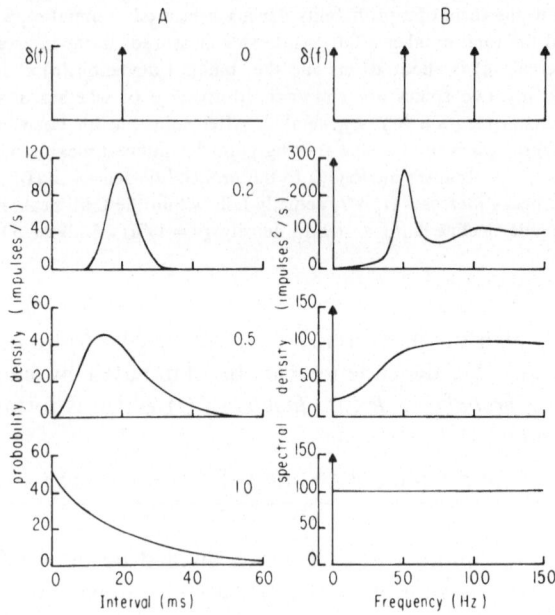

Fig. 9.14. (A) A range of interval histograms from a completely regular discharge (top) to an exponential distribution (bottom). The coefficient of variation ($\sigma/\mu$) is indicated in each part and increases from 0 to 1. All histograms have been calculated from gamma distributions (Equation 7.8) with $k = \infty$, 25, 4, and 1 respectively, and $r$ adjusted so that the mean interval was always 20 ms. (B) The corresponding power spectra. For the regular discharge, peaks of equal amplitude are seen at 50 Hz and integer multiples of this value. With increasing variability the peaks become less prominent until for the exponential distribution the spectral density is flat. The arrows represent Dirac $\delta$ functions either in time $\delta(t)$ or frequency $\delta(f)$. If the nerve impulse has a DC component, a $\delta$ function will occur at 0 Hz in parts of the figure.

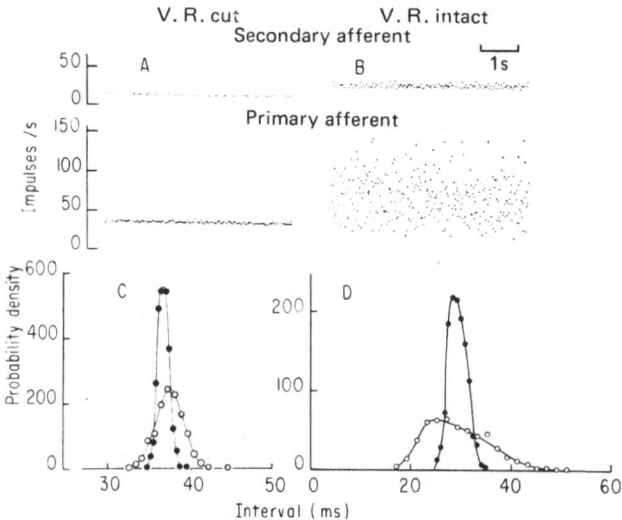

Fig. 9.15. The instantaneous firing rates of primary and secondary muscle spindle afferents are shown (A) without and (B) with spontaneous activity in γ-motoneurons. The activity in γ-motoneurons was eliminated by cutting the ventral roots (V.R.) which supply the motor fibers to the soleus muscle in the cat. Interval histograms are shown (C) without and (D) with ventral roots intact for primary (O) and secondary (●) afferents recorded simultaneously. In this preparation (different from that in A and B) the mean firing rates of the two were similar. (From Matthews and Stein, 1969b.)

physical processes. It can be specified by a single parameter, the mean rate at which events occur ($\nu = 1/\mu = 1/\sigma$). An exponential distribution is a special case ($k = 1$) of a gamma distribution. All the curves in Fig. 9.14A are gamma distributions [Equation (7.8)] with different values of the parameter $k$ (see also Fig. 7.7A). The third example shown ($k = 4$) could arise if four quanta or four synaptic potentials are needed for a nerve cell to discharge. An exact gamma distribution would only be found if no decay occurred between quanta or synaptic potentials [i.e., the nerve cell behaved as a perfect integrator rather than the leaky integrator of Equation (9.3)]. The range of distributions shown in Fig. 9.14 spans those commonly found in a number of nerve cells (Stein, 1965), although interval histograms with a coefficient of variation greater than one have also been described (Smith and Smith, 1965).

Figure 9.15 shows the discharge of primary and secondary muscle spindle afferents (A) without and (B) with spontaneous activity in γ-motoneurons. Note that in both (A) and (B), the primary afferent is more variable than the secondary afferent. The spontaneous activity in γ-motoneurons not only increases the mean firing rate, but also produces a considerable increase in variability. This increased variability presumably arises from the sensitivity of the afferents to the small contractions and relaxations that will be taking place in the intrafusal muscle fibers during this spontaneous activity. The greater variability of the primary afferents follows from their greater sensitivity to these small fluctuations. This variability does limit the ability of the endings to signal muscle length, as will be discussed later. This difference between primary and secondary afferents can also be seen in the interval histograms of Fig. 9.15C and D, in which the mean interval of both endings was quite similar. In both cases, the dispersion of the primary afferent (open circles) about the mean is much greater than that of the secondary afferent (closed circles).

Up to this point, only single interspike intervals have been considered. Another type of histogram is the *autocorrelogram*, which considers not only the time between successive spikes, but between any pair of spikes irrespective of the number of spikes in between (Fig. 9.13C). As discussed in Chapter 7, the probability density $f_2(t)$ for the time between two quanta (or for the total duration of two interspike intervals) can be determined from a convolution integral if the probability density for single quanta or intervals is known

$$f_2(t) = \int_0^t f_1(t - u)f_1(u)\,du \qquad (9.19)$$

Similarly, for $k$ interspike intervals

$$f_k(t) = \int_0^t f_1(t - u)f_{k-1}(u)\,du \qquad (9.20)$$

These formulas assume that successive interspike intervals are independent of one another, and form what is known as a *renewal process*. The name derives from one of its early applications in the study of replacement strategies for parts in industrial processes. In a renewal process, the time to failure of a new part is independent of the time to failure of the part it replaces.

The autocorrelogram considers intervals with all values of $k$, and the ordinate is often referred to as the *renewal density*, when normalized as in Fig. 9.13C. Formally, the renewal density is given by

$$r(t) = \sum_{k=1}^{\infty} f_k(t) \qquad (9.21)$$

In the autocorrelogram, peaks are seen corresponding to the time of one interval, two intervals, etc. Eventually the histogram flattens out to a level corresponding to the mean rate of firing in impulses/s or, in the industrial application, the mean rate for replacement of a given part.

A renewal process has no "memory" in that one interval cannot affect the next. Although successive intervals are often independent, significant serial correlations between intervals have been observed for muscle spindle afferents (Matthews and Stein, 1969b), second-order sensory fibers on which muscle spindles synapse (Jansen *et al.*, 1966), and motoneurons (Andreassen, 1978). Just as the Poisson process can be considered as a simple type of renewal process, so the renewal process can be considered as a simple type of *Markov process*. In a Markov process, the duration of one interspike interval can be influenced by the "state" of the neuron at that time, which is usually specified by the previous interspike interval. There is a well-developed theory of Markov processes (Cox and Miller, 1965), but these processes will not be considered further here.

Another common extension of this type of statistical analysis is to consider the relationship between spikes from two nerve cells, or between a series of stimulus pulses and a single nerve cell. In Fig. 9.16, the times from each stimulus to the occurrence of later spikes were measured and accumulated in what is known as a *poststimulus time histogram*. This histogram can be normalized as before by dividing the numbers in each bin by the total number of stimuli and the bin width. The units are then impulses/s and the poststimulus time histogram gives the modulation in the discharge rate produced by the stimulus. One could also include prestimulus times, but if an external stimulus is applied only random fluctuations will be observed in the prestimulus portion of the histogram.

In considering the relationship among two spike trains, perhaps neuron 1 affects neuron 2, neuron 2 affects neuron 1, or both neurons are affected by a third neuron. To decide among these possibilities, the *cross correlogram* of Fig. 9.16 may be useful. Times to the right of zero often represent effects of neuron 2 on neuron 1, whereas times to the left of zero represent effects of neuron 1 on nueron 2. Other neurons can provide common inputs which may be both excitatory and inhibitory and may

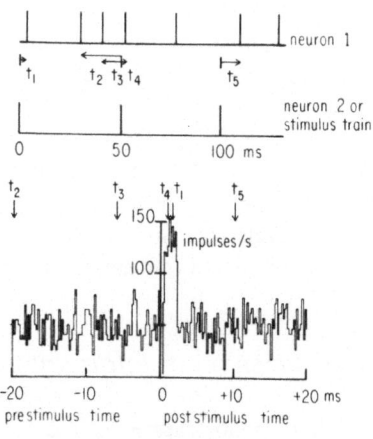

Fig. 9.16. In a poststimulus time histogram, the times from each stimulus to later spikes (e.g., $t_1, t_4, t_5$) are measured and plotted as shown. By dividing the ordinate by the total number of stimuli and the bin width, the results are expressed in impulses/s. Changes or modulation in firing rate can then be measured with respect to the prestimulus period. The same type of measurement can be used to determine the correlation between firing patterns in the two neurons and the result is then referred to as a cross-correlogram.

extend on both sides of the origin. Thus, the cross correlogram can be quite complex, but methods have been developed for determining the patterns of synaptic connections which might produce a given cross correlogram (e.g., Gerstein and Perkel, 1972). In general, even if the autocorrelograms of both neurons are also known, more than one pattern of connections could give rise to a particular set of histograms.

In considering the relationship between one neuron and a second, or between one neuron and a stimulus train, we have returned to considering input–output relationships. As pointed out earlier, the cross correlation function can be Fourier transformed to give the cross spectrum, which is closely related to the frequency–response function of a neuron. Previously, an analog input (a length change) generated a digital output (a series of sensory spikes) or a digital input (spikes in a motoneuron) generated an analog output (tension changes). In a cross correlogram or poststimulus time histogram, both inputs and outputs are digital events. Similarly, both inputs and outputs could be analog signals. The same methods and the same concepts apply.

The form for the spectrum of a series of pulses may not be familiar to many readers, and should be considered briefly (see Cox and Miller, 1965, for further details). According to the Fourier theorem, any waveform can be approximated by the sum of various sine or cosine waves, even a regular train of impulses occurring at a rate $\nu$. However, to fit a train of impulses at $t = 0$, $\mu, 2\mu, 3\mu$, etc., requires the sum of a large number of cosine waves with the frequencies $0, 2\pi\nu, 4\pi\nu, 6\pi\nu$, etc., where $\nu = 1/\mu$ as before. Each of these waves has positive contributions at the times of nerve impulses $(0, \mu, 2\mu, \text{etc.})$, but they can have either positive or negative contributions at other times, which will result in cancellation of the different waves being added. Thus, for the regular pulse train in Fig. 9.14, the spectrum will consist of a regular series of peaks at the firing rate and all multiples of it. At the other extreme, for the exponential interval distribution of the Poisson process, the spectrum will be constant with a spectral density equal to the mean rate at all frequencies. For the intermediate gamma distributions, the spectrum contains little low frequency power, an initial peak or two at approximately the mean firing rate, and its first few multiples. The spectrum then becomes flat at higher frequencies at a level equal to the mean firing rate. As with the cross spectrum and the cross-correlation function, the spectrum of a single pulse train and its autocorrelation function are related by the Fourier transform.

## Transmission of Information

The mean rate $\nu$ or the total number of impulses generated over a period of time is often considered as the information-containing parameter in a nerve impulse train. Thus, a higher rate of nerve impulses in a motoneuron produces a greater tension, and a stronger sensory stimulus generally causes a higher rate of impulses in a sensory neuron. Other parameters of the pattern may also be important in that one impulse may potentiate or depress the response to a second impulse, as was seen in the case of synaptic transmission (Chapter 7) or muscular contraction (Chapter 8). Other possible neural codes have been considered (Perkel and Bullock, 1968), but the mean firing rate seems to be the most important single parameter for transmission of information.

The measurement of information transmission can be formalized using a theoretical framework first introduced by Shannon (1948). A method for applying this theory to nerve cells under constant conditions is shown in Fig. 9.17. The nerve cell may have a minimum spontaneous level of firing $v_{min}$ and a stimulus applied for a time $t$ increases the firing rate to a value of $v$ impulses/s. In a time $t$ on average $x = vt$ impulses will be fired. The variable $x$ is determined by the stimulus and considered as the input to a communications channel in Fig. 9.17. On a given trial, some number $y$ impulses may be produced, and there are three sets of probabilities. $p(x)$ is the probability that the stimulus on any given trial was $x$, $p(y)$ is the probability that the response on any trial was $y$, and $p(y/x)$ is the conditional probability that the response was $y$, given that the stimulus was $x$.

Information is closely related to the uncertainty of a probability distribution or the entropy (a measure of randomness in thermodynamics). The uncertainty in the response depends on the number of different possible outputs and their relative probabilities according to the formula

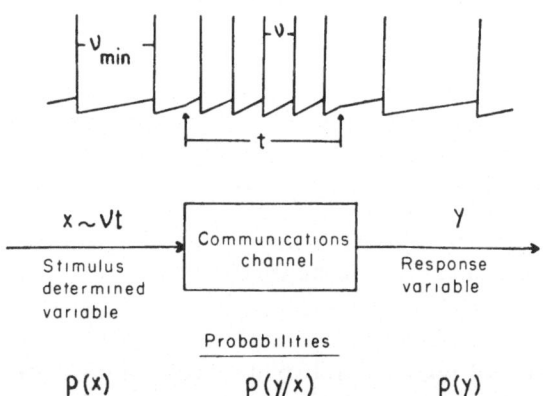

Fig. 9.17. Transmission of information by an idealized spike train. The analysis involves the probabilities of variables associated with a communications channel, as indicated in the text. (From Stein, 1970.)

(Ash, 1965)

$$H(Y) = -\sum_j p(y_j) \log_2 p(y_j) \qquad (9.22)$$

The capital $Y$ is used for the set $\{y_j\}$ and the summation is over all possible values of $j$. This formula will not be derived here, but the rationale for the logarithmic term and the use of logarithms to the base 2 can be understood from a simple example. Assume that a neuron fires from zero to $N - 1$ impulses with equal probability. Then, $p(y_j) = 1/N$ for all $j$ and Equation (9.22) reduces to

$$H(Y) = \log_2 N \qquad (9.23)$$

If only 0 or 1 impulse is fired, $H(Y) = 1$ bit, where *bit* is an abbreviation meaning one *bi*nary digi*t* or one binary choice. In other words, one yes or no answer or binary choice could eliminate all uncertainty. If $N = 4$, two answers are needed to eliminate all uncertainty (were less than 2 impulses generated? If so, was one impulse generated?). With $N = 8$, three answers or choices are required, and so on. Thus, the possible amount of information that could be transmitted depends on the uncertainty that can be removed, which varies as the logarithm to the base 2 of the number of equiprobable choices or categories.

The amount of uncertainty that is actually removed depends on the conditional probabilities. For a given value of $x$, the uncertainty in the output is given by

$$H(Y/x) = -\sum_j p(y_j/x) \log_2 p(y_j/x) \qquad (9.24)$$

and considering all possible $x$ in the set $X$, we have

$$H(Y/X) = \int_{x_{\min}}^{x_{\max}} p(x) H(Y/x)\, dx \qquad (9.25)$$

The *information* $I$ transmitted by a nerve cell is the reduction in uncertainty it produces when considered as an information channel. This reduction is the difference between the uncertainty in the output $H(Y)$ and the uncertainty even when the input is known $H(Y/X)$:

$$I = H(Y) - H(Y/X) \qquad (9.26)$$

Finally, the maximum information capacity or *channel capacity* $C$ that the nerve cell offers as a communications channel is

$$C = \max_{p(x)} I \qquad (9.27)$$

In principle, all possible probability distributions $p(x)$ must be considered to find the maximum information transmitted. This is difficult to do experimentally, but various schemes have been developed which permit the channel capacity to be estimated. The ability of a few sensory fibers in the cat to transmit information about steady inputs has been tabulated (Stein, 1970). With the exception of the secondary muscle spindle afferent, none of the receptors was able to transmit more than about three bits, which corresponds to distinguishing eight different categories. The values were quite close to the "magic number 7" mentioned earlier (Miller, 1956) for the number of categories human observers can distinguish about a single stimulus variable.

This agreement is probably fortuitous since there will normally be a number of nerve cells responding in parallel which could increase the information capacity. Some of this information will then be lost in synaptic transmission through the many stages needed for human decision making. The values for steady inputs will also be well below the true information capacity of a nerve cell, because nerve cells respond preferentially to changing stimuli and adapt to maintained ones. Stein and French (1970) and Eckhorn *et al.* (1976) have considered methods for analyzing information transmission by nerve cells with changing stimuli.

The reason for the variation in the information capacity of different nerve cells is the difference in the variability of their discharge. The high value for the secondary muscle spindle afferent is simply a consequence of its low variability. Why are some nerve cells variable? Are they less well designed, or is there some reason for their being "noisier"? A possible answer to this question is given in Fig. 9.18. This figure shows the response to a sinusoidal input for an electrical analog to the leaky integrator model of a nerve cell (Fig. 9.1B). The subthreshold depolarization leads to spikes which occur at three times over two cycles of a sine wave. This pattern repeats itself indefinitely so that a poststimulus time histogram shows three sharp peaks at particular phases of the sinusoidal cycle (Fig. 9.19A). This phenomenon is known as *phase-locking* and represents a severe distortion of the sinusoidal input. Addition of a second input consisting of white noise in increasing amounts (Fig. 9.19B, C, and D) disrupts the phase-locked pattern and produces a poststimulus time histogram with a clear sinusoidal character (see fitted curve using dots).

Fig. 9.18. A neural analog discharging at 100 pulses/s was stimulated with a 67 Hz sine wave. The resultant voltage changes in an RC integrator (Equation (9.3), $\tau = 10$ ms) are shown in the middle trace. Output pulses (top trace) occur when a threshold voltage is reached and resets the integrator. This figure shows several sweeps which superimposed exactly with three pulses occurring for every two cycles of the applied sinusoid. (From Stein, 1970.)

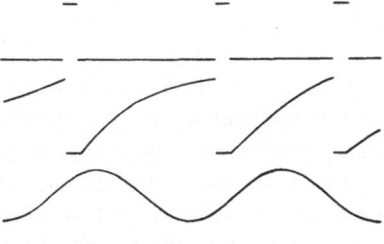

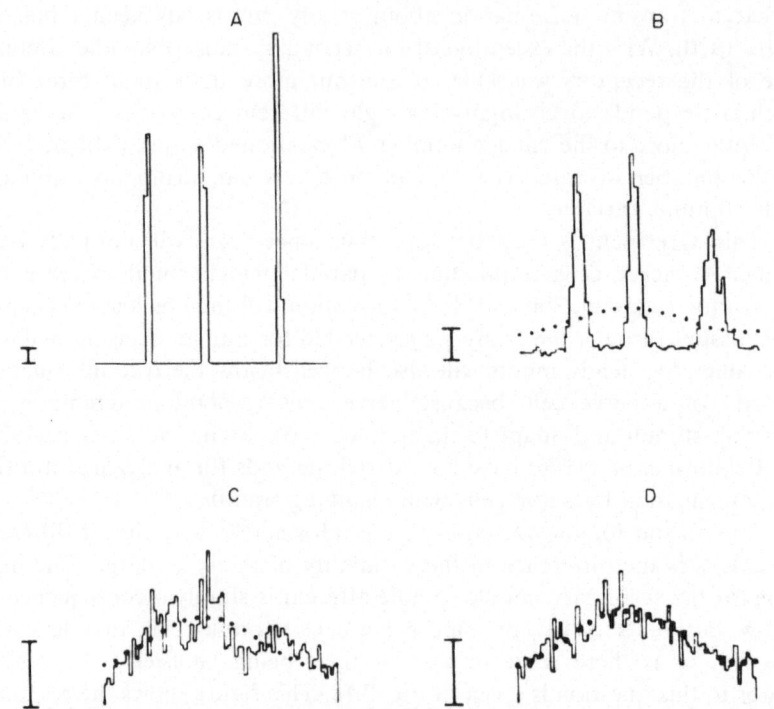

Fig. 9.19. Poststimulus time histograms for the response of a neural analog to the sinusoid shown in Fig. 9.18 with increasing amounts of noise added (A–D). The histogram covers exactly one full cycle (and is sometimes called a *cycle histogram*) with the peak of the applied sine wave in the middle. The bars represent a modulation of 100 pulses/s and the dots give the best fitting sine wave for the responses. Note the increasingly sinusoidal character of the response produced by adding more noise. (From Stein, 1970.)

Thus, although a regularly firing cell such as the secondary muscle spindle afferent can transmit more information about steady lengths of the muscle, the greater variability of the primary afferent can reduce distortion in the details of rapidly varying stimuli. The increased variability produced by γ-motoneurons may further increase the ability of muscle spindle afferents to faithfully signal details of rapidly changing inputs. The hypothesis has recently been confirmed experimentally by Inbar *et al.*, 1979). Finally, cochlear neurons (Kiang *et al.*, 1965) are among the most variable of sensory neurons (their discharge is close to a Poisson process). These neurons respond to the very rapidly changing auditory signals and the very randomness of their discharge may permit us to distinguish the same note played by a violin, an oboe, or a trumpet. Thus, rather than being neural noise, variability may be just another parameter that the nervous system uses in optimizing the design of a sensory neuron for a particular task.

## Problems

9.1. (a) Determine the frequency–response function for Equation (9.9) and give an expression for the amplitude and phase from Equations (9.14) and (9.15).
(b) What are the limiting values for gain and phase at low ($z_0$) and high ($z_\infty$) frequencies?
(c) Show that at a frequency $\omega = b^{-1}$, $|z^*(\omega)| = 2^{1/2}z_0$ and $\angle z^*(\omega) = 45°$.

9.2. A new species, *Biologia edmontonia*, has been discovered and Fig. 9.12A shows some of the first recordings from this species. In response to a brief tap to a tendon such that a muscle is extended 1 mm for 0.1 s, a muscle afferent fiber discharges nerve impulses at the following times: 38, 68, 95, 123, 153, 187, 230, and 296 ms.

(a) Give a mathematical expression which approximately fits the response of the afferent (see the curve of "instantaneous firing rate" in Fig. 9.12).
(b) Show that the transfer function for this receptor is

$$z^*(s) = \frac{10^3 s}{(s + 10)^2 (1 - e^{-0.1s})}$$

(c) Plot the frequency–response function of this afferent.
(d) How does it differ from that of other known muscle afferents?

9.3. The response of the slowly adapting stretch receptor to a step input is well fitted by a curve of the form $y(t) = at^{-k}$; $0 < k < 1$ (Brown and Stein, 1966). What will the transfer function for this receptor be? What will the gain and phase of the response be to sinusoids of different frequencies if the receptor behaves linearly?

9.4. Within a certain range the pumping of an electrogenic pump will be increased in proportion to the firing rate $v$ (also the influx of new ions) and decreased by its current pumping rate $p$ (also the efflux of ions which previously flowed in). Thus, $dp/dt = av - bp$, where $a$ and $b$ are constants.

(a) If the firing rate $v$ is proportional to the net current produced by the stimulus input ($i$) less that produced by the pump $p$ [i.e., $v = c(i - p)$], where $c$ is a constant, show that the transfer function $z^*(s) = v^*(s)/i^*(s) = c(s + b)/(s + ca + b)$.
(b) This is a classical form for a *high-pass filter* (the response at high frequencies is greater than at low frequencies). What are the limiting values $z_0$ and $z_\infty$?
(c) If $ca \gg b$, what are the "break frequencies" at which $|z^*(\omega)| = 2^{1/2}z_0$ and $|z^*(\omega)| = z_\infty/2^{1/2}$?
(d) Find values of $a$, $b$, and $c$ which would give a good fit to the frequency–response curve of a passive muscle spindle secondary afferent (Fig. 9.11B). Note that a good fit does not imply the mechanism underlying the frequency–response curve of this afferent is an electrogenic pump. Any mechanism which produces this form of high-pass filter could give as good a fit.

9.5. (a) Write expressions for $z_r$ and $z_i$ from Equation (9.13) in terms of the functions $x_r$, $x_i$, $y_r$, and $y_i$ in Table 9.3. What will the gain and phase of the frequency–response function be in terms of these functions?
(b) Write expressions for the coherence function $\gamma^2(\omega)$ in Table 9.3. Show that for a single data set $\gamma^2(\omega) = 1$. If several data sets are summed, the estimates of $S_{xx}(\omega)$ and $S_{yy}(\omega)$ will add since they are real positive functions. Cancellation can occur for $S_{xy}(\omega)$, so $0 \leqslant \gamma^2(\omega) \leqslant 1$ [see Bendat and Piersol (1971) for a proof].

# 10

# *Control of Movement*

In this chapter, the properties of nerve cells, muscle cells, and the synapses that link them will be synthesized to study whole neuromuscular systems and the way they control movement. Many of the same principles can be used to study other parts of the nervous system, but the control of movement will be emphasized as a logical extension of the material dealt with in previous chapters. First, the basic reflexes will be summarized, together with some elementary principles of reflex organization. Then some of these ideas will be quantified to analyze how the peripheral neuromuscular system functions in the control of movement. Finally, pattern generators in the central nervous system will be described which interact with the peripheral structures.

## *Reflexes*

The simplest and probably the most studied reflex connection is the monosynaptic pathway from primary muscle spindle afferents to $\alpha$-motoneurons, which is illustrated in Fig. 10.1. This reflex is strongest to motoneurons of the muscle in which the muscle spindle lies. Weaker connections are also found to motoneurons of muscles which act together with that muscle (*synergists*). Mendell and Henneman (1971) found that single primary muscle spindle afferents from medial gastrocnemius muscle in the cat make synaptic connection with virtually all the motoneurons of that muscle. However, only about two-thirds of motoneurons to a synergist, lateral gastrocnemius, had synaptic connections from medial gastrocnemius muscle afferents.

In a more extensive study, Scott and Mendell (1976) confirmed that synergists generally receive a lower frequency of synaptic inputs than the muscle in which the afferent originates. The studies were carried out by triggering a signal averager from the spikes of a muscle spindle afferent

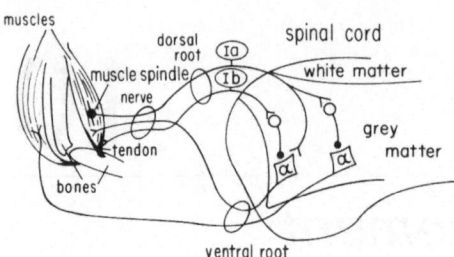

Fig. 10.1 Schematic representation of some simple reflex connections from group I muscle afferents onto α-motoneurons (α) to the same muscle and its antagonists. The sensory fibers enter the grey matter of the spinal cord at the dorsal or back surface. Their cell bodies are found in an enlargement of the dorsal root known as the dorsal root ganglion. The motoneurons have their cell bodies in the spinal cord and send their fibers out the ventral root on the other surface of the spinal cord to join the sensory fibers in a muscle nerve. Ia = primary muscle spindle afferent, Ib = Golgi tendon organ, ○- = interneuron, < = excitatory synaptic connection, ● = inhibitory synaptic connection, ●- = muscle spindle.

(referred to as *spike-triggered averaging*) while holding the muscle stretched to a constant length. As a result of averaging the intracellularly recorded potentials in an α-motoneuron, very small signals can be detected (down to 1 μV; Stauffer *et al.*, 1976). This same technique has been used to verify that secondary muscle spindle afferents also have some monosynaptic connections (Kirkwood and Sears, 1974), although the percentage is less (about half showed EPSP's, of which two-thirds were of monosynaptic latency; Stauffer *et al.*, 1976). The average strength of the connections is also weaker (30 μV compared to 65 μV; Stauffer *et al.*, 1976). Synchronous stimulation of the muscle nerve at a strength which will excite all primary afferents causes the discharge of most motoneurons. By potentiating transmission with repetitive stimulation, all motoneurons can be caused to fire [see Clamann *et al.* (1974) and their references to earlier literature]. Because of the weaker connections, few motoneurons in synergist muscles will fire, and the weaker effects of secondary muscle spindles alone would presumably not discharge many motoneurons.

These weaker synaptic connections can facilitate other excitatory inputs to the motoneuron. From experiments combining vibration (which mainly excites primary afferents) and stretch (which excites both primary and secondary afferents), Matthews (1969) suggested dividing the effects of stretch into two components. The *phasic-stretch reflex* or *tendon jerk*, which results from brief inputs such as a tendon tap, is almost entirely the effect of primary muscle spindle afferents. The tendon jerk is merely a mechanical method for producing a nearly synchronous burst of impulses in all primary muscle spindle afferents. The *static-stretch reflex*, which is the response to maintained stretch, results in part from the effect of secondary afferents. With the long period of time available, these effects may be mediated by polysynaptic connections as well as the relatively weak monosynaptic inputs from secondary afferents.

Muscle spindle afferents are the only sensory receptors which have a direct monosynaptic connection to α-motoneurons. Golgi tendon organs have disynaptic connections via a small inhibitory interneuron, as shown in Fig. 10.1. Stimulation of the tendon organs will therefore excite the interneuron, which will in turn inhibit the α-motoneurons in the same muscle. This pathway might be involved in the *clasp-knife reflex*, in which the muscle gives suddenly when the force in it reaches high levels, much as the blade in a pocket knife suddenly swings shut when a certain force is applied to it. However, Golgi tendon organs can be excited by quite small active contractions in a few motoneurons (Houk and Henneman, 1967), and are probably involved in the continuous regulation of muscle force, not merely in a protective function at high force levels (Rymer *et al.*, 1979).

Muscle spindle afferents also have inhibitory connections via inter-neurons to motoneurons of *antagonist* muscles (i.e., muscles which oppose the action of the muscle in which the spindles lie). Thus, at the same time as they excite a synergist group of muscles, they produce *reciprocal inhibition* in groups of muscles which would oppose their action (Eccles *et al.*, 1956). This connection is one example of reciprocal reflexes which are quite common in the nervous system. Golgi tendon organs often have a reciprocal excitatory effect on antagonistic motoneurons (Watt *et al.*, 1976).

In some movements a limb is held rigid by contracting two antagonis-tic groups of muscles. The presence of the inhibitory interneuron provides an additional location where the reciprocal reflex pathway can be modi-fied. In recent years, a wide variety of synaptic influences has been found onto Ia inhibitory interneurons (i.e., interneurons which are excited mono-synaptically by primary or group Ia muscle spindle afferents and which inhibit motoneurons). These interneurons may serve as a final common pathway for inhibitory influences (Hultborn, 1972, 1977), in analogy to the role of α-motoneurons (proposed by Sherrington, 1906) as a final common pathway for diverse excitatory and inhibitory reflex effects. The pathways from muscle receptors to γ-motoneurons are much weaker and contain mixtures of excitation and inhibition at longer latencies (Ellaway and Trott, 1978).

Cutaneous afferents can have various reflex effects, the best known of which is the *flexor* or *withdrawal reflex*. If a painful stimulus is applied to the skin, the animal will withdraw its limb from the stimulus. The reflex pathway may involve a variety of sensory fibers which have been loosely grouped under the term *flexor-reflex afferents* (Eccles and Lundberg, 1959). Although this term has proven useful operationally, it must be used with caution for several reasons. Firstly, all the afferents which are grouped in this way may not have similar reflex connections. Secondary muscle spindle afferents were included among flexor reflex afferents for many years before the secondary afferents in extensor muscles were shown to excite the extensors rather than the flexors. Secondly, the flexor reflex has at least two phases, an early and a late reflex response (Jankowska et al., 1967), and the same afferents or the same pathways may not be involved in both phases. The late phase may be produced, for example, by high-threshold afferents in response to strong, noxious stimuli, while the early response could be produced by lower threshold afferents. The early response would then represent a reflex adjustment of the limb, which might be carried out during a normal movement (see, for example, Duysens and Stein, 1978), rather than a withdrawal from a noxious stimuli. Finally, even the early flexion reflex is multisynaptic, and the pattern of stimulation or other inputs can modify the response at several points in the pathway. Sherrington (1906) reviewed evidence that even the sign of the reflex can be altered. A reflex extension is seen if a noxious stimulus is applied to the skin at a point where an extension, rather than a flexion, is required to withdraw the limb from the stimulus.

Rather than a single unified response, the flexor reflex is a general term referring to a range of reflex effects from a variety of afferents. The extra synaptic delays are presumably required so that an appropriate reflex response can be produced to any spatial and temporal pattern of afferent discharges. The reflex responses are not confined to the stimulated limb, for adjustments will be needed in the other limbs to prevent the animal from falling if one limb is withdrawn. The general term given to the adjustments in the opposite or *contralateral* limb is the *crossed-extensor reflex*. This multisynaptic-reflex response will again vary with the particular afferents stimulated, and the particular response in the *ipsilateral* limb (i.e., the limb stimulated). The reflex responses extend not only to $\alpha$-moto-neurons, but also to the smaller $\gamma$-motoneurons (Hunt, 1951). In contrast, the effects from muscle receptors are mainly confined to $\alpha$-motoneurons, as mentioned earlier.

Reflexes also affect higher centers to control the position of the neck, head, and eyes in space (Magnus, 1924). The nature of these pathways will not be considered here, although the interested reader may wish to consult a recent, thorough account of postural reflexes (Roberts, 1978). The response of a limb to muscle stretch can also involve *long-loop reflex pathways*. Although the precise pathways involved are not known with certainty, some responses are probably mediated by a fast pathway to the somatosensory cortex and then to the motor cortex and/or the red nucleus. These reflex pathways have also been recently and thoroughly reviewed (Desmedt, 1978a). Again, the extra synapses in these pathways permit greater flexibility in the response, and the ability of other inputs to "gate" the reflex response (Evarts and Tanji, 1974). Descending inputs from the brain can also modify the spinal reflex circuitry in ways which will be considered later.

## Size Principle

Although the reflex effects described above can be observed for motoneurons of various sizes, the magnitude of the synaptic potentials varies systematically with motoneuronal size. This trend has been studied most thoroughly for the monosynaptic connections between primary muscle spindle afferents and α-motoneurons. As mentioned earlier (in the section *Reflexes*), nearly all of the primary afferents in a given muscle have monosynaptic connections to each motoneuron projecting to that muscle.

The compound synaptic potential produced by maximally stimulating primary afferents in the muscle nerve increases systematically with the input resistance of the motoneuron's cell body [data from Burke (1968) has been replotted as Fig. 10.2A]. The input resistance varies inversely with the conduction velocity and hence the size of the motor axon (Fig. 10.2B). More quantitative treatment of the data (Stein and Bertoldi, 1980) suggests that large and small motoneurons receive approximately the same number of Ia connections. However, the same synaptic conductance change will produce a larger EPSP as a result of the larger input resistance of the smaller motoneurons. Thus, small motoneurons will be recruited first by increasing stretch of the muscle, and larger motoneurons will be recruited systematically according to size.

This orderly pattern of recruitment has been called the *size principle* by Henneman (1974a, b) who states, "It is the size of a motor neuron that

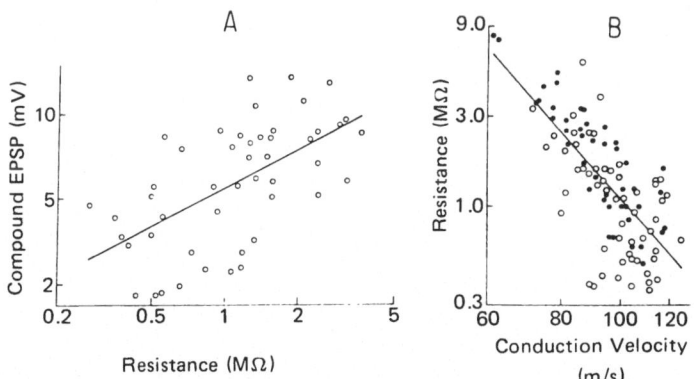

Fig. 10.2. (A) The compound EPSP evoked by stimulating primary muscle spindle afferents varies with the size of the motoneuron, as measured by the input resistance of its soma during intracellular recording. (B) The input resistance varies inversely with the conduction velocity of a motor axon, which is a measure of axon diameter and is presumably correlated with overall cell size. (See further discussion in the text and in Stein and Bertoldi, 1980.) Data are replotted on logarithmic scales and fitted with straight lines according to a least mean square criterion from (A) Burke (1968) and (B) Kernell (1966, filled circles) and Burke (1967, open circles).

determines its threshold and relative excitability. As a consequence, the motor units of a muscle can be fired in only one particular order as determined by the sizes of their neurons." Note that this order of recruitment to natural or synaptic inputs is just the inverse of the stimulation order with electrical stimuli. Electrical stimuli excite the largest axons preferentially (Chapter 6).

The size principle has several attractive features. As discussed in Chapter 8, smaller motoneurons innervate fewer and smaller muscle fibers than larger motoneurons. Thus, at low force levels these motoneurons will produce fine gradations of force, while large motoneurons will be available to produce the larger forces for high levels of contraction. These same gradations will apply to reflex and voluntary contractions since the orderly pattern of recruitment applies also to voluntary contractions in human subjects (Milner-Brown *et al.*, 1973). To determine the contractile force generated by individual motor units, the motor unit potential was recorded by a selective needle inserted into a muscle (the first dorsal interosseus) of the hand. Spike triggered averaging was used to extract the component of the overall force which was correlated with the discharge of single motor units. A nearly linear relationship was observed between the twitch tension generated by a motoneuron and the threshold force level of a voluntary contraction at which it became active (Fig. 10.3).

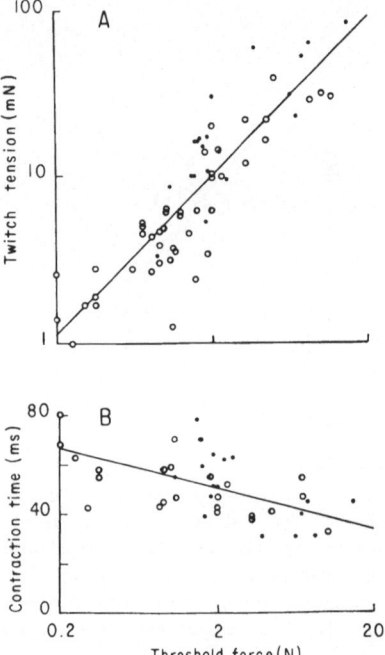

Fig. 10.3. Twitch tension (A) and contraction time (B) of single motor units in human first dorsal interosseus muscles during voluntary contractions. The values were obtained by spike-triggered averaging from unit EMG potentials recorded with a selective needle electrode in the muscle. (Data from Milner-Brown *et al.*, 1973.)

Another attractive feature of the size principle is that it will utilize the chemical capabilities of the motor units effectively. The small motoneurons are connected to slowly contracting, fatigue resistant muscle fibers (Fig. 10.3B). Thus, if a particular posture must be maintained for long periods of time at a modest force level, this posture will mainly involve slow, small motor units which have the abundant oxidative capacity suited to the task. If somewhat more phasic activities are required such as walking, the faster contracting, fatigue resistant motor units will be recruited to assist (Burke *et al.*, 1976). These motor units have both oxidative and glycolytic capabilities, so that the extra force can be generated and an even pace maintained for fairly long periods of time. Finally, for the most vigorous movements, such as running and jumping, the fast-twitch fatigable motor units can be activated. These units have mainly glycolytic capabilities which permit them to produce the extreme forces required, but only for brief periods of time until their glycogen supplies are depleted.

Finally, having a single order of recruitment should simplify the function of the brain in controlling movements. For any given level of signal descending to the spinal cord, an appropriate number of motoneurons would be recruited with suitable properties to carry out the desired movement. If approximately the same *number* of muscle spindle afferents connect synaptically to motoneurons independent of size (Kuno and Miyahara, 1969; Stein and Bertoldi, 1980), the density of afferent terminals and the synaptic potentials from this source will vary inversely with size (Burke, 1973), and the relative excitability to a wide variety of inputs follows as a natural consequence.

However attractive the size principle seems, a number of questions remain. How does the size principle arise during development? Are large motor units genetically large, or do they develop into large motor units after a motoneuron manages to innervate a large number of muscle fibers? Can the size principle redevelop in an adult after nerve injury? Milner-Brown *et al.* (1974) were unable to demonstrate an orderly recruitment of motor units following nerve section and resuture of the ulnar nerve in human subjects. However, Bagust and Lewis (1974) and Gordon and Stein (1980) have demonstrated a rematching of the nerve and muscle components of single motor units in resutured cat nerves. Is the size principle universal? Do we inevitably have to recruit slow motor units before fast ones, even in a large, brief movement where this would seem unnecessary? Recent evidence indicates that the orderly recruitment of motor units applies to human ballistic movements (Desmedt and Godeaux, 1977), but not to the rapid alternating movements when a cat shakes its paw (Smith *et al.*, 1980). Reports of exceptions or possible exceptions to the size principle persist (Grimby and Hannerz, 1968; Wyman *et al.*, 1974, Buller *et al.*, 1978). These questions will not be considered further here, but there are full discussions in recent volumes (Vrbova *et al.*, 1978; Desmedt, 1980).

## Feedback Systems

To the extent that a graded input from the central nervous system produces a graded output from a group or pool of motoneurons, the properties of these reflex pathways can be analyzed as a whole. General methods are available which are known by the term *systems analysis*. These methods have been considered to derive the frequency response or transfer functions of muscle receptors (Chapter 9) and muscles (Chapter 8). Now, these functions will be combined with the properties of reflex pathways to analyze a complete sensory feedback system. Linear systems will be considered here, but the analysis can be extended to consider nonlinearities which will always be present to some extent (Stein and Oğuztöreli, 1976; Marmarelis and Marmarelis, 1978).

The assumption of linearity produces major simplifications in the analysis of complex systems. For example, if two linear systems are in series (Fig. 10.4A), both systems can be described completely by their transfer functions, which are denoted $g^*(s)$ and $h^*(s)$ in terms of the Laplace transform variable $s$. As discussed in the last chapter, the Laplace transform of the output from the first system $y^*(s)$ can be expressed simply in terms of the Laplace transform of the input $x^*(s)$ and the transfer function,

$$y^*(s) = x^*(s)\, g^*(s) \tag{10.1}$$

Similarly, for the second system,

$$z^*(s) = y^*(s)h^*(s) \tag{10.2}$$

Combining these two equations,

$$z^*(s) = x^*(s)\, g^*(s)h^*(s) \tag{10.3}$$

In other words, the transfer function of the complete system is just the product of the transfer functions for the individual systems. This ability to combine systems in simple algebraic fashion becomes increasingly important as the complexity of the systems increases. Consider one further example, the simple feedback system shown in Fig. 10.4B. The circle on the left is a simple summing point, and the output of this point, $e$, which is known as the error signal, is the difference between the desired input $i$ and the feedback signal $f$. From this diagram, one can easily show that the output of the system is related to the input by the following equation.

$$o^*(s) = \frac{i^*(s)g^*(s)}{1 + g^*(s)h^*(s)} \tag{10.4}$$

Proof of this relation is left as a problem (Problem 10.2). The product

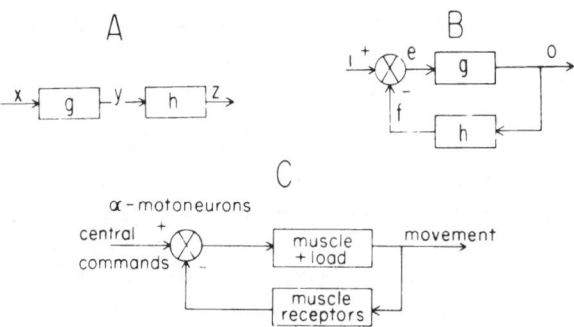

Fig. 10.4. Linear systems can be combined in various ways: (A) The systems g and h are connected in series. (B) The systems g and h are connected in a simple feedback loop via the summing point (⊗), which takes the difference between the input i and the feedback signal f. The resulting error signal e is amplified to produce an output o. All variables indicated can be considered either as functions of time t or the Laplace variable s. (C) A block diagram of a feedback system involving a muscle and its load, together with feedback from muscle receptors and their associated reflex pathways. The properties of muscles and loads have been considered in Chapter 8 (Fig. 8.13A) and those of muscle receptors in Chapter 9 (Fig. 9.10). Together they modify commands from the brain or other centers to produce appropriate movements.

$g^*(s)h^*(s)$ is particularly important in feedback systems and is referred to as the *loop gain* (i.e., the total gain around the feedback loop). The response of the system will be quite different, depending on the value of loop gain. When $g^*(s)h^*(s) \ll 1$, Equation (10.4) predicts the properties of the system will be determined solely by the *forward gain* $g^*(s)$, i.e.,

$$o^*(s) = i^*(s)\,g^*(s) \tag{10.5}$$

However, if $g^*(s)h^*(s) \gg 1$, the properties of the system will be determined solely by the *feedback gain* $h^*(s)$, namely

$$o^*(s) = i^*(s)/h^*(s) \tag{10.6}$$

Finally, if the value of $g^*(s)h^*(s)$ is near 1, or in particular near $-1$ for some value of $s$, the output will be very large for that value of $s$. To illustrate how a loop gain of $-1$ might arise, consider a sinusoidal input of frequency $\omega$, so that $s = j\omega$ can be substituted in Equation (10.4). The loop gain $g^*(j\omega)h^*(j\omega)$ will generally be a complex number, but there may be values of $\omega$ where a phase lag of 180° is found around the feedback loop. Then the product $g^*(j\omega)h^*(j\omega)$ will be a negative number. If that number is near $-1$, the denominator in Equation (10.4) will be near 0, and the output $o^*(s)$ can reach very large values. This is one possible explanation for the large tremors which can occur under some physiological and pathological conditions, although there are alternate explanations (Desmedt, 1978a; Stein and Lee, 1980).

## Stability and Oscillations: Physiological Tremor

To pursue this point further, the boxes in the feedback system can be identified with elements of a physiological system (Fig. 10.4C). The summing point on the left represents $\alpha$-motoneurons which sum inputs from a wide range of pathways in the central nervous system as well as reflex pathways to serve as a "final common pathway" (Sherrington, 1906). Note that this simple representation ignores the time-dependent properties of motoneurons (Baldissera et al., 1978). The forward gain corresponds to the muscle and its load, which together determine the output. The feedback gain corresponds to the properties of the muscle receptors, and in particular the muscle spindle primary afferents because of their strong monosynaptic connections to the $\alpha$-motoneurons. In addition, pure time delays will arise from conduction delays, synaptic delays, and excitation–contraction delays. Together, these delays contribute to the latency of a reflex pathway (i.e. the total time from the application of a perturbation to a muscle until the onset of a reflexly-generated force to resist that perturbation).

If the delay is 25 ms, the effect will be as shown in curve A of Fig. 10.5. A time delay will not affect the loop gain of a reflex pathway, but will introduce a phase lag which increases linearly with frequency. The reason for the curvature in Fig. 10.5 is simply that frequency is plotted on a logarithmic scale, as is customary in a Bode plot (see Chapter 9). If a muscle is included whose gain declines, with the second power of frequency at high frequencies (see Chapter 8), the response will be as shown by curve B in Fig. 10.5. Note that the muscle properties also introduce additional phase lags. A pure time delay of 25 ms would produce a phase lag of 180° (a half-cycle) at a frequency of 20 Hz. With the additional phase lags of the muscle, 180° phase lag is now seen at 7.5 Hz. Some of these phase lags can be canceled by the phase lead produced by the muscle spindle afferents as a result of their velocity sensitivity.

Curve C gives the overall response of the system after including the properties of muscle spindles. At higher frequencies the cancellation of phase lags is no longer sufficient and a phase lag of 180° is observed at 13.5 Hz. This is close to the tremor frequency found in cat muscles (Lippold, 1970; Nichols et al., 1978), which is somewhat higher than the oscillations at approximately 10 Hz observed in man. The high-pass characteristics of the muscle receptors (i.e., gain increasing with frequency) can also offset the low-pass characteristic of the muscles to some extent, as first pointed out by Poppele and Terzuolo (1968). In curve C, the gain is now quite constant up to about 10 Hz. Even at 13.5 Hz (the frequency at which there will be 180° phase lag), the gain is still higher in the presence of muscle receptors than at 7.5 Hz in the absence of muscle receptors (curve B).

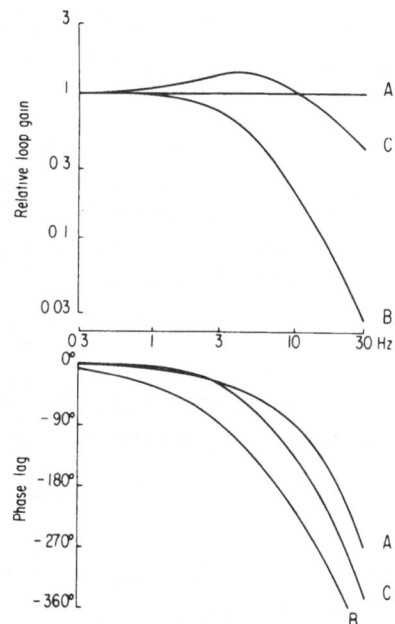

Fig. 10.5. Computed gain and phase for a feedback loop with properties based on the stretch reflex from mammalian muscles. Further explanation of the curves is given in the text and in Stein (1974).

Thus, muscle receptors contribute to the tendency for oscillations at high frequencies, at the same time as they resist perturbations at lower frequencies. The stability of the muscle at higher frequencies depends on the low-pass characteristics of muscles described previously. Muscles simply do not respond well to rapidly varying inputs, but rather produce a partially fused tension output. The curves of loop gain are normalized with respect to the low frequency values, since the absolute value of loop gain has been difficult to measure in biological systems. Usually a length change is imposed on a muscle, and the response of the muscle is measured as a tension change while the muscle is held at constant length (*isometric* conditions). The input–output relation then has the units of stiffness (newtons (N) of force per mm of stretch).

In order to measure loop gain, both input and output should be measured in the same units, so that the ratio of the two is dimensionless. Houk *et al.* (1970) managed to do this by stimulating small ventral root filaments which produced a small tension increase under isometric conditions. The response was a reflexly-induced tension decrease and the magnitude of this decrease could be measured before and after cutting the dorsal roots to eliminate the reflex response. The measured gain for soleus muscle in decerebrate cats was only 0.2–0.8, but they noted that this value mainly measures the gain of the tension control system via feedback from Golgi tendon organs. Because of the isometric conditions used, length feedback via muscle spindles would be minimized. Furthermore, the inhib-

itory reflexes from Golgi tendon organs may be depressed in the decerebrate cat, so their values probably underestimate the loop gain found
in intact animals.

The stability of the system can also be affected by the way in which a
muscle is loaded. Fig. 8.13A shows a muscle attached to a load which in
general contains a mass $M$, a stiffness $k_e$, and a viscosity or damping $D$.
This load itself constitutes a second-order system and is a classical example
of a mechanical oscillator. The tendency for oscillation depends on the
damping ratio (D'Azzo and Houpis, 1966)

$$\zeta = \frac{D}{(k_e M)^{1/2}}$$

In the absence of the viscosity or damping $D$, the mechanical system would
oscillate at a natural frequency $\omega_n$ given by

$$\omega_n = (k_e / M)^{1/2} \tag{10.8}$$

All real systems contain some damping, which causes the oscillation to
decay in response to a brief perturbation. As long as the damping ratio
$\zeta \leqslant 1$, the response is given by

$$Ae^{-pt} \sin(\omega_d t) \tag{10.9}$$

where $\omega_d = \omega_n (1 - \zeta^2)^{1/2}$ and $p = \omega_n \zeta$. Proof of these well-known relations
is left as a problem (Problem 10.3). As the damping ratio is increased, the
tendency for oscillation is decreased and the frequency of oscillation will
decrease somewhat.

Connecting such a system to a muscle or an intact limb can clearly
increase the tendency for oscillation or tremor. This is particularly true if
the natural frequency of the mechanical oscillator is close to the frequency
at which physiological tremor in the limb occurs naturally. Even attaching
a spring to a limb (Joyce and Rack, 1974) or to a partially isolated muscle
(Nichols et al., 1978) can increase the tendency for oscillation dramatically
because of the interaction of its elasticity with the mass of the limb and the
muscle properties. Since the muscle itself represents a second-order system,
connecting loads which also constitute a second-order system will produce
a fourth-order system. Oğuztöreli and Stein (1975) showed mathematically
that the resulting system can have at most one frequency of oscillation,
and that any oscillations must decay with time in the absence of reflex
feedback, as in Equation (10.9). With the addition of reflex pathways,
these decaying oscillations can be converted to maintained oscillations
which may occur at the frequency of the mechanical system, the frequency
of the reflex system, or some intermediate frequency. The tendency for
oscillation will be greatest when the natural frequencies of the mechanical
and reflex systems are close together.

One feature which distinguishes reflexly induced oscillations from others which may be generated wholly within the central nervous system (see next section on pattern generators) is the fact that the amplitude and frequency of reflexly generated oscillations will be sensitive to external loading. Such oscillations can also be reset to a high degree by brief perturbations applied to a limb (Stein *et al.*, 1978). Further details on tremor and factors which affect it can be found in several recent reviews ( Desmedt, 1978b; Stein and Lee, 1980).

## Pattern Generation

The oscillations found in physiological tremor are unwanted, while those in many pathological tremors are quite bothersome. However, our existence depends on several lower frequency oscillations, such as those of the heart, the respiratory system, the limbs during locomotion, and the jaws during chewing movements. These oscillations have sometimes been thought to arise from chains of reflexes (e.g., Mott and Sherrington, 1895). For example, pressure receptors, skin receptors, and joint receptors, which are all active while a limb is bearing weight, could give rise to a reflex flexion of the limb. Flexing the limb would stretch the extensor muscles and produce a stretch reflex. The resulting reflex extension could again bring the limb in contact with the ground. Although such reflexes can modify ongoing patterns of locomotion (Grillner, 1975; Duysens and Stein,

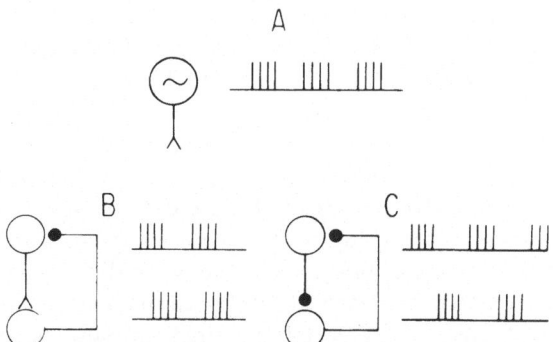

Fig. 10.6. (A) A cell with an oscillating membrane potential may produce bursts of impulses in isolation (Kandel, 1976) or drive another cell to produce such bursts (Pearson and Fourtner, 1975). Cells which do not produce bursts in isolation can do so when connected (B) in a negative feedback loop, or (C) with mutual inhibition. Patterns can be produced with varying degrees of overlap and phase lags between the bursts in each cell. Other synaptic inputs (not shown) may turn the bursts in any of the cells on and off or change the number of spikes in a burst.

1978), other mechanisms are clearly needed to turn locomotion on or off and to regulate the speed of locomotion. In standing, weight is placed on a limb without producing a reflex flexion and the gait of locomotion is switched abruptly from a walk to a trot as speed is increased, despite a smooth change in reflex feedback.

Evidence has been accumulating in recent years for the existence of central oscillators capable of producing appropriate patterns for these rhythmic movements, even in the absence of sensory feedback (Dellow and Lund, 1971; Grillner, 1975; Wyman, 1977). The nature of the neural circuits comprising these pattern generators or central oscillators has been more difficult to determine. In general, central oscillators can be of two types: the first is driven by single cells or groups of cells, each of which is capable of producing the pattern on its own. This idea is analogous to the ability of single heart cells to produce rhythmic contractions when isolated, and such rhythmic cells are generally referred to as *pacemakers* or *bursters*, if the pattern consists of bursts of impulses separated by pauses (Fig. 10.6A). The second possibility is that the oscillation arises from the synaptic connections between cells, none of which can produce oscillations in isolation. It is generally assumed that vertebrate pattern generators consist of such oscillatory networks, although individual pacemakers and bursters have been identified and studied in some detail in invertebrate systems (Kandel, 1976). However, recent evidence suggests that single vertebrate neurons can produce bursts (Spencer, 1977), probably by dendritic mechanisms (Wong *et al.*, 1979).

Where an oscillator contains groups of individual pacemakers, synaptic connections may still be important in strengthening and synchronizing the oscillation, much as the connections between cells in the heart or in the gut (El-Sharkawy *et al.*, 1978) lead to a single unified oscillation. Because of the possible death or damage of individual cells, some redundancy is important. In contrast to the heart, but not to smooth muscle of the intestine, neural oscillators often require processes which are much slower than the time needed to generate single impulses. Thus, bursts of impulses are recorded in particular phases of the oscillation, since the behavioral event (e.g., walking or breathing) occurs with a period of seconds, rather than milliseconds. One exception to this rule is the flight of insects (Wilson and Waldron, 1968), which occurs at rates such that elevator and depressor motoneurons to the wing may only fire a single impulse per cycle.

Slow processes are widely distributed among nerve cells (see Chapter 9), for example, in the *adaptation* of nerve cells to maintained inputs. Adaptation may be due to the accumulation of a permeability change following successive impulses or to the turning on of an electrogenic sodium pump as a result of the sodium influx during successive impulses. To the extent that these processes (a) increase immediately following each nerve impulse, (b) sum linearly, and (c) decay exponentially, adaptation occurs with a smooth, often exponential time course (Kernell, 1968; Sokolove, 1972; Stein *et al.*, 1974). A delay in turning on (for example, if the onset of the adaptation process also has an exponential time course) or some nonlinearity in the process can lead to oscillatory responses. The two may be related; if there is a threshold nonlinearity, in that the adaptation process does not turn on until after a certain number of impulses occurs, time delays will be introduced. Including time constants for the turning on and turning off of an adaptation mechanism in a simple neural model leads to a third-order differential equation which gives the possibility of decaying or well-maintained oscillations (Stein *et al.*, 1974).

Bursters in the mollusc *Aplysia* show slow sodium and potassium currents which turn on and off with a time course of seconds or even minutes (Kandel, 1976). Pacemaker neurons have been discovered in insect systems which do not generate impulses at all. These nonspiking interneurons produce continuous, slow oscillations in potential which lead synaptically to bursts of impulses in motoneurons (Pearson and Fourtner, 1975).

The simplest kind of network oscillator contains two interconnected neurons, neither of which can oscillate on its own. The greatest tendency for oscillation will occur when the signs of the connection are different so that they form a negative feedback loop (Fig. 10.6B). This is analogous to the oscillations produced by the negative feedback of reflex pathways except that the pathways are restricted to the central nervous system. Typical examples are the inhibition produced by Renshaw cells (Renshaw, 1946) back onto $\alpha$-motoneurons which excite them, or of cortical cells back onto thalamic cells which excite them (Andersen and Andersson, 1968). The latter mechanism has been suggested as a mechanism for the $\alpha$-rhythm found in the cerebral cortex, but this explanation has been disputed (Elul, 1972; Lippold, 1973).

Another pattern of connections is mutual inhibition, which has been suggested for many years as a basis for the oscillations between flexor and extensor motoneurons in locomotion (Brown, 1911, 1914), inspiratory and expiratory motoneurons in respiration (Burns and Salmoiraghi, 1960), and levator and depressor motoneurons in flight (Wilson and Waldron, 1968). Inhibitory connections do not occur directly between motoneurons (Wyman, 1977), so interest has centered on mutually inhibitory interneurons.

Evidence for mutual inhibition in the respiratory system as a basis for breathing is weak or negative (Wyman, 1977), but Miller and Scott (1977) have suggested mutual inhibition between Ia inhibitory interneurons as a basis for the oscillations of the spinal locomotor generator. The Ia inhibitory interneurons are the link between the Ia or primary muscle spindle afferents and the inhibition of *antagonist* motoneurons via a disynaptic pathway (see the section *Reflexes*). The Ia inhibitory interneurons to both flexors and extensors are known to receive a large number of connections from descending systems and also to mutually inhibit each other (Hultborn, 1972, 1977). The presence of descending inputs from the neurons responsible for regulating the locomotory patterns is important. Locomotion can be turned on and off and regulated in speed by stimulation of a midbrain location, which has become known as the *mesencephalic locomotory center* (Shik and Orlovsky, 1976). Brain stem centers also appear to be important in the regulation of rhythms involved in breathing and chewing.

The inclusion of the Ia inhibitory interneurons as a key element in Miller and Scott's model for the control of locomotion is interesting since it suggests that afferent input should be able to modify locomotion. Various effects have been described by many authors (reviewed by Grillner, 1975), but cutaneous rather than muscle afferents are most effective. The strength of reflex connections can also be modulated considerably during locomotion (Forssberg *et al.*, 1975). The pressure on the foot as a result of supporting the weight of the body during the stance phase of locomotion leads to a reflex extension, rather than the flexion reflex which might be expected. If these same afferent fibers are stimulated during the swing phase of the step, then increased flexion is produced (Duysens and Stein, 1978). Thus, the sign as well as the magnitude of the reflex response is altered during each step cycle, a phenomenon referred to as *reflex reversal* (Forssberg *et al.*, 1975).

Miller and Scott (1977) also included a mutual inhibition of Renshaw cells in their model (Ryall *et al.*, 1971), as well as feedback inhibition from the Renshaw cells onto the Ia inhibitory interneurons (Hultborn *et al.*, 1971). The importance of these connections for the spinal locomotor generator remains to be clearly demonstrated experimentally and there is evidence to the contrary (Menzies *et al.*, 1978). Finally, it is not clear how the reflex effects discussed above interact with the generator. Thus, a number of important and intriguing questions remain to be answered at the level of neuronal systems, as well as at the cellular and membrane levels. Another indication of this fact is that the final section in this book has been written in the form of a question.

## What Is Controlled?

The output of a neural oscillator or other central pattern generator is a train of impulses directed to the appropriate muscles. What features of the muscle is the central nervous system controlling with these impulses? The answer to this question is not as simple as it might seem initially, since the available circuitry could be used to control at least three different variables.

Table 10.1. Representation of Pathways for Controlling Force, Length, or Velocity [a]

| Motoneuron | Muscle fiber | Sensory fiber | Feedback path | Controlled variable |
|---|---|---|---|---|
| α ⟶ β ⟵ | Extrafusal ⟶ | Golgi tendon organ (Ib) ⟶ | Disynaptic inhibition | Force |
| Dynamic γ ⟶ | Nuclear bag ⟶ intrafusal | Primary muscle ⟶ spindle (Ia) | Tendon jerk | Velocity |
| Static γ ⟵ | Nuclear chain ⟶ intrafusal | Secondary muscle ⟶ spindle (II) | Static-stretch reflex | Length |

[a] See text for explanation

The top line in Table 10.1 shows how inputs directly to α-motoneurons could be used to control the force output of a muscle via its extrafusal muscle fibers. The actual force would be sensed by Golgi tendon organs and fed back via their disynaptic inhibitory connections to the α-motoneurons. This pathway may be involved in the clasp-knife reflex seen at high force levels, but is also involved in continuous regulation of force (see the section *Reflexes*). If the muscle becomes fatigued and the force output drops, the inhibition fed back to the α-motoneurons will be decreased. This *disinhibition* will produce increased firing in the motoneurons and hence increased force to compensate for the fatigue. The gain of this pathway is not very great in the decerebrate cat, but it could produce useful compensation for some degree of fatigue (Houk et al., 1970). There is also evidence that the firing of pyramidal tract cells from the motor cortex is correlated with the force output of motoneurons (Evarts, 1966), so that the desired force could be signaled by this pathway.

Similarly, the bottom line in Table 10.1 shows how activity in static γ-motoneurons could be used to control muscle length. Because there are connections also to nuclear bag intrafusal muscle fibers and primary muscle spindle afferents, both types of muscle spindle sensory fiber will be involved in the control of length. The existence of two distinct types of nuclear bag intrafusal muscle fibers (see Chapter 9) has been omitted from the table, but this simplification will not affect the flow of information shown. A role for γ-motoneurons in the servo-control of muscle length was first proposed by Merton (1953) and has generated controversy and experimentation ever since. Whether the strength of the static stretch reflex is strong enough to control muscle length has been discussed repeatedly (Matthews, 1972; Stein, 1974). One corollary of the original hypothesis is that activity should occur in static γ-motoneurons prior to activity in α-motoneurons. This does happen in some experimental situations, but often activity occurs at about the same time in both types of motoneuron (Vallbo, 1971).

A third control pathway arises from the dynamic γ-motoneurons, and in some muscles from β-motoneurons (fibers which generally innervate both slow, extrafusal and nuclear bag intrafusal fibers; Barker *et al.*, 1977). The effect of this stimulation is to sensitize the muscle spindle to phasic inputs such as a tendon jerk. Following section of the spinal cord, static γ-motoneurons become silent, but there is still activity in some dynamic γ-motoneurons (Alnaes *et al.*, 1965). This activity may well contribute to the brisk tendon jerks observed in chronic spinal animals and man, which often give rise to damped oscillations rather than a single contractile response. Furthermore, stimulation of some higher centers particularly increases the activity of dynamic γ-motoneurons (reviewed by Matthews, 1972), but the role of a separate velocity control system remains speculative.

Although the most rapid movements must be controlled mainly by α-motoneurons since there is no time for useful feedback, most studies agree that movements generally involve *coactivation* of α- and γ-motoneurons (Matthews, 1972). This coactivation, which is also referred to as *α–γ linkage* (Granit, 1970), has a number of attractive features. Firstly, it eliminates some of the delays involved in the reflex pathways, since activity in α-motoneurons will get the movement going quickly. Secondly, rather than having to control a movement completely, the reflex pathways merely assist in correcting for deviations from the desired movement. This has been referred to as *servo-assisted control* (Matthews, 1972; Stein, 1974) or *conditional feedback* (Houk, 1972), since the muscle receptors need only be activated when there is a deviation from the desired movement.

The idea of servo-assisted movements or conditional feedback is illustrated in Fig. 10.7 for several possible conditions. In each condition α- and static γ-motoneurons are activated to produce a given movement. In the first condition, the load is unexpectedly heavy, so that little or no shortening is produced by the extrafusal muscle fibers. Nonetheless, shortening of the poles of the intrafusal fiber can occur with a stretching of the center. Thus, the sensory endings of the muscle spindle afferent are stretched, the rate of afferent impulses increases, and the resultant stretch reflex will assist in overcoming the heavy load. In the second condition of normal load, both extrafusal and intrafusal muscle fibers contract in parallel so that the net effect on the sensory receptors will be small. The third condition is an unexpectedly light load in which the extrafusal fibers shorten very rapidly. The shortening of the poles of the slower intrafusal muscle fiber will not keep pace and the central region will be unloaded. A pause in the discharge of the sensory discharge from the muscle receptor will remove excitation from the α-motoneurons via the stretch reflex pathway.

The degree of α–γ coactivation can be varied in mammalian systems where most motoneurons to intrafusal fibers are distinct from those to

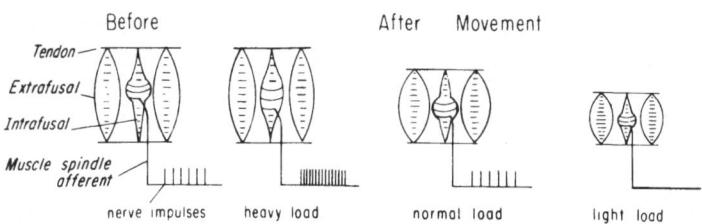

Fig. 10.7. Schematic representation of the effect of movements produced by coactivation of α- and γ-motoneurons on the discharge of a muscle spindle afferent when the load is heavier than, lighter than, or as expected. The horizontal lines indicate the striation spacing of the contractile portions of extrafusal and intrafusal muscle fiber while the coils around the central, less contractile region of the intrafusal fiber represent the stretch of the afferent endings.

extrafusal fibers (the β-motoneurons are the exception to the rule). The proportion of activity in each type of motoneuron could therefore be adjusted to suit different situations or different loads. The presence of both dynamic and static γ-motoneurons also provides the possibility of selecting either the velocity or the final length which is desired for a given movement. This increased flexibility is desirable with the wide range of movements required by mammals, including man.

The coactivation of α- and γ-motoneurons has led to the suggestion that the ratio of the length of a muscle to its force can be regulated, rather than the length or force individually (Houk, 1979). The ratio of force to length has the dimensions of stiffness, and the stiffness of the reflex pathways will add to the stiffness a muscle possesses at rest or during activity. Muscle alone can have quite complex properties (Chapter 8). A short stretch will produce a considerable amount of tension, but then bonds will rupture and the muscle force declines. Reflexes can compensate for these properties of muscle and make it a more linear, and hence a simpler element to control (Nichols and Houk, 1976). Whether this compensation is the primary role of reflexes remains far from certain, but it is an interesting and testable hypothesis.

Finally, one could argue that the ratio of force to velocity (i.e., viscosity) might also be controlled through the use of α- and dynamic γ-motoneurons. Although there is little evidence for this possibility, the fact remains that any or all of five variables: force, velocity, length, stiffness, or viscosity, could be controlled. There is not enough evidence to decide among these possibilities, but my hunch is that with $10^{12}$ neurons and a considerable choice of motoneurons and peripheral receptors, most if not all of these variables can be used to control particular movements as required. We certainly have the ability to just balance the force exerted by a load, to move a limb with a constant velocity accurately to a given final length, to put the right spring or stiffness in our step to cushion a fall, and perhaps to provide viscous drag to damp out undesired oscillations. Whether this is done by changing the variables controlled by the nervous system can only be determined by further experimental and theoretical studies on the biophysical aspects of nerve and muscle.

## Problems

10.1. If all parts of a neuron were scaled uniformly, what relation would be expected between somatic input resistance and axonal conduction velocity? The slope of the best-fitting straight line in the log–log plot of Fig. 10.2B is between $-3$ and $-4$. How might the discrepancy between prediction and experiment be explained? [See Stein and Bertoldi (1980) for a discussion of this problem.]

10.2. Prove Equation (10.4) using the notation indicated in Fig. 10.4B.

10.3. Suppose the muscle is detached from the load in Fig. 8.13 and a forcing function $x(t)$ is applied to the load to produce a displacement $y(t)$.

(a) Show that the differential equation for the system is given by

$$k_e\, y + D\, dy/dt + M\, d^2y/dt^2 = x(t)$$

(b) Show that the transfer function $g^*(s) = y^*(s)/x^*(s)$ of this system is

$$g^*(s) = \frac{g_0}{1 + 2\zeta s/\omega_n + (s/\omega_n)^2}$$

(c) The corresponding frequency response is obtained by substituting $s = j\omega$. Show that the magnitude of the frequency response has a peak at $\omega = \omega_n(1 - 2\zeta^2)^{1/2}$ for $\zeta < 2^{-1/2}$.

(d) If a Dirac delta function $\delta(t)$ is applied to this system and $\zeta < 1$, show that the response is given by Equation (10.9).

(e) What will the response be if $\zeta = 1$ or $\zeta > 1$?

# Solutions to Problems

2.1. From Equation (2.7), where $S_1 = K^+$, $S_2 = Ca^{++}$, $Z_1 = 1$, $Z_2 = 2$,

$$\frac{[Ca^{++}]_o}{[Ca^{++}]_i} = \left( \frac{[K^+]_o}{[K^+]_i} \right)^2$$

or $[Ca^{++}]_i / [Ca^{++}]_o = 100$.

2.2. Equilibrium constant $K = 1$, since the chemical form of the substances is unchanged, so

$$1 = \frac{[K^+]_i [Cl^-]_i}{[K^+]_o [Cl^-]_o} \quad \text{or} \quad \frac{[K^+]_o}{[K^+]_i} = \frac{[Cl^-]_i}{[Cl^-]_o}$$

2.3. From approximate charge neutrality, $[K^+]_o = [Cl^-]_o = x$, $[K^+]_i = [Cl^-]_i + [A^-] = y$. The osmotic pressure on the outside is

$$[K^+]_o + [Cl^-]_o = 2x$$

The osmotic pressure on the inside is

$$[K^+]_i + [Cl^-]_i + [A^-] = 2y$$

From the Gibbs–Donnan equilibrium condition,

$$\frac{[K^+]_o}{[K^+]_i} = \frac{[Cl^-]_i}{[Cl^-]_o} = \frac{x}{y} = \frac{y - [A^-]}{x}$$

Thus, $x^2 = y^2 - y[A^-]$ or

$$x = y \left[ (1 - [A^-]/y)^{1/2} \right] < y, \quad \text{if } [A^-] > 0$$

2.4.

$$Q = 0.1\text{–}1.0 \ \mu C/cm^2 \text{ or } (1\text{–}10) \times 10^{-12} \text{ moles}/cm^2$$

$$\text{Area} = 4\pi r^2 = 4(3.14)(5 \times 10^{-4})^2 = 6 \times 10^{-7} \text{ cm}^2$$

$$\text{Volume} = 4\pi r^3/3 = 4(3.14)(5 \times 10^{-4})^3/3 = 5 \times 10^{-10} \text{ cm}^3 = 5 \times 10^{-13} \text{ L}$$

Charge required to generate the membrane potential is

$$[(1-10) \times 10^{-12}](6 \times 10^{-7}) = (6-60) \times 10^{-19} \text{ moles of ions}$$

In the volume of a cell, this represents $1-10 \ \mu M$, which is negligible compared to $100-200$ mM.

2.5.   The flow of the solute $S$ is $J_s = (J_v + J_D)C_s$ moles/s, where $J_v + J_D$ (in L/s) represents the total flow of solute volume ($J_v$ is the flow of the bulk solution and $J_D$ is the flow of solute relative to the solution) and $C_s$ is the molar concentration (moles/L) of the solute.

2.6.   If $J_v = 0$ from Equation (2.12) for all values of $\Delta p$ and $\Delta C$, then $L_p = L_{pD} = 0$. From Onsager's relation $L_{Dp} = 0$, so Equation (2.13) becomes $J_D = L_D RT\Delta C$. Thus, a single constant $L_D$ is sufficient to describe the flow of substances.

3.1.   From the equations given in Problem 3.1

$$C_m = [C] + [SC] + [I][C]/K_I \quad \text{or} \quad [C] = \frac{C_m - [SC]}{1 + [I]/K_I}$$

Using the notation of Fig. 3.1, Equation (3.2) will still hold, and after substituting for $[C]$ from the equation above and rearranging, we find

$$[SC]\left\{(b+d) + \frac{a[S]_o}{1 + [I]/K_I}\right\} = \frac{a[S]_o C_m}{1 + [I]/K_I}$$

Substituting this equation into Equation (3.1) gives

$$J_s = \frac{ab[S]_o C_m}{a[S]_o + (b+d)(1 + [I]/K_I)}$$

The desired equation follows from substituting the definitions of $M = bC_m$ and $K_m = (b+d)/a$.

3.2.   From the assumptions in Problem 3.2,

$$S + C \underset{d}{\overset{a}{\rightleftharpoons}} SC \overset{b}{\longrightarrow} C + S$$

$$I + C \underset{f}{\overset{e}{\rightleftharpoons}} IC$$

$$S + IC \underset{d}{\overset{a}{\rightleftharpoons}} ISC$$

$$I + SC \underset{f}{\overset{e}{\rightleftharpoons}} ISC$$

where $K_I = f/e$, $K_m = (b+d)/a \approx d/a$ and

$$C_m = [C] + [SC] + [IC] + [ISC]$$

$$= [C] + [SC] + [I]([C] + [SC])/K_I$$

Solving for $[C]$ gives

$$[C] = \frac{C_m}{1 + [I]/K_i} - [SC]$$

In the steady state the rate of formation of the complex $[SC]$ must equal the rate of breakdown, so

$$a[S]_o[C] + f[ISC] = (b + d + e[I])[SC]$$

After substituting for $[C]$ and $[ISC] = [I][SC]/K_i$ and simplifying,

$$a[S]_o\left(\frac{C_m}{1 + [I]/K_i} - [SC]\right) = (b + d)[SC]$$

Solving for $[SC]$ and substituting the result in Equation (3.1) gives the desired result.

3.3. From Equation (3.5),

$$\frac{1}{J_s} = \frac{1}{M}\left(1 + \frac{K_m}{[S]_o}\right)$$

The form of the equation is a straight line in which the intercept on the $y$ axis (when $1/[S]_o = 0$) is $1/M$, and the intercept on the $x$ axis (when $1/J_s = 0$) is $-K_m$. Changing $I$ and $K_m$ will not affect the $y$ intercept in Problem 3.1 or the $x$ intercept in Problem 3.2. Increasing $I$ (or decreasing $K_i$) will make the $x$ intercept more negative in Problem 3.1 and the $y$ intercept less positive in Problem 3.2.

3.4. If $r = b/c \gg 1$ and $K_d$, $[S]_o$, and $[S]_i$ have similar orders of magnitude, the equation in Problem 3.6, which is derived from Equation (3.11), reduces to

$$J_s = \frac{cC_m K_d([S]_i - [S]_o)}{K_d([S]_i + [S]_o) + 2[S]_i[S]_o}$$

i.e., transport is limited by the rate $c$ at which the free carrier moves across the membrane. On the other hand, if $[S]_i \gg K_d$ and $[S]_o \gg K_d$, this equation reduces further to

$$J_s = \frac{cC_m K_d([S]_i - [S]_o)}{2[S]_i[S]_o}$$

which is limited by the fact that $([S]_i - [S]_o)$ must be smaller than $[S]_i$, if $[S]_i > [S]_o$ (or $[S]_o$, if $[S]_o > [S]_i$) and the ratio $K_d/[S]_o$ (or $K_d/[S]_i$) $\ll 1$.

3.5.  If $b = c$, then $r = 1$, and Equation (3.11) becomes

$$J_s^{\text{in}} = \frac{bC_m[S]_o}{2(K_d + [S]_o)}$$

which is identical to Equation (3.5) if we let $M = bC_m/2$ and $K_m = K_d$. Similarly, if $c \gg b$, $r = b/c \ll 1$, so Equation (3.11) reduces to

$$J_s^{\text{in}} = \frac{bC_m[S]_o K_d}{(K_d + [S]_o)K_d + (K_d + [S]_i)K_d}$$

under conditions when $K_d$, $[S]_o$, $[S]_i$ have similar orders of magnitude. However, the Michaelis–Menten equation was derived under conditions where the backward reaction was negligible because $[S]_i \ll K_d$. Then, this equation becomes

$$J_s^{\text{in}} = \frac{bC_m[S]_o}{2K_d + [S]_o}$$

which has the desired form, if we let $M = bC_m$ and $K_m = 2K_d$.

3.6.  If movement across the membrane is rate limiting, as assumed in the text, then the inward movement of the substance $S$ is given by an analogous equation to Equation (3.11), namely

$$J_s^{\text{out}} = \frac{bC_m[S]_i(K_d + r[S]_o)}{(K_d + [S]_i)(K_d + r[S]_o) + (K_d + [S]_o)(K_d + r[S]_i)}$$

and the net flux is given by the difference between this equation and Equation (3.11).

$$J_s^{\text{net}} = \frac{bC_m K_d([S]_i - [S]_o)}{(K_d + [S]_o)(K_d + r[S]_i) + (K_d + [S]_i)(K_d + r[S]_o)}$$

Expanding the terms on the bottom gives the desired form of the equation. This equation will reduce to Fick's law if $[S]_i < K_d$ and $[S]_o < K_d$ and to the mobile carrier equation if $r = 1$. To prove the latter result, note that the mobile carrier equation (3.6) can be rearranged in the form

$$J_s = \frac{MK_m([S]_i - [S]_o)}{(K_m + [S]_o)(K_m + [S]_i)}$$

4.1.  The two straight lines will cross when $E_k = V$ in Fig. 4.2 or

$$[K^+]_o = \frac{q[Na^+]_o[K^+]_i}{[K^+]_i + q[Na^+]_i}$$

Rearranging this equation to solve for $q$ yields

$$q = \frac{[K^+]_o [K^+]_i}{[Na^+]_o [K^+]_i - [Na^+]_i [K^+]_o}$$

4.2. From the diagram, the voltage at any point in the membrane will be (see also the notation of Fig. 4.1)

$$V = (V_m + W)(1 - x/d)$$

$$f(V) = \frac{d}{\int_0^d e^{k(V_m + W)(1 - x/d)} \, dx} = \frac{k(V_m + W)}{e^{k(V_m + W)} - 1}$$

Substituting this equation into Equation (4.6) gives the desired result. If $V_m = E_s$,

$$V_m = \frac{1}{k} \ln \frac{[S]_o}{[S]_i}$$

from the Nernst equation (4.8), or after taking exponentials of both sides and rearranging,

$$[S]_o = [S]_i e^{kV_m}$$

Thus, the term in brackets on the right side of the equation in the first part of this problem will be 0, and so $J_s = 0$.

For $V_m \gg 0$, $J_s$ reduces to

$$J_s = k(V_m + W)P_s[S]_i e^{-kW}$$

since, for example, $e^{k(V_m + W)} \gg 1$. For $V_m \ll 0$,

$$J_s = k(V_m + W)P_s[S]_o$$

The slope of $J_s$ with respect to $V$ is obtained by differentiation. Thus

$$dJ_s/dV = \begin{cases} kP_s[S]_i e^{-kW}, & V \gg 0 \\ kP_s[S]_o, & V \ll 0 \end{cases}$$

The curve will be outwardly rectifying if the slope for $V \gg 0$ is greater than for $V \ll 0$; i.e.,

$$kP_s[S]_i e^{-kW} > kP_s[S]_o$$

or $e^{-kW} > [S]_o/[S]_i$. Taking natural logarithms of both sides gives

$$-kW > \ln([S]_o/[S]_i) = kE_s$$

from Equation (4.8). After rearranging and simplifying, the condition for outward rectification is $W + E_s < 0$. Similarly, a necessary requirement for a linear relation is that the slopes are equal when

$V_m \gg 0$ and $V_m \ll 0$, or $W + E_s = 0$. Inward rectification requires the slope for $V_m \ll 0$ to be greater than for $V_m \gg 0$, or $W + E_s > 0$.

4.3.  If movement of carrier is rate limiting, then

$$J_s^{\text{in}} = b_i [SC]_o$$

and Equation (3.9) becomes

$$[C]_i + r_o [SC]_i = [C]_o + r_i [SC]_o$$

Substituting from Equation (3.7) yields

$$[SC]_i (r_o + K_d/[S]_i) = [SC]_o (r_i + K_d/[S]_o)$$

and Equation (3.8) becomes

$$C_m = [SC]_i (1 + K_d/[S]_i) + [SC]_o (1 + K_d/[S]_o)$$

Eliminating $[SC]_i$ between these two equations and solving for $[SC]_o$ gives the desired result.

4.4.  The corresponding equation for $J_s^{\text{out}}$ will be, by symmetry,

$$J_s^{\text{out}} = \frac{b_o C_m [S]_i (K_d + r_i [S]_o)}{(K_d + [S]_o)(K_d + r_o [S]_i) + (K_d + [S]_i)(K_d + r_i [S]_o)}$$

and the ratio $J_s^{\text{out}}/J_s^{\text{in}}$ is given by

$$\frac{J_s^{\text{out}}}{J_s^{\text{in}}} = \frac{b_o [S]_i (K_d + r_i [S]_o)}{b_i [S]_o (K_d + r_o [S]_i)}$$

$$= \frac{e^{k(V_m - E_s)}(cK_d + b_i [S]_o)}{(cK_d + b_i [S]_o e^{k(V_m - E_s)})}$$

after substituting from the Nernst equation (4.8). Since $e^{k(V_m - E_s)} > 1$ when $V_m > E_s$, and the other quantities in this equation are positive, the ratio of the quantities in the brackets is less than or equal to 1, so

$$J_s^{\text{out}}/J_s^{\text{in}} \leqslant e^{k(V_m - E_s)}$$

The equation for $J_s^{\text{out}}/J_s^{\text{in}}$ can be rearranged in the form

$$\frac{J_s^{\text{out}}}{J_s^{\text{in}}} = \frac{cK_d + b_i [S]_o}{cK_d e^{-k(V_m - E_s)} + b_i [S]_o}$$

Since $e^{-k(V_m - E_s)}$ will be less than 1 when $V_{,,} > E_s$, it follows easily that $J_s^{\text{out}}/J_s^{\text{in}} \geqslant 1$.

If $n$ molecules of a substance are transported by the carrier,

$$J_s^{in} = b_i[S_n C]_o$$

If the major forms of the carrier are as free carrier or combined with $n$ molecules of the substance, and these are in a steady state,

$$C_m = [C]_i + [C]_o + [S_n C]_i + [S_n C]_o$$

$$[C]_i + r_o[S_n C]_i = [C]_o + r_i[S_n C]_o$$

and

$$K_d = \frac{[C]_i[S]_i^n}{[S_n C]_i} = \frac{[C]_o[S]_o^n}{[S_n C]_o}$$

The proof proceeds as previously to give

$$J_s^{in} = \frac{b_i C_m[S]_o^n(K_d + r_o[S]_i^n)}{(K_d + [S]_o^n)(K_d + r_o[S]_i^n) + (K_d + [S]_i^n)(K_d + r_i[S]_o^n)}$$

Similarly, the ratio $J_s^{out}/J_s^{in}$ becomes

$$\frac{J_s^{out}}{J_s^{in}} = \frac{b_o[S]_i^n(K_d + r_i[S]_o^n)}{b_o[S]_o^n(K_d + r_o[S]_i^n)}$$

where $b_o = b_i e^{nkV_m}$ and $[S]_o = [S]_i e^{kV_m}$ as before. Thus,

$$\frac{J_s^{out}}{J_s^{in}} = \frac{e^{nk(V_m - E_s)}(cK_d + b_i[S]_o^n)}{(cK_d + b_i[S]_o e^{nk(V_m - E_s)})}$$

from which the desired inequalities follow. Taking into account intermediate species of carrier between $C$ and $S_n C$ (e.g., $SC$, $S_2 C$) further complicates the equations but does not affect the overall inequality.

4.5. From the results in Problems 4.3 and 4.4, it follows that

$$J_s^{net} = J_s^{out} - J_s^{in}$$

$$= \frac{C_m c K_d(b_o[S]_i - b_i[S]_o)}{(K_d + [S]_o)(cK_d + b_o[S]_i) + (K_d + [S]_i)(cK_d + b_i[S]_o)}$$

For $V \gg 0$, $b_o$ becomes very large, $b_i$ very small, and

$$J_s^{net} = \frac{C_m c K_d}{K_d + [S]_o}$$

For $V \ll 0$, $b_i$ becomes very large, $b_o$ very small, and

$$J_s^{\text{net}} = \frac{-C_m c K_d}{K_d + [S]_i}$$

Saturation will also occur for asymmetrical membranes as long as $b_o$ increases without bound for large, positive values of $V$ and $b_i$ increases without bound for large, negative values of $V$. If the carrier $C$ is negatively charged but the complex $SC$ is uncharged, Equation (3.9) becomes

$$c_o[C]_i + b[SC]_i = c_i[C]_o + b[SC]_o$$

The inward flux is identical in form to the equation in Problem 4.3 except $r_o = b/c_o$ and $r_i = b/c_i$. The net flux becomes

$$J_s^{\text{net}} = \frac{C_m b K_d([S]_i c_i - [S]_o c_o)}{(K_d + [S]_o)(c_o K_d + b[S]_i) + (K_d + [S]_i)(c_i K_d + b[S]_o)}$$

When $c_i$ increases without bound, $J_s^{\text{net}}$ saturates at

$$J_s^{\text{net}} = \frac{C_m b[S]_i}{K_d + [S]_i}$$

and when $c_o$ increases without bound $J_s^{\text{net}}$ saturates at

$$J_s^{\text{net}} = \frac{-C_m b[S]_o}{K_d + [S]_o}$$

4.6.   The inward flux is identical in form to the equation in Problem 4.3 except that $r_o = b_o/c_o$ and $r_i = b_i/c_i$. From symmetry, the outward flux can be written by exchanging the subscripts $i$ and $o$, and the net flux follows from subtraction of the inward from the outward flux.

 If the membrane is symmetrical,

$$c_o = ce^{kV_m/2}, \qquad c_i = ce^{-kV_m/2}, \qquad b_o = be^{kV_m}, \qquad b_i = be^{-kV_m}$$

Thus,

$$J_s^{\text{net}} = \frac{K_d C_m bc([S]_i e^{kV_m/2} - [S]_o e^{-kV_m/2})}{(K_d + [S]_o)(cK_d e^{kV_m/2} + be^{kV_m}[S]_i) + (K_d + [S]_i)(cK_d e^{-kV_m/2} + be^{-kV_m}[S]_o)}$$

For sufficiently large, positive values of $V_m$,

$$J_s^{\text{net}} \propto e^{-kV_m/2}$$

which approaches zero, while for sufficiently large, negative values of $V_m$,

$$J_s^{net} \propto e^{kV_m/2}$$

which also approaches zero.

The reason for this result is that with large potential differences across the membrane, both $C^+$ and $SC^{++}$ will accumulate at the negative edge of the membrane and there will be little net transport. If either is uncharged, it will continue to shuttle across the membrane irrespective of the electric field.

4.7. Substituting in the equation given in Problem 4.6 and dividing top and bottom by $cK_d$ gives

$$J_s^{net} = \frac{C_m b(e^{k_{sc}V_m/2}e^{-k_c V_m/2}[S]_i - e^{-k_{sc}V_m/2}e^{k_c V_m/2}[S]_o)}{(K_d + [S]_o)(e^{k_c V_m/2} + \lambda[S]_i e^{k_{sc}V_m/2}) + (K_d + [S]_i)(e^{-k_c V_m/2} + \lambda[S]_o e^{-k_{sc}V_m/2})}$$

The denominator is in the desired form and the numerator can be brought into the desired form since

$$e^{k_{sc}V_m/2}e^{-k_c V_m/2} = e^{kV_m/2}$$

and from the Nernst equation,

$$([S]_o/[S]_i)^{1/2} = e^{kE_s/2}$$

When $V_m = E_s$, we have $\sinh(0) = 0$ and $J_s = 0$.

5.1. If $I_{Rh}$ is the rheobasic current and $V_{Th}$ is the threshold voltage, then from Equation (5.3),

$$V_{Th} = V_r + I_{Rh}R$$

or $I_{Rh} = (V_{Th} - V_r)/R$. For other currents greater than the rheobasic value, the time to reach threshold will be

$$V_{Th} = V_R + IR(1 - e^{-t/\tau})$$

which can be rearranged in the desired form. For $t \ll \tau$, the equation becomes, after expanding the exponential in a power series and neglecting higher order terms,

$$I = I_{Rh}\tau/t$$

or the charge $Q = It = I_{Rh}\tau$ is constant.

5.2. The ratio of currents will be

$$r = \frac{I_{s_2}}{I_{s_1}} = \frac{J_{s_2}^{net}}{J_{s_1}^{net}} = \frac{J_{s_2}^{out} - J_{s_2}^{in}}{J_{s_1}^{out} - J_{s_1}^{in}}$$

If the movement in the two directions is independent and the internal concentration of $s$ does not change, $J_{s_2}^{out} = J_{s_1}^{out}$. Dividing top and bottom by $J_{s_2}^{out}$ gives

$$r = \frac{1 - \left(J_{s_2}^{in}/J_{s_2}^{out}\right)}{1 - \left(J_{s_1}^{in}/J_{s_1}^{out}\right)} = \frac{1 - e^{-k(V_m - E_{s_2})}}{1 - e^{-k(V_m - E_{s_1})}}$$

from Ussing's flux ratio test in Chapter 4. The second part follows easily since

$$\Delta I = I_{s_2} - I_{s_1} = I_{s_1}(r - 1)$$

5.3.   The differential equation can be rearranged and integrated

$$\int \frac{dm}{\alpha - (\alpha + \beta)m} = \int dt$$

yielding

$$-\frac{1}{\alpha + \beta}\left(\ln\left[\alpha - (\alpha + \beta)m\right] - \ln\left[\alpha - (\alpha + \beta)m_o\right]\right) = t$$

where $m_o$ is the value of $m$ at $t = 0$. Taking exponentials of both sides gives

$$\alpha - (\alpha + \beta)m = (\alpha - (\alpha + \beta)m_o)e^{-t/(\alpha + \beta)}$$

Rearranging to solve for $m$, we find

$$m = m_\infty + (m_o - m_\infty)e^{-t/\tau_m}$$

where $m_\infty = \alpha/(\alpha + \beta)$ is the steady state value of $m$ (as $t \to \infty$) and $\tau_m = (\alpha + \beta)^{-1}$ is the time constant with which $m$ approaches the steady state value. If $\alpha/\beta = e^{k(V - E)}$ where $k = ZF/RT$, then

$$m_\infty = \frac{\alpha}{\alpha + \beta} = \frac{1}{1 + \beta/\alpha} = \frac{1}{1 + e^{-k(V - E)}}$$

For the time constant $\tau_m$ to be maximum, $d\tau/dV = 0$. If $\tau = (\alpha + \beta)^{-1}$, then

$$\frac{d\tau}{dV} = \frac{d(\alpha + \beta)/dV}{(\alpha + \beta)^2} = 0$$

If $\alpha = \alpha_o e^{k(V - E)/2}$, $\beta = \alpha_o e^{-k(V - E)/2}$ from Equation (5.9) and $\alpha$, $\beta > 0$ for all finite values of $V$, then for $d\tau/dV = 0$,

$$\frac{d(\alpha + \beta)}{dV} = \frac{\alpha_o k}{2}\left[e^{k(V - E)/2} - e^{-k(V - E)/2}\right] = 0$$

which is only true when $V = E$.

5.4.   (a) $m$; (b) $n$, $h$; (c) $h$, $n$; (d) $n$, $h$.

5.5. (a) Rearranging the differential equation and substituting from Equation (5.3) gives

$$\mu \, dU/dt + U = IR(1 - e^{-t/\tau})$$

Taking Laplace transforms gives

$$\mu s U^* + U^* = IR\left(\frac{1}{s} - \frac{1}{s + 1/\tau}\right)$$

where transformed variables are indicated by the superscript * and are functions of the Laplace variable $s$. Solving for $U^*$ gives

$$U^* = \frac{IR}{\mu\tau}\left[\frac{1}{s(s + 1/\tau)(s + 1/\mu)}\right]$$

Taking the inverse Laplace transform yields the desired result.
(b) From the result in part (a) and Equation (5.3),

$$V - U = U_o = -IR\left[e^{-t/\tau} + \left(\frac{\tau}{\mu - \tau}\right)e^{-t/\tau} - \left(\frac{\mu}{\mu - \tau}\right)e^{-t/\mu}\right]$$

Simplifying the right-hand side of this equation and solving for $I$ gives the desired result.
(c) If $V - U$ reaches its maximum at time $t$,

$$d(V - U)/dt = \frac{IR\mu}{\mu - \tau}\left(\frac{1}{\tau}e^{-t/\tau} - \frac{1}{\mu}e^{-t/\mu}\right) = 0$$

This can only be true if

$$\frac{1}{\tau}e^{-t/\tau} = \frac{1}{\mu}e^{-t/\mu}$$

which, after rearranging and taking logarithms, gives the desired result.

6.1. If the resistivity of the external medium is $\rho_e$ the longitudinal resistance $R_l$ of a length $l$ is

$$R_l = \rho_e l \Big/ \left[\pi(b/2)^2 - \pi(a/2)^2\right]$$

$$= \frac{4\rho_e l}{\pi(b^2 - a^2)}$$

The resistivity in the radial direction $R_r$ is

$$R_r = \frac{\rho_e}{l}\int_{a/2}^{b/2}\frac{dr}{2\pi r}$$

$$= \frac{\rho_e}{2\pi l}\ln(b/a)$$

Substituting the above equations into the condition that $R_r \ll R_l$ and simplifying gives the desired result.

6.2.  If the resistance of a unit membrane is $R$, the resistance $R_m$ of a layer of membrane with radius $r$ covering an internode of length $l$ is

$$R_m = \frac{R}{2\pi rl}$$

The capacitance $C_m$ of the same internode, if a layer of membrane has a capacitance $C$ per unit area, is

$$C_m = 2\pi rl\, C$$

so the product $\tau = R_m C_m$ is independent of $r$.

6.3.  (a) From Equation (6.20),

$$\frac{\partial V}{\partial T} = \frac{r_a \lambda}{2(\pi T)^{1/2}}\, e^{-T} e^{-X^2/4T}\left(-\frac{1}{2T} - 1 + \frac{X^2}{4T^2}\right)$$

For $\partial V/\partial T = 0$, the quantity in brackets on the right-hand side of this equation must be zero. Thus, $X^2 = 4T^2 + 2T$ or

$$X = \pm 2T\left[(8T-1)/8T\right]^{1/2} \sim \pm 2T \qquad \text{for } T > 1.$$

(b) For $X = 2T$

$$V = \frac{r_a \lambda}{(2\pi X)^{1/2}}\, e^{-X}$$

(c) $V(X=1) = (V_o/X^{1/2})e^{-X} = 1$ mV. Thus, $V_o = e$ mV and

$$V(X=3) = e^{-2}/3^{1/2} = 78\ \mu V < 100\ \mu V$$

6.4.  From Equation (6.23),

$$\frac{V_{x=0}}{I} = \frac{r_a \lambda}{2} = \left(\frac{\rho R}{\pi^2 a^3}\right)^{1/2}$$

after substituting from the definitions of $r_a$ and $\lambda$, and simplifying.

6.5.  From Equations (6.2) and (6.25),

$$\frac{\partial^2 V}{\partial x^2} = j(r_a + r_e)$$

Since $j = \pi aJ$ and $J = C\partial V/\partial t + V/R$, this equation becomes

$$\frac{R}{\pi a(r_a + r_e)}\frac{\partial^2 V}{\partial x^2} = RC\frac{\partial V}{\partial t} + V$$

This equation is identical in form to Equation (6.6) if $\tau = RC$ and

$$\lambda^2 = \frac{R}{\pi a(r_a + r_e)}$$

To obtain $\lambda$ in the required form substitute

$$r_a = \frac{4\rho_a}{\pi a} \quad \text{and} \quad r_e = \frac{4\rho_e}{\pi(b^2 - a^2)}$$

and simplify.

6.6. If $V_p$ is the peak of a monophasic action potential of duration $t$ and conduction velocity $v$, the diphasic potential of Equation (6.28) will have a positive peak of magnitude $V_p$ and a negative peak of magnitude $V_p$ if $\delta > v(t - t_p)$, where $t_p$ is the time at which the monophasic action potential reaches its peak ($t_p \leqslant t$). Thus, the peak-to-peak value of $V_2$ will be twice that of $V_1$. Equation (6.29) can be rewritten

$$V_3(x, \delta) = \frac{1}{2} \{ V_1(x - \delta) + V_1(x + \delta) \} - V_1(x)$$

so if $\delta > v(t - t_p)$, $V_3$ will have two positive phases of magnitude $V_p/2$ separated by a negative phase of magnitude $V_p$. Thus, the peak-to-peak value of $V_3$ will be 1.5 times that of $V_1$.

6.7. From Equation (6.31),

$$\frac{\partial V_2}{\partial x} = \frac{q\sigma}{(x^2 + y^2)^{3/2}} \left( 1 - \frac{3x^2}{x^2 + y^2} \right)$$

$\partial V_2/\partial x$ will only be zero when the quantity in brackets on the right side of this equation is zero; i.e., when

$$y = \pm 2^{1/2}x$$

and the distance between the two peaks in the $x$-direction is $2^{1/2}y$. Substituting the condition for maximal values into Equation (6.31) gives

$$V_2 = \frac{\pm 2q\delta}{3^{3/2}y^2} \propto y^{-2}$$

7.1.

$$V = 4/3 \, \pi r^2 = (4/3)\pi(25 \times 10^{-9})^3 = 6.54 \times 10^{-23} \text{ m}^3 = 6.54 \times 10^{-20} \text{ L}$$

$$\text{molarity} = \frac{6000}{6 \times 10^{23} \times 6.54 \times 10^{-20}} = 0.15 \text{ M}$$

A molecule such as ACh, if dissociated into positive and negative ions, will contribute $2 \times 0.15 = 0.3$ osmoles of osmotic pressure.

7.2. Starting with the binomial distribution [Equation (7.10)], note that

$$\lim_{n \to \infty} \frac{n!}{(n - k)!} = n(n - 1) \cdots (n - k + 1) = n^k$$

and

$$\lim_{n \to \infty} (1 - p)^{n-k} = (1 - \mu/n)^{n-k} = 1 - \mu \frac{(n - k)}{n} + \cdots$$

$$= 1 - \mu + \cdots = e^{-\mu}$$

Substituting these results and the fact that $\mu = np$ into Equation (7.10) gives the Poisson distribution [Equation (7.9)].

The variance of the Poisson distribution is

$$\sigma^2 = \sum_{k=0}^{n} (k - \mu)^2 g(k) = \sum_{k=0}^{\infty} (k^2 - 2k\mu + \mu^2)\left( \frac{\mu^k e^{-\mu}}{k!} \right)$$

Consider the following sums

$$\sum_{k=0}^{\infty} \frac{\mu^k}{k!} = e^\mu \qquad \text{(see, for example, Selby, 1975)}$$

$$\sum_{k=0}^{\infty} \frac{k\mu^k}{k!} = \mu \sum_{k=1}^{\infty} \frac{\mu^{k-1}}{(k - 1)!} = \mu \sum_{j=0}^{\infty} \frac{\mu^j}{j!} = \mu e^\mu$$

$$\sum_{k=0}^{\infty} \frac{k^2 \mu^k}{k!} = \mu \sum_{k=1}^{\infty} \frac{k\mu^{k-1}}{(k - 1)!} = \mu \frac{d}{d\mu}\left[ \sum_{k=1}^{\infty} \frac{\mu^k}{(k - 1)!} \right]$$

$$= \mu \frac{d}{d\mu}\left( \mu \sum_{j=0}^{\infty} \frac{\mu^j}{j!} \right) = \mu \frac{d}{d\mu}(\mu e^\mu)$$

$$= \mu e^\mu (1 + \mu)$$

Thus,

$$\sigma^2 = \mu(1 + \mu) - 2\mu^2 + \mu^2 = \mu$$

For the binomial distribution, we start from the binomial theorem

$$(x + y)^n = \sum_{k=0}^{n} \binom{n}{k} x^k y^{n-k}$$

where $\binom{n}{k} = n!/k!(n - k)!$. Thus, the sum

$$\sum_{k=0}^{n} \binom{n}{k} p^k (1 - p)^{n-k} = (p + 1 - p)^n = 1$$

is independent of $p$ and

$$\frac{d}{dp}\left[ \sum_{k=0}^{n} \binom{n}{k} p^k (1 - p)^{n-k} \right] = 0$$

After performing the differentiation and simplifying, we have

$$\sum_{k=0}^{n} \binom{n}{k} (k - np) p^{k-1} (1 - p)^{n-k-1} = 0$$

Multiplying both sides of this equation by $p(1 - p)$ gives

$$\sum_{k=0}^{n} \binom{n}{k}(k - np)p^k(1 - p)^{n-k} = 0$$

and the mean is then

$$\mu = \sum_{k=0}^{n} kf(k) = \sum_{k=0}^{n} k\binom{n}{k}p^k(1 - p)^{n-k}$$

$$= np \sum_{k=0}^{n} \binom{n}{k}p^k(1 - p)^{n-k} = np$$

To obtain the variance, differentiate again,

$$\frac{d}{dp}\left[\sum_{k=0}^{n} \binom{n}{k}(k - np)p^k(1 - p)^{n-k}\right] = 0$$

or, after simplification,

$$\sum_{k=0}^{n} \binom{n}{k}[(k - np) - np(1 - p)]p^k(1 - p)^{n-k} = 0$$

The variance is then

$$\sigma^2 = \sum_{k=0}^{n} (k - \mu)^2 g(k) = \sum_{k=0}^{n} (k - np)^2 \binom{n}{k}p^k(1 - p)^{n-k}$$

$$= np(1 - p)$$

7.3.  From Ohm's law,

$$I = g_s(V - E_s) + g_m(V - E_m) = 0$$

If $g_s = 0$

$$V = V_m = E_m$$

With nonzero values of $g_s$,

$$V = \frac{g_s E_s + g_m E_m}{g_s + g_m}$$

$$\Delta V = V - V_m = \frac{g_s(E_s - E_m)}{g_s + g_m}$$

The membrane conductance corresponds to the Michaelis constant in this example and the change in voltage, $\Delta V = (E_s - E_m)/2$, is half of its maximum value, when $g_s = g_m$.

7.4.  The surface area of the neuromuscular junction $= 10^9/10^4 = 10^5$ $\mu m^2 = 0.1$ mm$^2$. The total surface area $= \pi dl = \pi \times 0.075 \times 40 = 9.4$ mm$^2$. The fraction of surface covered by the neuromuscular junction is just over 1%. The rate of synthesis would be $10^3 \times 9.4 \times 10^6 \sim 10^{10}$ receptors/week $\sim 10^6$ receptors/min.

8.1.   From the definition of $\delta(x)$, we consider the limit, as $\Delta x \to 0$, of

$$\alpha = \begin{cases} \alpha_0/\Delta x, & 0 \leqslant x \leqslant \Delta x \\ 0, & \text{elsewhere} \end{cases}$$

In the range $0 \leqslant x \leqslant \Delta x$, $\alpha \gg \beta$, $B \sim 0$, and Equation (8.7) becomes

$$v \, dA/dx = \alpha(1 - A)$$

Taking Laplace transforms

$$svA^* = \alpha/s - \alpha A^*$$

assuming $A = 0$ for $x = 0$. Solving for $A^*$,

$$A^* = \frac{\alpha}{s} \left( \frac{1}{\alpha + sv} \right)$$

and

$$A = \frac{\alpha}{v} \int_0^x e^{-\alpha x/v} \, dx$$

$$= 1 - e^{-\alpha x/v}$$

In particular,

$$\lim_{\Delta x \to 0} A(\Delta x) = 1 - e^{-\alpha_0/v}$$

For $x > \Delta x$, $\alpha = 0$ and Equation (8.7) becomes

$$v \, dA/dx = -\beta A$$

Taking Laplace transforms and using the initial condition for $A^*(\Delta x)$ above gives

$$\lim_{\Delta x \to 0} A^* = \frac{(1 - e^{-\alpha_0/v})}{sv + \beta}$$

Taking inverse transforms gives the desired result.
        To solve for $B$ take Laplace transforms of Equation (8.8)

$$svB^* = \beta A^* - \gamma B^*$$

or

$$B^* = \frac{\beta A^*}{sv + \gamma} = \frac{\beta v(1 - e^{-\alpha_0/v})}{(sv + \gamma)(sv + \beta)}$$

Taking inverse Laplace transforms gives the desired form of $B$.

The total number of bonds formed will be, from Equation (8.8)

$$n \propto (1 - e^{-\alpha_0/v}) \int_0^\infty \left[ e^{-\beta x/v} + \left( \frac{\beta}{\beta - \gamma} \right)(e^{-\gamma x/v} - e^{-\beta x/v}) \right] dx$$

$$\propto \frac{1 - e^{-\alpha_0/v}}{\beta - \gamma} \int_0^\infty \left[ \beta e^{-\gamma r/v} - \gamma e^{-\beta x/v} \right] dx$$

Integrating this expression gives the desired result.
To find the velocity where this number is maximum,

$$\frac{dn}{dv} \propto 1 - e^{-\alpha_0/v} - \frac{\alpha_0 e^{-\alpha_0/v}}{v} = 0$$

or

$$e^{-\alpha_0/v} = \frac{1}{1 + \alpha_0/v}$$

This transcendental equation has two solutions: (1) when $v = 0$, $n = 0$, a minimum, and (2) when $v \to +\infty$, $n \propto \alpha_0(\gamma + \beta)/\beta\gamma$, a maximum.

8.2. (a) Taking Laplace transforms of Equation (6.15) and substituting from Equation (6.13) gives

$$I_a^* = -\left( \frac{1}{r_a \lambda} \right) \frac{\partial V^*}{\partial X}$$

$$= \frac{(s + 1)^{1/2}}{r_a \lambda} \left( A^* \exp\left[ -X(s + 1)^{1/2} \right] - B^* \exp\left[ X(s + 1)^{1/2} \right] \right)$$

From symmetry about $X = 0$, $A^* = B^*$. Thus,

$$I_a^* = \frac{2(s + 1)^{1/2}}{r_a \lambda} A^* \sinh\left[ -X(s + 1)^{1/2} \right]$$

At $X = K$, $I_a = -I$ (and at $X = -K$, $I_a = I$),

$$A^* = \frac{r_a \lambda I^*}{2(s + 1)^{1/2} \sinh\left[ K(s + 1)^{1/2} \right]}$$

Then, from Equations (6.19) and (6.13),

$$Z^* = \frac{V^*}{I^*} = \frac{r_a \lambda \{ \exp\left[ -X(s + 1)^{1/2} \right] + \exp\left[ X(s + 1)^{1/2} \right] \}}{2(s + 1)^{1/2} \sinh\left[ K(s + 1)^{1/2} \right]}$$

which gives the desired result after substituting from the definition of the cosh function.

(b) If $I = C \, \partial V / \partial t$, where $I$ is a constant, then after taking Laplace transforms,

$$I^* = I/s = C[sV^* - V(0)]$$

and

$$V^* = \frac{r_a \lambda I^* \cosh\left[X(s+1)^{1/2}\right]}{(s+1)^{1/2}\sinh\left[K(s+1)^{1/2}\right]}$$

from part (a). The initial value theorem states that

$$\lim_{t \to 0} f(t) = \lim_{s \to \infty} sf^*(s)$$

so

$$V(0) = \lim_{s \to \infty} \frac{r_a \lambda I \cosh\left[X(s+1)^{1/2}\right]}{(s+1)^{1/2}\sinh\left[K(s+1)^{1/2}\right]}$$

$$= \lim_{s \to \infty} \frac{r_a \lambda I \exp\left[X(s+1)^{1/2}\right]}{(s+1)^{1/2}\exp\left[K(s+1)^{1/2}\right]} = 0$$

since $x \leqslant K$. Similarly,

$$\lim_{t \to 0} \frac{dV}{dt} = \lim_{s \to \infty} s[sV^* - V(0)]$$

$$= \lim_{s \to \infty} \frac{sr_a \lambda \exp\left[(X-K)(s+1)^{1/2}\right]}{(s+1)^{1/2}}$$

$$= \begin{cases} \infty, & |X| = K \\ 0, & |X| < K \end{cases}$$

and

$$C = \frac{I}{dV/dt} = \begin{cases} 0, & |X| = k \\ \infty, & |X| < k \end{cases}$$

(c) The transverse tubules open out at their two ends to the surface membrane. As an action potential spreads along the surface membrane, current will be injected at the two ends of the cable-like structure of the tubules, as shown in this problem. At the point of injection, the membrane will charge up instantaneously (with no capacity), but other parts of the transverse tubules will not charge up at all until later times. Thus, with high frequency signals, only the surface membrane will contribute to the overall capacitance of the system.

8.3. (a) From Equation (8.16)

$$F(v + b) = (F_0 + a)b - a(v + b)$$

$$F = \frac{F_0 b - av}{v + b}$$

When $v = v_{max}$, $F = 0$ and $b = av_{max}/F_0$. Thus,

$$F = \frac{a(v_{max} - v)}{v + av_{max}/F_0}$$

Algebraic manipulation of this equation yields Equation (8.17).

(b) Expanding Fenn and Marsh's equation in a power series gives

$$F = W_0 \left[ 1 - av + \frac{(av)^2}{2} - \frac{(av)^3}{6} + \cdots \right] - kv$$

Expanding Hill's equation above gives

$$F = \frac{F_0(1 - \alpha v)}{1 + v/b} = F_0(1 - \alpha v)\left[ 1 - \frac{v}{b} + \left( \frac{v}{b} \right)^2 - \left( \frac{v}{b} \right)^3 + \cdots \right]$$

where $\alpha = a/F_0 b$. Zero-order terms can be matched if $W_0 = F_0$.

First-order terms can be matched if $a + k/W_0 = \alpha + 1/b$. Second-order terms can also be matched if $a^2/2 = \alpha/b + 1/b^2$. These three equations determine the three parameters in each model, and higher-order terms cannot be matched without some special constraint. For example, the third-order terms will match only if

$$\frac{a^3}{6} = \frac{1}{b^3} + \frac{\alpha}{b^2} = \frac{1 + \alpha b}{b^3}$$

Dividing this condition by the condition for second-order terms gives $a/3 = 1/b$. Similar division of the condition for second-order terms by that for first-order terms gives

$$\frac{a^2}{2(a + k/W_0)} = \frac{1}{b}$$

Clearly, all conditions can only hold in the special case where $a = k/W_0$.

8.4. (a) The equation for the fraction $n$ of bonds in the attached state is

$$v \, dn/dx = \alpha(1 - n) - \beta n$$

(b) If we are considering a *contraction* velocity $v$ (i.e., $v < 0$) no bonds will be formed until $x$ reaches $h$ from the right. Then, in the range $0 \leqslant x \leqslant h$,

$$v \, dn/dx = \alpha_1 x/h - (\alpha_1 + \beta_1)xn/h \qquad 0 \leqslant x \leqslant h$$

Let $\gamma = -hv$, where $\gamma > 0$ since $v < 0$. Then,

$$\frac{\gamma dn}{(\alpha_1 + \beta_1)n - \alpha_1} = x \, dx$$

and after integrating from $x$ to $h$

$$\frac{-\gamma}{\alpha_1 + \beta_1} \ln\left[\frac{(\alpha_1 + \beta_1)n - \alpha_1}{\alpha_1}\right] = \frac{h^2 - x^2}{2}$$

since $n(h) = 0$. Taking exponentials of both sides and rearranging

$$n = \frac{\alpha_1}{\alpha_1 + \beta_1}\left[1 - e^{[(x^2 - h^2)(\alpha_1 + \beta_1)]/2\gamma}\right]$$

For $x < 0$, the differential equation in (a) becomes

$$dn/dx = kn$$

where $k = -\beta_2/v$ and $k > 0$. Then

$$n = n_0 e^{-kn}$$

where, from above,

$$n_0 = \frac{\alpha_1}{\alpha_1 + \beta_1}\left[1 - e^{-h^2(\alpha_1 + \beta_1)/2\gamma}\right]$$

(c)

$$F \propto \int_{-\infty}^{\infty} xn \, dx = \int_{-\infty}^{0} xn \, dx + \int_{0}^{h} xn \, dx$$

$$\int_{-\infty}^{0} xn \, dx = n_0 \int_{-\infty}^{0} xe^{-kn} = -n_0/k^2$$

$$\int_{0}^{h} xn \, dx = \frac{\alpha_1}{\alpha_1 + \beta_1}\left[\frac{h^2}{2} - e^{-h^2(\alpha_1 + \beta_1)/2\gamma}\int_{0}^{h} xe^{x^2(\alpha_1 + \beta_1)/2\gamma} \, dx\right]$$

$$= \frac{\alpha_1}{\alpha_1 + \beta_1}\left\{\frac{h^2}{2} - \left(\frac{\gamma}{\alpha_1 + \beta_1}\right)\left[1 - e^{-h^2(\alpha_1 + \beta_1)/2\gamma}\right]\right\}$$

Finally,

$$F \propto \frac{\alpha_1}{\alpha_1 + \beta_1}\left\{\frac{h^2}{2} - \left(\frac{\gamma}{\alpha_1 + \beta_1} + \frac{1}{k^2}\right)\left[1 - e^{-h^2(\alpha_1 + \beta_1)/2\gamma}\right]\right\}$$

8.5. (a) $Z^*(s) \sim Z_0 s_0^2/s^2$,    $s \gg s_0$
(b) $Z^*(s) \sim Z_0$,    $s \ll s_0$
The two approximations above are equal if and only if $s = s_0$.
(c) The Laplace transform of $Z(t) = e^{-t/\tau_1} - e^{-t/\tau_2}$, where $\tau_1$ and $\tau_2$

are real, positive constants, is

$$Z^*(s) = \frac{\tau_1}{1 + s\tau_1} - \frac{\tau_2}{1 + s\tau_2} = \frac{\tau_1 - \tau_2}{(1 + s\tau_1)(1 + s\tau_2)}$$

Clearly, this expression can only agree with the form given in the problem if the quadratic expression there can be factored. The roots will be

$$s = -2\zeta/s_0 \pm \left(4\zeta^2/s_0^2 - 4/s_0^2\right)^{1/2}$$

$$= -\frac{2}{s_0}\left[\zeta \pm (\zeta^2 - 1)^{1/2}\right]$$

(d) There will be two real time constants $\tau_1$ and $\tau_2$ where

$$\tau_1 = \frac{2}{s_0}\left[\zeta - (\zeta^2 - 1)^{1/2}\right] \quad \text{and} \quad \tau_2 = \frac{2}{s_0}\left[\zeta + (\zeta^2 - 1)^{1/2}\right]$$

only if $\zeta^2 > 1$. For $\tau_1, \tau_2$ to be positive and distinct, $\zeta > 1$.

9.1. (a) From Equation (9.9),

$$z^*(s) = y^*(s)/x^*(s) = a(1 + sb)$$

$$\text{Amplitude} = |z^*(\omega)| = |a(1 + j\omega b)| = a(1 + \omega^2 b^2)^{1/2}$$

$$\text{Phase} = \angle z^*(\omega) = \tan^{-1}(\omega b)$$

(b)

$$z_0 = |z^*(0)| = a, \qquad \angle z^*(0) = 0$$

$$z_\infty = |z^*(\infty)| = \infty, \qquad \angle z^*(\infty) = 90°$$

(c)

$$|z^*(b^{-1})| = a2^{1/2} = z_0 2^{1/2}; \qquad \angle z^*(b^{-1}) = \tan^{-1}(1) = 45°$$

9.2. (a) From the data given, the instantaneous frequency can be calculated and compared to a function $y(t) = 10^3 t e^{-10t}$

| Impulse number | 1 | 2 | 3 | 4 | 5 | 6 | 7 | 8 |
|---|---|---|---|---|---|---|---|---|
| Time (ms) | 36 | 68 | 95 | 123 | 153 | 176 | 230 | 296 |
| Inst. freq. | 26.3 | 33.3 | 37.0 | 35.7 | 33.3 | 29.4 | 23.3 | 15.2 |
| $10^3 t e^{-10t}$ | 26.0 | 34.4 | 36.7 | 36.0 | 33.1 | 28.8 | 23.1 | 15.3 |

(b) The transform of the input function

$$x^*(s) = \frac{1}{s} - \frac{e^{-0.1s}}{s} = \frac{1}{s}(1 - e^{-0.1s})$$

since the response to a step lasting 0.1 s represents the sum of a step of infinite duration starting at $t = 0$ $[x^*(s) = 1/s]$ and a negative step starting at $t = 0.1$ s $[x^*(s) = e^{-0.1s}/s]$. The function $y^*(t)$ is obtained by transforming the $y(t)$ for part (a) and is given by

$$y^*(s) = \frac{10^3}{(s + 10)^2}$$

The transfer function is then obtained from the equation

$$z^*(s) = y^*(s)/x^*(s)$$

(c) The frequency–response function is (see diagram)

$$z^*(\omega) = \frac{10^3}{(j\omega + 10)^2(1 - e^{-0.1j\omega})}$$

which has a magnitude

$$|z^*(\omega)| = \frac{10^3}{(\omega^2 + 100)\left[(1 - \cos 0.1\omega)^2 + \sin^2 0.1\omega\right]^{1/2}}$$

$$= \frac{250\omega}{(\omega^2 + 100)|\sin 0.05\omega|}$$

using the relation $e^{jx} = \cos x + j \sin x$ and $\sin x/2 = [(1 - \cos x)/2]^{1/2}$. The phase is

$$\angle z^*(\omega) = \pi/2 - 2 \tan^{-1}(0.1\omega) - \tan^{-1}\left[\frac{\sin(0.1\omega)}{1 - \cos(0.1\omega)}\right]$$

$$= \pi/2 - 2 \tan^{-1}(0.1\omega) - \tan^{-1}\left(\frac{2\sin(0.1\omega)}{\sin^2(0.05\omega)}\right)$$

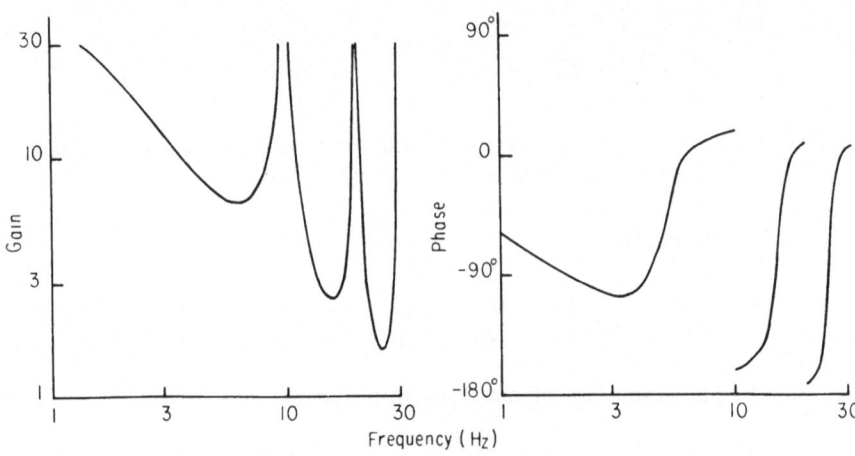

(d) The magnitude or gain of the frequency–response function goes to $\infty$ for $\omega = 2\pi f = 0$, $20\pi$, $40\pi$, etc. In addition, the response mainly shows phase lags and a decreasing gain with increases in frequency (except for the resonance peaks). Most muscle receptors show phase leads and increasing gain over a range of frequencies.

9.3. $y^*(s) = s^{k-1}\Gamma(1 - k)$; $x^*(s) = s^{-1}$; $z^*(s) = y^*(s)/x^*(s) = s^k\Gamma(1 - k)$; and $\Gamma(n) = \int_0^\infty e^{-x}x^{n-1}\,dx$ is the gamma function. The gain is $|z^*(\omega)| = \omega^k\Gamma(1 - k)$ and the phase is $\angle z^*(\omega) = k\pi/2$.

9.4. (a) Transforming the equations given

$$sp^* - p_0 = av^* - bp^*$$

$$v^* = c(i^* - p^*)$$

where $p_0$, the pumping rate at rest ($v = 0$), is taken to be zero, if $p$ is the increase in pumping rate with activity. Then,

$$p^* = av^*/(s + b)$$

and substituting this equation into the second equation above and rearranging gives the desired result.

(b) $z_0 = z^*(0) = cb/(ca + b)$; $z_\infty = Z^*(\infty) = c$. If $ca \gg b$

$$|z^*(\omega)| \sim \frac{c(\omega^2 + b^2)^{1/2}}{\left[\omega^2 + (ca)^2\right]^{1/2}}, \qquad z_0 \sim b/a$$

(c)

$$|z^*(b)| = 2^{1/2}b/a = 2^{1/2}z_0$$

$$|z^*(ca)| = c/2^{1/2} = Z_\infty/2^{1/2}$$

(d) From Fig. 9.11B, $z_0 = b/a \sim 100$, and $z_\infty = c \sim 3000$. The lower breakpoint occurs just above 3 Hz $= 6\pi$ radians/s so $b \sim 20$ and $a \sim 20/100 = 0.2$. The upper breakpoint occurs just below 100 Hz $= 200\pi$ radians/s, so $ca \sim 600$ and again $a \sim 0.2$.

9.5. (a) From Table 9.3, omitting ($\omega$) for compactness,

$$z_r = \frac{x_r y_r + x_i y_i}{x_r^2 + x_i^2}; \qquad z_i = \frac{x_r y_i + x_i y_r}{x_r^2 + x_i^2}$$

$$|z^*(\omega)| = (z_r^2 + z_i^2)^{1/2} = \frac{(x_r^2 + x_i^2)(y_r^2 + y_i^2)}{x_r^2 + x_i^2}$$

$$= \frac{|y^*(\omega)|}{|x^*(\omega)|}$$

$$\angle z^*(\omega) = \tan^{-1}(z_i/z_r) = \tan^{-1}\left(\frac{x_r y_i - x_i y_r}{x_r y_r + x_i y_i}\right)$$

(b)

$$\gamma^2(\omega) = \frac{|S_{xy}(\omega)|^2}{S_{xx}(\omega)S_{yy}(\omega)} = |z^*(\omega)|^2 \frac{S_{xx}(\omega)}{S_{yy}(\omega)}$$

$$= \frac{|y^*(\omega)|^2}{|x^*(\omega)|^2} \frac{|x^*(\omega)|^2}{|y^*(\omega)|^2} = 1$$

10.1.  The somatic conductance should vary as somatic membrane area which will vary as the square of the linear dimensions. Thus, the somatic resistance $r_s \propto d^{-2}$. The discrepancy could arise in the assumptions of uniform scaling, or that the resistance $R_s$ per unit membrane area is independent of cell size.

10.2.  From the diagram, one can write

$$o^*(s) = g^*(s)e^*(s)$$

$$e^*(s) = i^*(s) - f^*(s)$$

$$f^*(s) = h^*(s)o^*(s)$$

Therefore,

$$o^*(s) = g^*(s)\left[i^*(s) - h^*(s)o^*(s)\right]$$

and algebraic manipulation yields the desired result.

10.3.  (a) The forces produced by the load must balance the applied force $x(t)$ so

$$k_e y + Dv + Ma = x(t)$$

where the velocity $v = dy/dt$, the acceleration $a = d^2y/dt^2$, and we define $y = 0$ at the equilibrium position in the absence of applied forces.
(b) Taking Laplace transforms

$$k_e y^* + sDy^* + s^2My^* = x^*$$

$$g^* = y^*/x^* = \frac{1}{k_e + sD + s^2M}$$

which converts to the desired form if $g_0 = 1/k_e$,

$$\omega_n = (k_e/M)^{1/2} \qquad \zeta = \frac{D}{2(k_eM)^{1/2}}.$$

(c)

$$g^*(\omega) = \frac{g_0}{1 - (\omega/\omega_n)^2 + j2\zeta\omega/\omega_n}$$

$$|g^*(\omega)| = \frac{g_0}{\left\{\left[1 - (\omega/\omega_n)^2\right]^2 + \left[2\zeta\omega/\omega_n\right]^2\right\}^{1/2}}$$

which will have a maximum when $[1 - (\omega/\omega_n)^2]^2 + [2\zeta\omega/\omega_n]^2$ has a minimum. Differentiating with respect to $w$, the condition for a minimum is

$$\left[1 - (\omega/\omega_n)^2\right]\omega + 2\zeta^2\omega = 0$$

$$(\omega^3/\omega_n^2) + \omega(2\zeta^2 - 1) = 0$$

Solutions of this equation occur for

$$\omega = 0, \pm(1 - 2\zeta^2)^{1/2}$$

However, it can be shown that $\omega = 0$ represents a maximum for the denominator or a minimum for $|g^*(\omega)|$. The maximum will only occur for a real frequency if $1 > 2\zeta^2$ or $\zeta < 2^{-1/2}$.

(d) If $g^*(s) = g_0/[1 + 2\zeta s/\omega_n + (s/\omega_n)^2]$, the quadratic term in the denominator will have zeros when

$$s = -\omega_n\left[\zeta \pm (\zeta^2 - 1)^{1/2}\right]$$

If $\zeta < 1$, the roots will be complex numbers, and we can rewrite $g^*(s)$ as

$$g^*(s) = \frac{g_0\omega n^2}{(s + p + j\omega_d)(s + p - j\omega_d)}$$

where $p = \omega n\zeta$ and $\omega_d = \omega_n(1 - \zeta^2)^{1/2}$. The inverse transform is then

$$g(t) = \frac{g_0\omega_n^2}{2j\omega_d} e^{-pt}(e^{j\omega_d t} - e^{-j\omega_d t})$$

The desired result is obtained by substituting $A = g_0\omega_n^2/\omega_d$ and using the definition $\sin(x) = (1/2j)(e^{jx} - e^{-jx})$.

(e) If $\zeta = 1$, then

$$g^*(s) = \frac{g_0\omega_n^2}{(s + \omega_n)^2} \quad \text{and} \quad g(t) = g_0\omega_n^2 t e^{-\omega_n t}$$

If $\zeta > 1$, the roots of the denominator in $g^*(s)$ are both real and distinct numbers, where

$$g^*(s) = \frac{g_0 \omega_n^2}{(s + p_1)(s + p_2)}$$

with $p_1 = \omega_n[\zeta + (\zeta^2 - 1)^{1/2}]$ and $p_2 = \omega_n[\zeta - (\zeta^2 - 1)^{1/2}]$. Then

$$g(t) = \frac{g_0 \omega_n}{2(\zeta^2 - 1)^{1/2}} (e^{-p_2 t} - e^{-p_1 t})$$

# References

Adrian, R. H. (1969). Rectification in muscle membrane. *Prog. Biophys. Mol. Biol.* **19**:341–369.

Adrian, R. H. (1978). Charge movement in the membrane of striated muscle. *Ann. Rev. Biophys. Bioengng.* **7**:85–112.

Adrian, R. H., Chandler, W. K., and Hodgkin, A. L. (1970). Voltage clamp experiments in striated muscle fibres. *J. Physiol.* **208**:607–644.

Adrian, R. H., Chandler, W. K., and Rakowski, R. F. (1976). Charge movement and mechanical repriming in skeletal muscle. *J. Physiol.* **254**:361–388.

Adrian, R. H. and Peachey, L. D. (1973). Reconstruction of the action potential of frog sartorius muscle. *J. Physiol.* **235**:103–131.

Agin, D. (1964). Hodgkin–Huxley equations: Logarithmic relation between membrane current and frequency of repetitive activity. *Nature* **201**:625–626.

Alnaes, E., Jansen, J. K. S. and Rudjord, T. (1965). Fusimotor activity in the spinal cat. *Acta Physiol. Scand.* **63**:197–212.

Amplett, G. W., Perry, S. W., Syska, H., Brown, M. D., and Vrbova, G. (1975). Cross innervation and the regulatory protein system of rabbit soleus muscle. *Nature* **257**:602–604.

Anderson, C. R. and Stevens, C. F. (1973). Voltage clamp analysis of acetylcholine produced end-plate current fluctuations at frog neuromuscular junction. *J. Physiol.* **235**:655–691.

Andersen P. and Andersson, S. A. (1968). *Physiological Basis of the Alpha Rhythm*. Appleton–Century–Croft, New York.

Andreassen, S. (1978). *Interval Patterns of Single Motor Units*. Ph.D. Thesis, Institute of Neurophysiology, University of Copenhagen, and Electronics Laboratory, Technical University of Denmark.

Armstrong, C. M. and Bezanilla, F. (1973). Currents related to movement of the gating particles of the sodium channels. *Nature* **242**:459–461.

Armstrong, C. M. and Bezanilla, F. (1977). Inactivation of the sodium channel. II. Gating current experiments. *J. Gen. Physiol.* **70**:567–590.

Armstrong, C. M., Bezanilla, F., and Rojas, E. (1973). Destruction of sodium conductance inactivation in squid axons perfused with pronase. *J. Gen. Physiol.* **62**:375–391.

Ash, R. (1965). *Information Theory*. Wiley, New York.

Ashley, C. C. and Moisescu, D. G. (1972). Model for the action of calcium in muscle. *Nature New Biol.* **237**:208–211.

Bagshaw, C. R., Eccleston, J. R., Eckstein, F., Goody, R. S., Gutfreund, H. and Trentham, D. R. (1974). The magnesium ion-dependent adenosine triphosphate of myosin. *Biochem. J.* **141**:351–364.

Bagust, J. and Lewis, D. M. (1974). Isometric contractions of motor units in self-reinnervated fast and slow twitch muscles of the cat. *J. Physiol.* **237**:91–102.

Baker, P.F., Hodgkin, A. L., and Ridgway, E. B. (1971). Depolarization and calcium entry in squid giant axons. *J. Physiol.* **218**:709–755.

Baker, P. F. and Willis, J. S. (1972). Binding of the cardiac glycoside ouabain to intact cells. *J. Physiol.* **224**:441–462.

Baldissera, F., Gustafsson, B., and Parmiggiani, F. (1978). Saturating summation of the afterhyperpolarization conductance in spinal motoneurones: A mechanism for "secondary range" repetitive firing. *Brain Res.* **146**:69–82.

Bangham, A. D. (1972). Lipid bilayers and biomembranes. *Ann. Rev. Biochem.* **41**:753–776.

Barker, D., Emonet-Dénand, F., Harker, D., Jami, L., and Laporte, Y. (1976). Distribution of fusimotor axons to intrafusal muscle fibres in cat tennuisimus spindles as determined by the glycogen depletion method. *J. Physiol.* **261**:49–69.

Barker, D., Emonet-Denand, F., Harker, D. W., Jami, L., and Laporte, Y. (1977). Types of intra- and extrafusal muscle fibre innervated by dynamic skeleto-fusimotor axons in cat peroneus brevis and tennuisimus muscles, as determined by the glycogen depletion method. *J. Physiol.* **266**:713–726.

Basmajian, J. V. (1976). *Muscles Alive: Their Functions Revealed by Electromyography*, 2nd ed. Williams and Wilkens, Baltimore.

Bawa, P. and Stein, R. B. (1976). The frequency response of human soleus muscle. *J. Neurophysiol.* **39**:788–793.

Beidler, L. M. (1961). Taste receptor stimulation. *Prog. Biophys. Biophys. Chem.* **12**:107–151.

Bendat, J. S. and Piersol, A. G. (1971). *Random Data: Analysis and Measurement Procedures.* Wiley, New York.

Bennett, M. R. and Pettigrew, A. G. (1976). The formation of neuromuscular synapses. In *The Synapse.* Cold Spring Harbor, New York, *Symp. Quant. Biol.* **40**:409–424.

Bennett, M. V. L., ed. (1974). *Synaptic Transmission and Neuronal Interaction.* Raven Press, New York.

Berg, D. K. and Hall, Z. W. (1975). Increased extrajunctional acetylcholine sensitivity produced by chronic post-synaptic neuromuscular blockade. *J. Physiol.* **244**:659–676.

Berthold, C.-H. (1978). Morphology of normal peripheral axons. In *Physiology and Pathobiology of Axons* (Wasman, S. G., ed.). Raven Press, New York, pp. 3–63.

Bessou, P., Laporte, Y., and Pages, B. (1968). Frequencygrams of spindle primary endings elicited by stimulation of static and dynamic fusimotor fibres. *J. Physiol.* **196**:47–63.

Betz, W. J. (1970). Depression of transmitter release at the neuromuscular junction of the frog. *J. Physiol.* **206**:629–644.

Birks, R. I. (1977). A long-lasting potentiation of transmitter release related to an increase in transmitter stores in a sympathetic ganglion. *J. Physiol.* **271**:847–862.

Blaschko, H. and Muscholl, E. (1972). Catecholamines. In *Hand. Exp. Pharmakol.*, Vol. 33. Springer-Verlag, Berlin.

Blaustein, M. P. and Hodgkin, A. L. (1969). The effect of cyanide on the efflux of calcium from squid axons. *J. Physiol.* **200**:497–527.

Blaustein, M. P., Ratzlaff, R. W., and Kendrick, N. K. (1978). The regulation of intracellular calcium in presynaptic terminals. In *Calcium Transport and Cell Function* (Scarpa, A. and Carafoli, E., eds.). *Ann. N.Y. Acad. Sci.* **307**:195–211.

Blinks, J. R., Rüdel, R., and Taylor, S. R. (1978). Calcium transients in isolated amphibian skeletal muscle fibres: detection with aequorin. *J. Physiol.* **277**:291–323.

Bloom, F. E. (1972). Electron microscopy of catecholamine-containing structures. In *Catecholamines* (Blashko, H. and Muscholl, E., ed.). Springer-Verlag, Berlin. *Hand. Exp. Pharmakol.* **33**:46–78.

Boistel, J. and Fatt, P. (1958). Membrane permeability change during inhibitory transmitter action in crustacean muscle. *J. Physiol.* **144**:176–191.

Boyd, I. A. and Kalu, K. U. (1979). Scaling factor relating conduction velocity and diameter for myelinated afferent fibres in the cat hind limb. *J. Physiol.* **289**:277–297.

Boyd, I. A. and Martin, A. R. (1956). The end-plate potential in mammalian muscle. *J. Physiol.* **132**:74–91.

Boyd I. A. and Ward, J. (1969). The response of isolated cat muscle spindles to passive stretch. *J. Physiol.* **200**:104–105P.

Boyd, I. A., Gladden, M., McWilliam, P. N., and Ward, J. (1977). Control of dynamic and static nuclear bag fibres and nuclear chain fibres by $\gamma$ and $\beta$ axons in isolated cat muscle spindles. *J. Physiol.* **265**:133–162.

Bretscher, M. S. (1973). Membrane structure: some general principles. *Science* **181**:622–629.

Brink, F. (1954). The role of calcium ions in neural processes. *Pharmacol. Rev.* **6**:243–298.

Britton, H. G. (1964). Permeability of the human cell to labeled glucose. *J. Physiol.* **170**:1–20.

Britton, H. G. (1966). Fluxes in passive, monovalent, and polyvalent carrier systems. *J. Theor. Biol.* **10**:28–52.

Brodwick, M. S. and Eaton, D. C. (1978). Sodium channel inactivation in squid axon is removed by high internal pH or tyrosine-specific reagents. *Science* **200**:1494–1496.

Brown, H. F. and Noble, S. J. (1969). Membrane currents underlying delayed rectification and pace-maker activity in frog atrial muscle. *J. Physiol.* **204**:717–736.

Brown, M. C. (1971). The responses of frog muscle spindles and fast and slow muscle fibres to a variety of mechanical inputs. *J. Physiol.* **218**:1–17.

Brown, M. C., Crowe, A., and Matthews, P. B. C. (1965). Observations on the fusimotor fibres of the tibialis posterior muscle of the cat. *J. Physiol.* **177**:140–159.

Brown, M. C., Engberg, I., and Matthews, P. B. C. (1967). The relative sensitivity to vibration of muscle receptors of the cat. *J. Physiol.* **192**:773–800.

Brown, M. C., Goodwin, G. M., and Matthews, P. B. C. (1969). After-effects of fusimotor stimulation on the response of muscle spindle primary afferent endings. *J. Physiol.* **205**:677–694.

Brown, M. C., Jansen, J. K. S., and Van Essen, D. (1976). Polyneuronal innervation of skeletal muscle in new-born rats and its elimination during maturation. *J. Physiol.* **261**:387–422.

Brown, M. C. and Stein, R. B. (1966). Quantitative studies on the slowly adapting stretch receptor of the crayfish. *Kybernetik* **4**:175–185.

Brown, T. G. (1911). The intrinsic factors in the act of progression in the mammal. *Proc. Roy. Soc. London B* **84**:308–319.

Brown, T. G. (1914). On the nature of the fundamental activity of the nervous centres: together with an analysis of the conditioning of rhythmic activity in progression, and a theory of the evolution of function in the nervous system. *J. Physiol.* **48**:18–46.

Bülbring, E. and Shuba, M. F. (1976). *Physiology of Smooth Muscle.* Raven Press, New York.

Buller, A. J., Eccles, J. C., and Eccles, R. M. (1960). Interactions between motoneurons and muscles in respect to the characteristic speed of their responses. *J. Physiol.* **150**:399–416.

Buller, N. P., Garnett, R., and Stephens, J. A. (1978). The use of skin stimulation to produce reversal of motor unit recruitment order during voluntary muscle contraction in man. *J. Physiol.* **277**:1–2P.

Burke, R. E. (1967). Motor unit types of cat triceps surae muscle. *J. Physiol.* **193**:141–160.

Burke, R. E. (1968). Group Ia synaptic input to fast and slow twitch motor units of cat triceps surae. *J. Physiol.* **196**:605–630.

Burke, R. E. (1973). On the central nervous system control of fast and slow twitch motor units. In *New Developments in EMG and Clinical Neurophysiology* (Desmedt, J. E., ed). Karger, Basel **3**:69–94.

Burke, R. E. and Edgerton, V. R. (1975). Motor unit properties and selective involvement in movement. *Exer. Sport Sci. Rev.* **3**:31–81.

Burke, R. E. and Tsairis, P. (1973). Anatomy and innervation ratios in motor units of cat gastrocnemius. *J. Physiol.* **234**:749–765

Burke, R. E., Levine, D. N., Salcman, M., and Tsairis, P. (1974). Motor units in cat soleus muscle: physiological, histochemical, and morphological characteristics. *J. Physiol.* **238**: 503–514.

Burke, R. E., Levine, D. N., Zajac, F. E., Tsairis, P., and Engel, W. K. (1971). Mammalian motor units: physiological-histochemical correlation in three types in cat gastrocnemius. *Science* **174**:709–712.

Burke, R. E., Rymer, W., and Walsh, J. V. Jr. (1976). Relative strength of synaptic input from short-latency pathways to motor units of defined type in cat medial gastrocnemius. *J. Neurophysiol.* **39**:447–458.

Burke, R. E., Rudomin, P., and Zajac, F. E., III (1977). The effect of activation history on tension production by individual muscle units. *Brain Res.* **109**:515–529.

Burns, B. D. and Salmoiraghi, G. C. (1960). Repetitive firing of respiratory neurones during their burst activity. *J. Neurophysiol.* **23**:27–46.

Burnstock, G. (1976). Do some nerve cells release more than one transmitter? *Neuroscience* **1**:239–248.

Bush, B. M. H. and Roberts, A. (1968). Resistance reflexes from a crab muscle receptor without impulses. *Nature* **218**:1171–1173.

Butz, E. and Cowan J. D. (1974). Transient potentials in dendritic systems of arbitrary geometry. *Biophys. J.* **14**:661–689.

Cangiano, A. (1973). Acetylcholine supersensitivity: the role of neurotrophic factors. *Brain Res.* **58**:255–259.

Capaldi, R. A. (1974). A dynamic model of cell membranes. *Sci. Amer.* **230**:26–33.

Caputo, C. (1978). Excitation and contraction processes in muscle. *Ann. Rev. Biophys. Bioengng.* **7**:63–84.

Chandler, W. K. and Meves, H. (1965). Voltage clamp experiments on internally perfused giant axons. *J. Physiol.* **180**:788–820.

Chandler, W. K. and Meves, H. (1970). Evidence for two types of sodium conductance in axons perfused with sodium fluoride solution. *J. Physiol.* **211**:653–678.

Chandler, W. K., Hodgkin, A. L. and Meves, H. (1965). The effect of changing the internal solution on sodium inactivation and related phenomena in giant axons. *J. Physiol.* **180**:821–836.

Chock, S. P., Chock, P. B. and Eisenberg, E. (1976). Pre-steady-state kinetic evidence for a cyclic interaction of myosin subfragment one with actin during the hydrolysis of adenosine 5′-triphosphate. *Biochemistry* **150**:3244–3253.

Chung, E. K. (1974). *Electrocardiography*. Harper and Row, New York.

Clamann, H. P., Gillies, J. D., Skinner, R. D. and Henneman, E. (1974). Quantitative measures of output of a motoneuron pool during monosynaptic reflexes. *J. Neurophysiol.* **37**:1328–1337.

Clark, J. and Plonsey, R. (1966). A mathematical evaluation of the core conductor model. *Biophys. J.* **6**:95–112.

Close, R. I. (1972). Dynamic properties of mammalian skeletal muscles. *Physiol. Rev.* **52**:129–147.

Cole, K. S. (1949). Dynamic electrical characteristics of the squid axon membrane. *Arch. Sci. Physiol.* **3**:253–258.

Cole, K. S. and Curtis, H. J. (1939). Electrical impedance of the squid giant axon during activity. *J. Gen. Physiol.* **22**:649–670.

Cole, K. S. and Moore, J. W. (1960). Potassium ion current in the squid giant axon. *Biophys. J.* **1**:1–14.

Connor, J. A. and Stevens, C. F. (1971). Prediction of repetitive firing behaviour from voltage clamp data on an isolated neurone soma. *J. Physiol.* **213**:31–53.

Constantin, L. L. (1975). Electrical properties of the transverse tubular system. *Fed. Proc.* **34**:1390–1394.

Cooke, J. D. and Quastel, D. M. J. (1973). Cumulative and persistent effects of nerve terminal depolarization on transmitter release. *J. Physiol.* **228**:407–434.

Cooper, R. D. and Smith, R. S. (1974). The movement of optically detectable organelles in myelinated axons of *Xenopus laevis*. *J. Physiol.* **242**:77–97.

Cooper, S. and Eccles, J. C. (1930). The isometric responses of mammalian muscles. *J. Physiol.* **69**:377–385.

Coppin, C. M. L. and Jack, J. J. B. (1972). Internodal length and conduction velocity of cat muscle afferent nerve fibers. *J. Physiol.* **222**:91–93P.

Cox, D. R. and Lewis, P. A. W. (1966). *Statistical Analysis of Series of Events*. Methuen, London.

Cox, D. R. and Miller, H. D. (1965). *The Theory of Stochastic Processes*. Methuen, London.

Cragg, B. and Thomas, P. K. (1961). Changes in conduction velocity and fibre size proximal to peripheral nerve lesions. *J. Physiol.* **157**:315–327.

Curtin, N. A., Gilbert, C., Kretzschmer, K. M., and Wilkie, D. R. (1974). The effect of the performance of work on total energy output and metabolism during muscular contraction. *J. Physiol.* **238**:455–472.

Cussons, P. D., Hulliger, M., and Matthews, P. B. C. (1977). Effects of fusimotor stimulation on the sensitivity of the muscle spindle secondary ending to small amplitude sinusoidal stretch. *J. Physiol.* **265**:30–31P.

Dale, H. H. (1935). Pharmacology and nerve endings. *Proc. Roy. Soc. Med.* **28**:319–332.

Danielli, J. F. and Davson, H. (1935). A contribution to the theory of permeability of thin films. *J. Cell. Comp. Physiol.* **5**:495–508.

Danielli, J. F. and Harvey, E. N. (1935). The tension at the surface of mackerel egg oil, with remarks on the nature of the cell surface. *J. Cell. Comp. Physiol.* **5**:483–494.

Davidson, N. (1976). *Neurotransmitter Amino Acids*. Academic Press, London.

D'Azzo, J. J. and Houpis, C. H. (1966). *Feedback Control System Analysis and Synthesis*. McGraw-Hill, New York.

Del Castillo, J. and Katz, B. (1955). Local activity at a depolarized nerve–muscle junction. *J. Physiol.* **128**:396–411.

Del Castillo, J. and Katz, B. (1956). Biophysical aspects of neuro-muscular transmission. *Prog. Biophys. Biophys. Chem.* **6**:121–170.

Dellow, P. G. and Lund, J. P. (1971). Evidence for central patterning of rhythmical mastication. *J. Physiol.* **215**:1–15.

De Robertis, E. D. P. and Bennett, H. S. (1955). Some features of the submicroscopic morphology of synapses in frog and earthworm. *J. Biophys. Biochem. Cytol.* **1**:47–58.

Desmedt, J. E. (1978a). Cerebral motor control in man: long-loop mechanisms. *Prog. Clin. Neurophysiol.*, Vol. 4, Karger, Basel.

Desmedt, J. E. (1978b). Physiological tremor, pathological tremors, and gradation of muscle force. *Prog. Clin. Neurophysiol.*, Vol. 5, Karger, Basel.

Desmedt, J. E. (1980). Recruitment patterns of motor units and the gradation of muscle force. *Prog. Clin. Neurophysiol.*, Vol. 9, Karger, Basel.

Desmedt, J. E. and Godeaux, E. (1977). Ballistic contractions in man: characteristic recruitment pattern of single motor units of the tibialis anterior muscle. *J. Physiol.* **264**:673–693.

DeVoe, R. D. (1974). Principles of cell homeostasis. In *Medical Physiology*, 13th ed. (Mountcastle, V. B., ed.). C. V. Mosby, St. Louis.

Devreotes, P. N. and Fambrough, D. M. (1976). Turnover of acetylcholine receptors in skeletal muscle. In *The Synapse*, Cold Spring Harbor, New York, *Symp. Quant. Biol.* **40**:237–251.

De Weer, P. and Geduldig, D. (1973). Electrogenic sodium pump in squid giant axon. *Science* **179**:1326–1328.

Dodge, F. A. and Rahamimoff, R. (1967). Cooperative action of calcium ions in transmitter release at the neuromuscular junction. *J. Physiol.* 193:419–432.

Dowben, R. M. (1969). *General Physiology: A Molecular Approach.* Harper and Row, New York.

Dudel, J. and Kuffler, S. W. (1961). Presynaptic inhibition at the crayfish neuromuscular junction. *J. Physiol.* 155:543–562.

Duysens, J. and Stein, R. B. (1978). Reflexes induced by nerve stimulation in walking cats with implanted cuff electrodes. *Exp. Brain Res.* 32:213–224.

Draper, M. H. and Weidmann, S. (1951). Cardiac resting and action potentials recorded with an intracellular electrode. *J. Physiol.* 115:74–94.

Eaton, D. C., Brodwick, M. S., Oxford, G. S., and Rudy, B. (1978). Argenine-specific reagents remove sodium channel inactivation. *Nature* 271:473–476.

Ebashi, S., Endo, M., and Ohtsuki, I. (1969). Control of muscle contraction. *Quart. Rev. Biophys.* 2:351–384.

Eccles, J. C. (1964). *The Physiology of Synapses.* Springer–Verlag, Berlin.

Eccles, J. C. and McIntyre, A. K. (1953). The effects of disuse and of activity on mammalian spinal reflexes. *J. Physiol.* 121:492–516.

Eccles, J. C., Katz, B., and Kuffler, S. W. (1942). Effect of eserine on neuromuscular transmission. *J. Neurophysiol.* 5:211–230.

Eccles, J. C., Fatt, P., and Landgren, S. (1956). The central pathway for the direct inhibitory action of impulses in the largest afferent nerve fibres to muscle. *J. Neurophysiol.* 126:524–562.

Eccles, R. M. and Lundberg, A. (1959). Synaptic actions in motoneurons by afferents which may evoke the flexion reflex. *Arch. Ital. Biol.* 97:199–221.

Eckhorn, R., Grüsser, O.-J., Kröller, J., Pellnitz, K., and Pöpel, B. (1976). Efficiency of different neuronal codes: information transfer calculations for three different neuronal systems. *Biol. Cybern.* 22:49–60.

Edwards, C., Lorkovic, H., and Weber, A. (1966). The effect of the replacement of calcium by strontium on excitation–contraction coupling in frog skeletal muscle. *J. Physiol.* 186:295–306.

Eigen, M. and Winkler, R. (1970). Alkali-ion carriers: dynamics and selectivity. In *The Neurosciences Second Study Program* (Schmitt, F. O., ed.). Rockefeller University Press, New York, pp. 685–696.

Eisenberg, E., Dobkin, L., and Keilly, W. W. (1972). Heavy meromyosin: evidence for a refractory state unable to bind to actin in the presence of ATP. *Proc. Nat. Acad. Sci. U.S.A.* 69:667–671.

Eisenberg, E. and Hill, T. L. (1978). A crossbridge model of muscular contraction. *Prog. Biophys. Mol. Biol.* 33:55–82.

Eisenman, G. (1962). Cation selective glass electrodes and their mode of operation. *Biophys. J.* 2:259–323.

Ellaway, P. H. and Trott, J. R. (1978). Autogenic reflex action onto gamma motoneurones by stretch of triceps surae in the decerebrated cat. *J. Physiol.* 276:49–66.

Elliott, G. F., Lowy, J., and Worthington, C. R. (1963). An x-ray and light-diffraction study of the filament lattice of striated muscle in the living state and in rigor. *J. Mol. Biol.* 6:295–305.

Elliott, G. F., Rome, E. M., and Spencer, M. (1970). A type of contraction hypothesis applicable to all muscles. *Nature* 226:417–420.

El-Sharkawy, T. Y., Morgan, K. G., and Szurszewski, J. H. (1978). Intracellular electrical activity of canine and human gastric smooth muscle. *J. Physiol.* 279:291–301.

Elul, R. (1972). The genesis of the EEG. *Int. Rev. Neurobiol.* 15:227–272.

Emonet-Denand, F., Laporte, Y., Matthews, P. B. C., and Petit, J. (1977). On the subdivision of static and dynamic fusimotor actions on the primary ending of the cat muscle spindle. *J. Physiol.* **268**:827–861.

Endo, M. (1977). Calcium release from the sarcoplasmic reticulum. *Physiol. Rev.* **57**:71–108.

Erlanger, J. and Gasser, H. S. (1937). *Electrical Signs of Nervous Activity.* University of Pennsylvania Press, Philadelphia.

Evarts, E. V. (1966). Pyramidal tract activity associated with a conditioned hand movement in the monkey. *J. Neurophysiol.* **29**:1011–1027.

Evarts, E. V. and Tanji, J. (1974). Gating of motor cortex reflexes by prior instruction. *Brain Res.* **71**:479–494.

Fambrough, D. M. (1970). Acetylcholine sensitivity of muscle fiber membranes: mechanism of regulation by motoneurons. *Science* **168**:372–373.

Fambrough, D. M., Drachman, D. B. and Satyamurti, S. (1973). Neuromuscular junction in myasthenia gravis: decreased acetylcholine receptors. *Science* **182**:293–295.

Fatt, P. and Ginsborg, B. L. (1958). The ionic requirements for the production of action potentials in crustacean muscle fibres. *J. Physiol.* **142**:516–543.

Fatt, P. and Katz, B. (1951). An analysis of the end-plate potential recorded with an intracellular electode. *J. Physiol.* **115**:320–370.

Fatt, P. and Katz, B. (1952). Spontaneous subthreshold activity at motor nerve endings. *J. Physiol.* **117**:109–128.

Fechner, G. T. (1966). *Elements of Psychophysics* (Adler, H. E., Tr.; Howes, D. H. and Boring, E. G., eds.). Holt, Rinehart and Winston, New York.

Feller, W. (1957). *An Introduction to Probability Theory and Its Applications*, 2nd ed. Wiley, New York.

Fenn, W. O. and Marsh, B. S. (1935). Muscular force at different speeds of shortening. *J. Physiol.* **85**:277–297.

Fernand, V. S. V. and Hess, A. (1969). The occurence, structure, and innervation of slow and twitch muscle fibres in the tensor tympani and stapedius of the cat. *J. Physiol.* **200**:547–554.

Ferreira, H. G., Harrison, F. A., Keynes, R. D., and Zurich, L. (1972). Ion transport across an isolated preparation of sheep rumen epithelium. *J. Physiol.* **222**:77–93.

Fiehn, W. and Peter, J. B. (1971). Properties of the fragmented sarcoplasmic reticulum from fast-twitch and slow-twitch muscles. *J. Clin. Invest.* **50**:570–573.

Finkelstein, A. (1972). Thin lipid membranes. A model for cell membranes. *Arch. Intern. Med.* **129**:229–240.

Fisz, M. (1963). *Probability Theory and Mathematical Statistics.* Wiley, New York.

Flitney, F. W. and Hirst, D. G. (1978). Crossbridge detachment and sarcomere 'give' during stretch of active frog's muscle. *J. Physiol.* **276**:449–465.

Fohlmeister, J. F., Poppele, R. E., and Purple, R. L. (1977). Repetitive firing: quantitative analysis of encoder behavior of slowly adapting stretch receptor of crayfish and eccentric cell of *Limulus. J. Gen. Physiol.* **69**:849–877.

Forsling, M. L. and Widdas, W. F. (1968). The effect of temperature on the competitive inhibition of glucose transfer in human erythrocytes by phenolphthalein, phloretin, and stilboestrol. *J. Physiol.* **194**:545–554.

Forssberg, H., Grillner, S., and Rossignol, S. (1975). Phase dependent reflex reversal during walking in chronic spinal cats. *Brain Res.* **85**:103–107.

Fox, C. F. (1972). The structure of cell membranes. *Sci. Amer.* **227**:31–38.

Fozzard, H. A. (1977). Heart: excitation–contraction coupling. *Ann. Rev. Physiol.* **39**:201–220.

Frankenhaeuser, B. (1962). Instantaneous potassium currents in myelinated nerve fibres of *Xenopus laevis. J. Physiol.* **160**:46–53.

Frankenhaeuser, B. (1963). A quantitative description of potassium currents in myelinated nerve fibres of *Xenopus laevis. J. Physiol.* **169**:424–430.

Frankenhaeuser, B. and Hodgkin, A. L. (1957). The action of calcium on the electrical properties of squid axons. *J. Physiol.* **137**:218–244.

Frankenhaeuser, B. and Huxley, A. F. (1964). The action potential in the myelinated nerve fibre of *Xenopus laevis* as computed on the basis of voltage clamp data. *J. Physiol.* **171**:302–315.

French, A. S. and Holden, A. V. (1971a). Frequency domain analysis of neurophysiological data. *Comp. Prog. Bio-med.* **1**:219–234.

French, A. S. and Holden, A. V. (1971b). Alias free sampling of neuronal spike trains. *Kybernetik* **8**:165–171.

French, A. S. and Stein, R. B. (1970). A flexible neural analog using integrated circuits. *I.E.E.E. Trans. Biomed. Engng.* **17**:248–253.

Frye, C. D. and Edidin, M. (1970). The rapid intermixing of cell surface antigens after formation of mouse–human heterokaryons. *J. Cell Sci.* **7**:319–335.

Gainer, H., ed. (1977). *Peptides in Neurobiology*. Plenum Press, New York.

Ganong, W. F. (1975). *Review of Medical Physiology*, 7th ed. Lange Medical Publications, Los Altos, Ca.

Garay, R. P. and Garrahan, P. J. (1973). The interaction of sodium and potassium with the sodium pump in red cells. *J. Physiol.* **231**:297–325.

Garrahan, P. J. and Glynn, I. M. (1967a). The stocheiometry of the sodium pump. *J. Physiol.* **192**:217–235.

Garrahan, P. J. and Glynn, I. M. (1967b). The incorporation of inorganic phosphate into adenosine triphosphate by reversal of the sodium pump. *J. Physiol.* **192**:237–256.

Gasser, H. S. (1950). Ummedullated fibers originating in dorsal root ganglia. *J. Gen. Physiol.* **33**:651–690.

Gerstein, G. and Perkel, D. H. (1972). Mutual temporal relationships among neuronal spike trains. Statistical techniques for display and analysis. *Biophys. J.* **12**:453–473.

Gibbs, C. L. (1978). Cardiac energetics. *Physiol. Rev.* **58**:174–254.

Glaser, M. and Singer, S. J. (1971). Circular dichroism and the conformations of membrane proteins. Studies with red blood cell membranes. *Biochemistry* **10**:1780–1787.

Goldman, D. E. (1943). Potential, impedance, and rectification in membranes. *J. Gen. Physiol.* **27**:37–60.

Goldstein, D. A. and Solomon, A. K. (1960). Determination of equivalent pore radius for human red cells by osmotic pressure measurement. *J. Gen. Physiol.* **44**:11–17.

González-Serratos, H. (1971). Inward spread of activation in vertebrate muscle fibres. *J. Physiol.* **212**:777–799.

Goodenough, D. A. (1976). The structure and permeability of isolated hepatocyte gap junctions. In *The Synapse*. Cold Spring Harbor, New York, *Symp. Quant. Biol.* **40**:37–44.

Goodwin, G. M., Hulliger, M., and Matthews, P. B. C. (1975). The effects of fusimotor stimulation during small amplitude streching on the frequency response of the primary ending of the mammalian muscle spindle. *J. Physiol.* **253**:175–206.

Gordon, A. M., Huxley, A. F. and Julian, F. J. (1966). The variation in isometric tension with sarcomere length in vertebrate muscle fibres. *J. Physiol.* **184**:170–192.

Gordon, T. and Stein, R. B. (1980). Rematching of nerve and muscle properties in cat motor units after reinnervation. In *Plasticity of Muscle* (Pette, D., ed.). De Gruyter, Berlin. pp. 283-296.

Gordon, T., Hoffer, J. A., Jhamandas, J., and Stein, R. B. (1980). Long-term effects of axotomy on neural activity during locomotion. *J. Physiol.* **303**: 243-263.

Gorter, E. and Grendel, F. (1925). Bimolecular layers of lipoids on chromocytes of blood. *J. Exp. Med.* **41**:439–443.

Granit, R. (1970). *The Basis of Motor Control*. Academic Press, London.

Gray, E. G. (1957). The spindle and extrafusal innervation of a frog muscle. *Proc. Roy. Soc. London B* **146**:416–430.

Gray, E. G. (1959). Axo-somatic and axo-dendritic synapses of the cerebral cortex: an electron-microscopic study. *J. Anat.* **93**:420–433.

Gray, J. A. B. and Malcolm, J. L. (1950). The initiation of nerve impulses by mesenteric Pacinian corpuscles. *Proc. Roy. Soc. London B* **137**:96–114.

Greaser, M. L. and Gergely, J. (1971). Reconstitution of troponin activity from three protein components. *J. Biol. Chem.* **246**:4226–4233.

Grillner, S. (1975). Locomotion in vertebrates: central mechanisms and reflex interaction. *Physiol. Rev.* **55**:247–304.

Grimby, L. and Hannerz, J. (1968). Recruitment order of motor units in voluntary contraction: changes induced by proprioceptive afferent activity. *J. Neurol. Neurosurg. Psychiat.* **31**:565–573.

Guidotti, G. (1972). Membrane proteins. *Ann. Rev. Biochem.* **41**:731–752.

Guyton, A. C. (1971). *Basic Human Physiology: Normal Function and Mechanisms of Disease.* W. B. Saunders, Philadelphia.

Hagiwara, S. and Naka, K. (1964). The initiation of spike potential in barnacle muscle fibres under low intracellular $Ca^{++}$. *J. Gen. Physiol.* **48**:141–162.

Hagiwara, S. and Nakajima, S. (1966). Differences in Na and Ca spikes as examined by application of tetrodotoxin, procaine, and manganese ions. *J. Gen Physiol.* **49**:793–806.

Hagiwara, S. and Oomura, Y. (1958). The critical depolarization for the spike in the squid giant axon. *Jap. J. Physiol.* **8**:234–245.

Hartshorne, D. J. and Pyun, H. P. (1971). Calcium binding by the troponin complex and the purification and properties of troponin A. *Biochim. Biophys. Acta* **229**:698–711.

Hasselbach, W. (1964). Structural and enzymatic properties of the calcium transporting membranes of the sarcoplasmic reticulum. *Ann. N.Y. Acad. Sci.* **137**:1041–1048.

Hauswirth, O., Noble, D. and Tsien, R. W. (1968). Adrenaline: mechanism of action on the placemaker potential in cardiac Purkinje fibers. *Science* **162**:916–917.

Hellam, D. C. and Podolsky, R. J. (1969). Force measurements in skinned muscle fibres. *J. Physiol.* **200**:807–819.

Hendler, R. W. (1971). Biological membrane ultrastructure, *Physiol. Rev.* **51**:66–97.

Henneman, E. (1974a). Peripheral mechanisms involved in the control of muscle. In *Medical Physiology*, 13th ed. (Mountcastle, V. B., ed.). C. V. Mosby Co., St. Louis, pp. 617–635.

Henneman, E. (1974b). Organization of the spinal cord. In *Medical Physiology*, 13th ed. (Mountcastle, V. B., ed.). C. V. Mosby Co., St. Louis, pp. 636–650.

Henneman, E. and Olson, C. B. (1965). Relations between structure and function in the design of skeletal muscles. *J. Neurophysiol.* **28**:581–598.

Hess, A. (1961). The structure of slow and fast extrafusal muscle fibres in the extraocular muscles and their nerve endings in guinea pigs. *J. Cell. Comp. Physiol.* **58**:63–80.

Heuser, J. E. and Reese, T. S. (1973). Evidence for recycling of synaptic vesicle membrane during transmitter release at the frog neuromuscular junction. *J. Cell Biol.* **57**:315–344.

Heuser, J. E., Reese, T. S. and Landis, D. M. D. (1974). Functional changes in frog neuromuscular junctions studied with freeze-fracture *J. Neurocytol.* **3**:109–131.

Heuser, J. E., Reese, T. S., Dennis, M. J., Jan, Y., Jan, L. and Evans, L. (1979). Synaptic vesicle exocytosis captured by quick freezing and correlated with quantal transmitter release. *J. Cell. Biol.* **81**:275–300.

Hill, A. V. (1936). Excitation and accomodation in nerve. *Proc. Roy. Soc. London B* **119**:305.

Hill, A. V. (1938). The heat of shortening and the dynamic constants of muscle. *Proc. Roy. Soc. London B* **126**:136–195.

Hill, A. V. (1949). The abrupt transition from rest to activity in muscle. *Proc. Roy. Soc. London B* **136**:399–420.

Hill, A. V. (1964). The effect of load on the heat of shortening of muscle. *Proc. Roy. Soc. London B* **159**:297–318.

Hill, A. V. (1965). *Trails and Trials in Physiology*. Arnold, London.

Hill, D. K. (1968). Tension due to interaction between sliding filaments in resting striated muscle. The effect of stimulation. *J. Physiol.* **199**:637–684.

Hill, T. L. (1974). Theoretical formalism for the sliding filament model of contraction of striated muscle. Part I. *Prog. Biophys. Mol. Biol.* **28**:268–339.

Hill, T. L. (1975). Theoretical formalism for the sliding filament model of contraction of striated muscle. Part II. *Prog. Biophys. Mol. Biol.* **29**:105–159.

Hille, B. (1970). Ionic channels in nerve membranes. *Prog. Biophys. Mol. Biol.* **21**:3–32.

Hille, B. (1971). The permeability of the sodium channel to organic cations in myelinated nerve. *J. Gen. Physiol.* **58**:599–619.

Hille, B. (1975). The receptor for tetrodotoxin and saxitoxin: a structural hypothesis. *Biophys. J.* **15**:615–619.

Hille, B. (1976). Gating in sodium channels of nerve. *Ann. Rev. Physiol.* **38**:139–152.

Hille, B. (1977). Local anaesthetics: hydrophilic and hydrophobic pathways for the drug-receptor reaction. *J. Gen. Physiol.* **69**:497–515.

Hodgkin, A. L. (1958). Ionic movements and electrical activity in giant nerve fibres. *Proc. Roy. Soc. London B* **148**:1–37.

Hodgkin, A. L. and Horowicz, P. (1960). Potassium contractures in single muscle fibres. *J. Physiol.* **153**:386–403.

Hodgkin, A. L. and Huxley, A. F. (1952). A quantitiative description of membrane current and its application to conduction and excitation in nerve. *J. Physiol.* **117**:500–544.

Hodgkin, A. L., Huxley, A. F. and Katz, B. (1952). Measurement of current–voltage relations in the membrane of the giant axon of *Loligo. J. Physiol.* **116**:424–448.

Hodgkin, A. L. and Keynes, R. D. (1955a). Active transport of cations in giant axons from *Sepia and Loligo. J. Physiol.* **128**:28–60.

Hodgkin, A. L. and Keynes, R. D. (1955b). The potassium permeability of a giant nerve fibre. *J. Physiol.* **128**:61–88.

Hodgkin, A. L. and Rushton, W. A. H. (1946). The electrical constants of a crustacean nerve fibre. *Proc. Roy. Soc. London B* **133**:444–479.

Hoffman, J. F. (1973). Molecular aspects of the $Na^+$, $K^+$-pump in red blood cells. In *Organization of Energy-Transducing Membranes* (Nakao, M. and Packer, C., eds.). University Park Press, Baltimore, pp. 9–21.

Holman, J. P. (1969). *Thermodynamics*. McGraw–Hill, New York.

Hornykiewicz, O. (1966). Dopamine (3-hydroxytyramine) and brain function. *Pharmacol. Rev.* **18**:925–964.

Houk, J. C. (1972). The phylogeny of muscular control configurations. In *Biocybernetics IV* (Drischel, H. and Dettmar, P., eds.). VEB Gustav Fischer Verlag, Jena, pp. 125–144.

Houk, J. C. (1979). Regulation of stiffness by skeletomotor reflexes. *Ann. Rev. Physiol.* **41**:99–114.

Houk, J. and Henneman, E. (1967). Responses of Golgi tendon organs to active contractions of the soleus muscle of the cat. *J. Neurophysiol.* **30**:466–481.

Houk, J., Singer, J. J. and Goldman, M. C. (1970). An evaluation of length and force feedback in decerebrate cats. *J. Neurophysiol.* **33**:784–811.

Hubbard, J. I. (1973). Microphysiology of vertebrate neuromuscular transmission. *Physiol. Rev.* **53**:674–723.

Hulliger, M., Matthews, P. B. C. and Noth, J. (1977). Effects of combining static and dynamic fusimotor stimulation on the response of the muscle spindle primary ending to sinusoidal stretching. *J. Physiol.* **267**:839–856.

Hultborn, H. (1972). Convergence on interneurones in the reciprocal Ia inhibitory pathway to motoneurones. *Acta Physiol. Scand.,* Sup. 375.

Hultborn, H. (1977). Transmission in the pathway of reciprocal Ia inhibition to motoneurones and its control during the tonic stretch reflex. *Prog. Brain Res.* **44**:235–254.

Hultborn, J., Jankowska, E. and Lindstrom, S. (1971). Recurrent inhibition from motor axon collaterals of transmission in the 1a inhibitory pathway to motoneurones. *J. Physiol.* **215**:591–612.

Hunt, C. C. (1951). The reflex activity of mammalian small-nerve fibres. *J. Physiol.* **115**:456–469.

Hunt, C. C. and Ottoson, D. (1975). Impulse activity and receptor potential of primary and secondary muscle spindle endings of isolated mammalian muscle spindles. *J. Physiol.* **252**:259–281.

Hursh, J. B. (1939). Conduction velocity and diameter of nerve fibers. *Amer. J. Physiol.* **127**:131–139.

Husmark, I. and Ottoson, D. (1971). Is the adaptation of the muscle spindle of ionic origin? *Acta Physiol. Scand.* **81**:138–140.

Hutter, O. F. (1961). Ion movements during vagus inhibition of the heart. In *Nervous Inhibition* (Florey, E., ed.). Pergamon Press, Oxford, pp. 114–124.

Huxley, A. F. (1957). Muscle structure and theories of contraction. *Prog. Biophys. Biophys. Chem.* **7**:255–318.

Huxley, A. F. (1959). Ion movements during nerve activity. *Ann. N.Y. Acad. Sci.* **81**:221–246.

Huxley, A. F. (1974). Review lecture: Muscular contraction. *J. Physiol.* **243**:1–43.

Huxley, A. F. and Neidergerke, R. (1954). Interference microscopy of living muscle fibres. *Nature* **173**:971–973.

Huxley, A. F. and Simmons, R. M. (1971). Proposed mechanism of force generation in striated muscle. *Nature* **233**:533–538.

Huxley, A. F. and Simmons, R. M. (1972). Mechanical transients and the origin of muscular force. Cold Spring Harbor, New York, *Symp. Quant. Biol.* **37**:669–680.

Huxley, H. E. (1971). The structural basis of muscular contraction. *Proc. Roy. Soc. London B* **178**:131–199.

Huxley, H. E. and Brown, W. (1967). The low-angle x-ray diagram of vertebrate striated muscle and its behavior during contraction and rigor. *J. Mol. Biol.* **30**:383–434.

Huxley, H. E. and Hanson, J. (1954). Changes in the cross-striations of muscle during contraction and stretch and their structural implications. *Nature* **173**:973–976.

Inbar, G., Madrid, J., and Rudomin, P. (1979). The influence of the gamma system on the cross-correlated activity of Ia muscle spindles and its relation to information transmission. *Neurosci. Lett.* **13**:73–78.

Infante, A. A. and Davies, R. E. (1962). Adenosine triphosphate breakdown during a single twitch of frog sartorius muscle. *Biochem. Biophys. Res. Comm.* **9**:410–415.

Jack, J. J. B., Noble, D. and Tsien, R.W. (1975). *Electrical Current Flow in Excitable Cells.* Clarendon Press, Oxford.

Jack, J. J. B. and Redman, S. J. (1971). An electrical description of the motoneurone, and its application to the analysis of synaptic potentials. *J. Physiol.* **215**:321–352.

Jankowska, E., Jukes, M. G. M., Lund, S., and Lundberg, A. (1967). The effect of DOPA on the spinal cord. 5. Reciprocal organization of pathways transmitting excitatory action to alpha motoneurones of flexors and extensors. *Acta Physiol. Scand.* **70**:369–388.

Jansen, J. K. S. and Rudjord, T. (1964). On the silent period and Golgi tendon organs of the soleus muscle of the cat. *Acta Physiol. Scand.* **62**:364–379.

Jansen, J. K. S., Nicolaysen, K. and Rudjord, T. (1966). Discharge patterns of neurons of the dorsal spinocerebellar tract activated by static extension of primary endings of muscle spindles. *J. Neurophysiol.* **29**:1061–1086.

Jansen, J. K. S., Njå, A., Ormstad, K., and Walløe., L. (1971). On the innervation of the slowly adapting stretch receptor of the crayfish abdomen. An electrophysiological approach. *Acta Physiol. Scand.* **81**:273–285.

Jewell, B. R. and Wilkie, D. R. (1958). An analysis of the mechanical components in frog's striated muscle. *J. Physiol.* **143**:515–540.

Joyce, G. C. and Rack, P. M. H. (1974). The effects of load and force on tremor at the normal human elbow joint. *J. Physiol.* **240**:375–396.

Joyce. G. C., Rack, P. M. H. and Westbury, D. R. (1969). The mechanical properties of cat soleus muscle during controlled lengthening and shortening movements. *J. Physiol.* **204**:461–474.

Julian, F. J. (1969). Activation in a skeletal muscle contraction model with a modification for insect fibrillar muscle. *Biophys. J.* **9**:547–570.

Julian, F. J. and Goldman, D. E. (1962). Effects of mechanical stimulation on some electrical properties of axons. *J. Gen. Physiol.* **46**:297–313.

Julian, F. J. Sollins, K. R. and Sollins, M. R. (1974). A model for transient and steady-state mechanical behavior of contracting muscle. *Biophys. J.* **14**:546–567.

Kandel, E. R. (1976). *Cellular Basis of Behavior: An Introduction to Behavioral Neurobiology.* W. H. Freeman, San Francisco.

Kandel, E. R., Frazier, W. T. and Coggeshall, R. E. (1967). Opposite synaptic actions mediated by different branches of an identifiable interneuron in *Aplysia. Science* **155**: 346–349.

Katchalsky, A. and Curran, P. F. (1965). *Nonequilibrium Thermodynamics in Biophysics.* Harvard University Press, Cambridge, Massachusetts.

Katz, A. M. (1970). Contractile proteins of the heart. *Physiol. Rev.* **50**:63–158.

Katz, B. (1939). The relation between force and speed in muscular contraction. *J. Physiol.* **96**:45–64.

Katz, B. (1949). Les constantes électriques de la membrane de muscle. *Arch. Sci. Physiol.* **3**:285–299.

Katz, B. and Miledi, R. (1967). A study of synaptic transmission in the absence of nerve impulses. *J. Physiol.* **192**:407–436.

Katz, B. and Miledi, R. (1969). Tetrodotoxin-resistant electrical activity in presynaptic terminals. *J. Physiol.* **203**:459–487.

Katz, B. and Miledi, R. (1970). Further study of the role of calcium in synaptic transmission. *J. Physiol.* **207**:789–801.

Katz, B. and Miledi, R. (1971). The effect of prolonged depolarization on synaptic transfer in the stellate ganglion of the squid. *J. Physiol.* **216**:503–512.

Katz, B. and Miledi, R. (1972). The statistical nature of the acetylcholine potential and its molecular components. *J. Physiol.* **224**:665–699.

Kedem, O. (1961). Criteria of active transport. In *Membrane Transport and Metabolism* (Kleinzeller, A. and Kotyk, A., eds.). Academic Press, New York, pp. 87–93.

Kedem, O. and Katchalsky, A. (1958). Thermodynamic analysis of the permeability of biological membranes to nonelectrolytes. *Biochim. Biophys. Acta* **27**:229–246.

Kedem, O. and Katchalsky, A. (1961). A physical interpretation of the phenomenological coefficients of membrane permeability. *J. Gen. Physiol.* **45**:143–179.

Kernell, D. (1966). Input resistance, electrical excitability, and size of ventral horn cells in cat spinal cord. *Science* **152**:1637–1640.

Kernell, D (1968). The repetitive impulse discharge of a simple neurone model compared to that of spinal motoneurones. *Brain Res.* **11**:685–687.

Keynes, R. D. and Rojas, E. (1974). Kinetics and steady-state properties of the charged system controlling sodium conductance in the squid giant axon. *J. Physiol.* **239**:393–434.

Kiang, N. Y. S., Watanabe, T., Thomas, E. C., and Clark, L. F. (1965). *Discharge Patterns of Single Fibers in the Cat's Auditory Nerve.* M.I.T. Res. Mon. No. 35, M.I.T. Press, Cambridge.

Kiloh, L. G., McComas, A. J. and Osselton, J. W. (1972). *Clinical Electroencephalography,* 3rd ed. Butterworths, London.

Kirkwood, P. A. and Sears, T. A. (1974). Monosynaptic excitation of motoneurones from secondary endings of muscle spindles. *Nature* **252**:243–244.

Knibestöl, M. and Vallbo, Å. B. (1980). Intensity of sensation related to activity of slowly adapting mechanoreceptive units in the human hand. *J. Physiol.* **300**:251–267.

Korn, E. D. (1966). Structure of biological membranes. *Science* **153**:1491–1498.

Krnjevic, K. (1974). Chemical nature of synaptic transmission in vertebrates. *Physiol. Rev.* **54**:418–540.

Krnjevic, K., Puil, E. and Werman, R. (1978a). EGTA and motoneuronal after-potentials. *J. Physiol.* **275**:199–223.

Krnjevic, K. Puil, E. and Werman, R. (1978b). Significance of 2, 4-dinitrophenol action on spinal motoneurones. *J. Physiol.* **275**:225–239.

Kuffler, S. W. (1954). Mechanisms for activation and motor control of stretch receptors in lobster and crayfish. *J. Neurophysiol.* **17**:558–574.

Kuffler, S.W. and Nicholls, J. G. (1976). *From Neuron to Brain.* Sinauer Associates, Sunderland, Massachusetts.

Kuffler, S. W. and Yoshikami, D. (1975a). The distribution of acetylcholine sensitivity at the post-synaptic membrane of vertebrate skeletal twitch muscles: iontophoretic mapping in the micron range. *J. Physiol.* **244**:703–730.

Kuffler, S. W. and Yoshikami, D. (1975b). The number of transmitter molecules in a quantum: an estimate from iontophoretic application of acetylcholine at the neuromuscular synapse. *J. Physiol.* **251**:465–482.

Kuffler, S. W. and Vaughan-Williams, E. M. (1953). Small-nerve junctional potentials. The distribution of small motor nerves to frog skeletal muscle, and the membrane characteristics of the fibres they innervate. *J. Physiol.* **121**:289–317.

Kuno, M. and Miyahara, J. T. (1969). Analysis of synaptic efficiency in spinal motoneurones from 'quantum' aspects. *J. Physiol.* **201**:479–493.

Kyte, J. (1975). Structural studies of sodium and potassium ion-activated adenosine triphosphatase. The relationship between molecular structure and the mechanism of active transport. *J. Biol. Chem.* **250**:7443–7449.

Ladbrooke, B. D., Williams, R. M., and Chapman, D. (1968). Studies on lecithin-cholesterol-water interactions by differential scanning calorimetry and x-ray diffraction. *Biochim. Biophys. Acta* **150**:333–340.

Lapicque, L. (1907). II. Recherches quantitatives sur l'excitation électrique des nerfs traitée comme une polarisation. *J. Physiol. Path. Gen.* **9**:622–635.

Lass, Y., Halevi, Y., Landau, E. M. and Gitter, S. (1973). A new model for transmitter mobilization in the frog neuromuscular junction. *Pflügers Arch.* **343**:157–163.

Läuger, P. (1972). Carrier-mediated ion transport. *Science* **178**:24–30.

Lea, E. J. A. (1963). Permeation through long narrow pores. *J. Theor. Biol.* **5**:102–107.

LeFevre, P. G. (1959). Molecular structural factors in competitive inhibition of sugar transport. *Science* **130**:104–105.

Lehman, W., Kendrick-Jones, J. and Szent-Gyorgi, A. (1972). Myosin-linked regulatory systems: comparative studies. Cold Spring Harbor, New York, *Symp. Quant. Biol.* **37**:319–330.

Lehninger, A. L. (1975). *Biochemistry*, 2nd ed. Worth Publishers, New York.

Lenman, J. A. R. and Ritchie, A. E. (1970). *Clinical Electromyography.* Pitman Medical, Bath.

Lennerstrand, G. (1968). Position and velocity sensitivity of muscle spindles in the cat. IV. Interaction between two fusimotor fibres converging on the same spindle ending. *Acta Physiol. Scand.* **74**:257–273.

Levi-Montalcini, R. and Angeletti, P. (1969). Nerve growth factor. *Physiol. Rev.* **48**:534–569.

Lewis, D. M., Bagust, J., Webb, S. N., Westerman, R. A. and Finol, H. J. (1977). Axon conduction velocity modified by reinnervation of mammalian muscle. *Nature* **270**:745–746.

Liley, A. W. (1956). The effects of presynaptic polarization on the spontaneous activity of the mammalian neuromuscular junction. *J. Physiol.* **134**:427–443.

Ling, G. N. (1962). *A Physical Theory of the Living State.* Blaisdell, New York.

Ling, G. and Gerard, R. W. (1949). The normal membrane potential of frog sartorius fibers. *J. Cell Comp. Physiol.* **34**:393–396.

Lippold, O. C. J. (1970). Oscillation in the stretch reflex arc and the origin of the rhythmical 8–12 c/s component of physiological tremor. *J. Physiol.* **206**:359–382.

Lippold, O. C. J. (1973). *The Origin of the Alpha Rhythm.* Churchill Livingstone, Edinburgh.

Llinás, R. R. (1977). Calcium and transmitter release in squid synapse. In *Soc. Neurosci. Symp.* (Cowan, W. M. and Ferendelli, J. A., ed.). Society for Neuroscience, Bethesda, Md., **2**:139–160.

Llinás, R. and Nicholson, C. (1971). Electrophysiological properties of dendrites and somata in alligator Purkinje cells. *J. Neurophysiol.* **34**:532–551.

Llinás, R. and Nicholson, C. (1975). Calcium role in depolarization–secretion coupling: an aequorin study in squid giant synapse. *Proc. Nat. Acad. Sci. U.S.A.* **72**:187–190.

Lodish, H. F. and Rothman, J. E. (1979). The assembly of cell membranes. *Sci. Amer.* **310**:48–63.

Loewenstein, W. R. (1974). Membrane transformations in neoplasia. In *Control Processes in Neoplasia* (Mehlman, M. A. and Hanson, R. W., eds.). Academic Press, New York.

Loewenstein, W. R. and Mendelson, M. (1965). Components of receptor adaptation in a Pacinian corpuscle. *J. Physiol.* **177**:378–397.

Lohmann, K. (1934). Weitere Untersuchungen über das Co-Ferment der Milchsäurebildung. *Biochem. Z.* **271**:278–279.

Lømo, T. and Rosenthal, J. (1972). Control of ACh sensitivity by muscle activity in the rat. *J. Physiol.* **221**:493–513.

Lømo, T. and Westgaard, R. H. (1975). Further studies on the control of ACh sensitivity by muscle activity in the rat. *J. Physiol.* **252**:603–626.

Lømo, T. and Westgaard, R. H. (1976). Control of ACh sensitivity in rat muscle fibers. In *The Synapse.* Cold Spring Harbor, New York. *Symp. Quant. Biol.* **40**:263–274.

Lorente de Nó, R. (1947). *A Study of Nerve Physiology.* Rockefeller Institute, New York.

Lowey, S., Slayter, H. S., Weeds, A. G. and Baker, H. (1969). Substructure of the myosin molecule. I. Subfragments of myosin by enzymic degradation. *J. Mol. Biol.* **42**:1–29.

Lucy, J. A. (1964). Globular lipid micelles and cell membranes. *J. Theor. Biol.* **7**:360–373.

Lymn, R. W. and Taylor, E. W. (1971). Mechanism of adenosine triphosphate hydrolysis by actomyosin. *Biochemistry* **10**:4617–4624.

Magazanik, L. G. and Vyskočil, F. (1970). Dependence of acetylcholine desensitization on the membrane potential of frog muscle fibre and on the ionic changes in the medium. *J. Physiol.* **210**:507–518.

Magleby, K. L. and Stevens, C. F. (1972a). The effect of voltage on the time course of end-plate currents. *J. Physiol.* **223**:151–171.

Magleby, K. L. and Stevens, C. F. (1972b). A quantitative description of end-plate currents. *J. Physiol.* **223**:173–197.

Magleby, K. L. and Zengel, J. E. (1975). A dual effect of repetitive stimulation on post-tetanic potentiation of transmitter release at the frog neuromuscular junction. *J. Physiol.* **245**:163–182.

Magleby, K. L. and Zengel, J. E. (1976). Augmentation: a process that acts to increase transmitter release at the frog neuromuscular junction. *J. Physiol.* **257**:471–494.

Magnus, R. (1924). *Korperstellung.* Springer, Berlin.

Malhotra, S. K. and van Harreveld, A. (1968). Molecular organization of the membranes of cells and cellular organelles. In *Biological Basis of Medicine* (Bittar, E. E. and Bittar, N., eds.). Academic Press, London, **1**:3–68.

Mallart, A. and Martin, A. R. (1967). An analysis of facilitation of transmitter release at the neuromuscular junction of the frog. *J. Physiol.* **193**:679–694.

Mallart, A. and Martin, A. R. (1968). The relation between quantum content and facilitation at the neuromuscular junction of the frog. *J. Physiol.* **196**:593–604.

Mannard, A. and Stein, R. B. (1973). Determination of the frequency response of isometric soleus muscle in the cat using random nerve stimulation. *J. Physiol.* 229:275–296.

Mannherz, H. G. and Goody, R. S. (1976). Proteins of contractile systems. *Ann. Rev. Biochem.* 45:427–465.

Marks, W. B. and Loeb, G. E. (1976). Action currents, internodal potentials, and extracellular records of myelinated mammalian nerve fibers derived from node potentials. *Biophys. J.* 16:655–668.

Marmarelis, P. Z. and Marmarelis, V. Z. (1978). *Analysis of Physiological Systems: The White Noise Approach.* Plenum Press, New York.

Matthews, B. H. C. (1933). Nerve endings in mammalian muscle. *J. Physiol.* 78:1–53.

Matthews, P. B. C. (1964). Muscle spindles and their motor control. *Physiol. Rev.* 44:219–288.

Matthews, P. B. C. (1969). Evidence that the secondary as well as the primary endings of the muscle spindles may be responsible for the tonic stretch reflex of the decerebrate cat. *J. Physiol.* 204:365–393.

Matthews, P. B. C. (1972). *Mammalian Muscle Receptors and Their Central Actions.* Arnold, London.

Matthews, P. B. C. and Stein, R. B. (1969a). The sensitivity of muscle spindle afferents to small sinusoidal changes in length. *J. Physiol.* 200:723–743.

Matthews, P. B. C. and Stein, R. B. (1969b). The regularity of primary and secondary muscle spindle afferent discharges and its functional implications. *J. Physiol.* 202:59–82.

Matthews, P. B. C. and Westbury, D. R. (1965). Some effects of fast and slow motor fibres on muscle spindles of the frog. *J. Physiol.* 178:178–192.

McAllister, R. E., Noble, D. and Tsien, R. W. (1975). Reconstruction of the electrical activity of cardiac Purkinje fibres. *J. Physiol.* 251:1–59.

McLaughlin, S. G. A., Szabo, G. and Eisenman, G. (1971). Divalent ions and the surface potential of charged phospholipid membranes. *J. Gen. Physiol.* 58:667–687.

McPhedran, A. M., Wuerker, R. B. and Henneman, E. (1965). Properties of motor units in a homogeneous red muscle (soleus) of the cat. *J. Neurophysiol.* 28:71–84.

Meech, R. W. (1978). Calcium-dependent potassium activation in nervous tissues. *Ann. Rev. Biophys. Bioengng.* 7:1–18.

Melzack, R. and Wall, P. D. (1965). Pain mechanisms: a new theory. *Science* 150:971–979.

Mendell, L. M. and Henneman E. (1971). Terminals of single Ia fibers: location, density, and distribution within a pool of 300 homonymous motoneurons. *J. Neurophysiol.* 34:171–187.

Mendell, L. M., Munson, J. B. and Scott, J. G. (1976). Alterations of synapses on axotomized motoneurones. *J. Physiol.* 255:67–79.

Menzies, J. E., Albert, C. P. and Jordan, L. M. (1978). Testing a model for the spinal locomotor generator. *Soc. Neurosci. Abstr.* 4:1219.

Merton, P. A. (1953). Speculations on the servo-control of movement. In *The Spinal Cord* (Wolstenholme, G. E. W., ed.). Churchill, London, pp. 247–255.

Miledi, R. and Potter, L. T. (1971). Acetylcholine receptors in muscle fibers. *Nature* 233:599–603.

Miller, G. A. (1956). The magical number seven plus or minus two. Some limits on our capacity for processing information. *Psych. Rev.* 6:63–81.

Miller, S. and Scott, P. D. (1977). The spinal locomotor generator. *Exp. Brain Res.* 30:387–403.

Milner-Brown, H. S., Stein, R. B. and Lee, R. G. (1974). The pattern of recruiting human motor units in neuropathies and motor neurone disease. *J. Neurol. Neurosurg. Psychiat.* 37:665–669.

Milner-Brown, H. S., Stein, R. B. and Yemm, R. G. (1973). The orderly recruitment of human motor units during voluntary isometric contractions. *J. Physiol.* 230:359–370.

Milsum, J. H. (1966). *Biological Control Systems Analysis.* McGraw-Hill, New York.

Monnier, A.-M. (1934). *L'Excitation Électrique des Tissues*. Hermann, Paris.

Moore, G. P., Perkel, D. H. and Segundo, J. P. (1966). Statistical analysis and functional interpretation of neuronal spike data. *Ann. Rev. Physiol.* **28**:493–522.

Moore, J. W. and Cole, K. S. (1963). Voltage clamp techniques. In *Physical Techniques in Biological Research* (Nastuk, W. L., ed.). Academic Press, New York, pp. 263–321.

Moore, J. W. (1962). *Physical Chemistry*, 3rd ed. Prentice Hall, Englewood Cliffs, New Jersey.

Morse, P. M. (1962). *Thermal Physics*. Benjamin, New York.

Mott, G. W. and Sherrington, C. S. (1895). Experiments upon the influence of sensory nerves upon movement and nutrition of the limbs. Preliminary communication. *Proc. Roy. Soc. London* **57**:481–488.

Mountcastle, V. B. (1974). *Medical Physiology*, 13th ed. C. V. Mosby, St. Louis.

Murray, J. M. and Weber, A. (1974). The cooperative action of muscle proteins. *Sci. Amer.* **251**:58–71.

Nachmanson, D. (1970). Proteins in excitable membranes. *Science* **168**:1059–1066.

Nagano, K., Fujihara, Y., Hara, Y. and Nakao, M. (1973). ATP as a modulator of $Na^+$, $K^+$-ATPase. In *Organization of Energy-Transducing Membranes* (Nakao, M. and Parker, L., eds.). University Park Press, Baltimore, pp. 47–61.

Nakajima, S. and Onodera, K. (1969a). Membrane properties of the stretch receptor neurones of crayfish with particular reference to mechanisms of sensory adaptation. *J. Physiol.* **200**:161–185.

Nakajima, S. and Onodera, K. (1969b). Adaptation of the generator potential in the crayfish stretch receptors under constant length and constant tension. *J. Physiol.* **200**:187–204.

Nakajima, S. and Takahashi, K. (1966). Post-tetanic hyperpolarization and electrogenic Na pump in stretch receptor neurone of crayfish. *J. Physiol.* **187**:105–127.

Nakao, M., Nakao, T., Ohta, H., Nagai, F., Kawai, K., Fujihara, Y., and Nagano, K. (1973). The $Na^+$, $K^+$-ATPase molecule. In *Organization of Energy-Transducing Membranes* (Nakao, M. and Parker, L., eds.). University Park Press, Baltimore, pp. 23–34.

Narahashi, T. (1963). Dependence of resting and action potentials on internal potassium in perfused squid giant axons. *J. Physiol.* **169**:91–115.

Neher, E. and Sakmann, B. (1976). Single channel currents recorded from membrane of denervated frog muscle fibres. *Nature* **260**:799–802.

Neher, E. and Stevens, C. F. (1977). Conductance fluctuations and ionic pores in membranes. *Ann. Rev. Biophys. Bioeng.* **6**:345–381.

Nichols, T. R. and Houk, J. C. (1976). The improvement in linearity and the regulation of stiffness that results from the actions of the stretch reflex. *J. Neurophysiol.* **39**:119–142.

Nichols, T. R., Stein, R. B., and Bawa, P. (1978). Spinal reflexes as a basis for tremor in the premammillary cat. *Can. J. Physiol. Pharmacol.* **56**:375–383.

Noble, D. (1962). A modification of the Hodgkin–Huxley equations applicable to Purkinje fibre action and pacemaker potentials. *J. Physiol.* **160**:317–352.

Noble, D. (1975). *The Initiation of the Heartbeat*. Clarendon Press, Oxford.

Noble, D. and Stein, R. B. (1966). The threshold conditons for initiation of action potentials by excitable cells. *J. Physiol.* **187**:129–162.

Noble, D. and Tsien, R. W. (1969). Outward membrane currents activated in the plateau range of potentials in cardiac Purkinje fibres. *J. Physiol.* **200**:205–231.

Obata, K., Takeda, K., and Shinozaki, H. (1970). Further study on pharmacological properties of the cerebellar-induced inhibition of Deiters neurones. *Exp. Brain Res.* **11**:327–342.

Ochs, S. (1974). Axoplasmic transport-energy metabolism and mechanism. In *The Peripheral Nervous System* (Hubbard, J. I., ed.). Plenum Press, New York, pp. 47–67.

Oğuztöreli, M. N. and Stein, R. B. (1975). An analysis of oscillations in neuromuscular systems. *J. Math. Biol.* **2**:87–105.

Oğuztöreli, M. N. and Stein, R. B. (1976). The effects of multiple reflex pathways on the oscillations in neuromuscular systems. *J. Math. Biol.* **3**:87–101.

Oğuztöreli, M. N. and Stein, R. B. (1977). A kinetic study of muscular contractions. *J. Math. Biol.* **5**:1–31.

Osborne, N. N. (1979). Is Dale's principle valid? *Trends in Neurosci.* **2**:73–75.

Ottoson, D. and Shepherd, G. M. (1971). Transducer properties and integrative mechanisms in the frog's muscle spindle. In *Handbook of Sensory Physiology* (Loewenstein, W. R., ed.). Springer-Verlag, Heidelberg, **1**:442–449.

Overton, E. (1899). Über die allgemeinen osmotischen Eigenschaften der Zelle, ihre vermutlichen Ursachen und ihre Bedeutung für die Physiologie. *Vjschr. Naturforsch. Ges. Zurich* **44**:88–135. Repr. in *Papers on Biological Membrane Structure* (Branton, D. and Park, R. R., eds.). Little Brown, Boston.

Paintal, A. S. (1966). The influence of diameter of medullated nerve fibres of cats on the rising and falling phases of the spike and its recovery. *J. Physiol.* **184**:791–811.

Paintal, A. S. (1967). A comparison of the nerve impulses of mammalian nonmedullated nerve fibres with those of the smallest diameter medullated fibres. *J. Physiol.* **193**:523–533.

Paintal, A. S. (1978). Conduction properties of normal peripheral mammalian axons. In *Physiology and Pathobiology of Axons* (Waxman, S. G., ed.). Raven Press, New York, pp. 131–144.

Payton, B. W., Bennett, M. V. L. and Pappas, G. D. (1969). Permeability and structure of junctional membranes at an electrotonic synapse. *Science* **166**:1641–1643.

Peachey, L. D. (1965). The sarcoplasmic reticulum and transverse tubules of the frog's sartorius. *J. Cell Biol.* **25**:209–231.

Pearson, K. G. and Fourtner, C. R. (1975). Nonspiking interneurons in walking system of the cockroach. *J. Neurophysiol.* **38**:33–52.

Pearson, K. G., Stein, R. B., and Malhotra, S. K. (1970). Properties of action potentials from insect motor nerve fibres. *J. Exp. Biol.* **53**:299–316.

Perkel, D. H. and Bullock, T. H. (1968). Neural coding. *Neurosci. Res. Prog. Bull.* **6**:221–348.

Peters, A., Palay, S. L., and Webster, H. de F. (1970). *The Fine Structure of the Nervous System*. Harper and Row, New York.

Plonsey, R. (1974). The active fiber in a volume conductor. *I.E.E.E. Trans. Biomed. Engng.* **21**:371–381.

Podolsky, R. J., Gulati, J., and Nolan, A. C. (1974). Contraction transients of skinned muscle fibers. *Proc. Nat. Acad. Sci. U.S.A.* **7**:1516–1519.

Podolsky, R. J. and Nolan, A. C. (1972). Muscle contraction transients, crossbridge kinetics and the Fenn effect. Cold Spring Harbor, New York, *Symp. Quant. Biol.* **37**:661–668.

Pollak, V. (1971). The waveshape of action potentials recorded with different types of electromyographic needles. *Med. Biol. Engng.* **9**:657–664.

Poppele, R. (1973). Systems approach to the study of muscle spindles. In *Control of Posture and Locomotion* (Stein, B., Pearson, K. G. Smith, R. S., and Redford, J. B., ed.). Plenum Press, New York, pp. 127–146.

Poppele, R. E. and Bowman, R. J. (1970). Quantitative description of linear behavior of mammalian muscle spindles. *J. Neurophysiol.* **33**:59–72.

Poppele, R. E. and Terzuolo, C. A. (1968). Myotatic reflex: its input–output relation. *Science* **159**:743–745.

Porter, K. R. and Bonneville, M. A. (1968). *Fine Structure of Cells and Tissues*, 3rd ed. Lea and Febiger, Philadelphia.

Prigogine, I. (1967). *Thermodynamics of Irreversible Processes*, 3rd ed. Wiley, New York.

Prosser, C. L. (1974). Smooth muscle. *Ann. Rev. Physiol.* **36**:503–535.

Pumphrey, R. J. and Young, J. Z. (1938). The rates of conduction of nerve fibres of various diameters in cephalopods. *J. Exp. Biol.* **15**:443–466.

Purves, D. (1975). Functional and structural changes in mammalian sympathetic neurons following interruption of their axons. *J. Physiol.* **252**:429–463.

Purves, D. and Njå, A. (1976). Effect of nerve growth factor on synaptic depression after axotomy. *Nature* **260**:535-536.

Purves, D. and Sakmann, B. (1974). The effect of contractile activity on fibrillation and extrajunctional acetylcholine sensitivity in rat muscle maintained in organ culture. *J. Physiol.* **237**:157-182.

Quastel, D. M. J., Hackett, J. T., and Cooke, J. D. (1971). Calcium: is it required for transmission secretion? *Science* **172**:1034-1036.

Rall, W. (1962). Electrophysiology of a dendritic neuron model. *Biophys. J.* **2 Sup.**:145-167.

Rall, W. (1969). Time constants and electrotonic length of membrane cylinders and neurons. *Biophys. J.* **9**:1483-1508.

Rang, H. P. and Ritchie, J. M. (1968). On the electrogenic sodium pump in mammalian non-myelinated nerve fibres and its activation by various external cations. *J. Physiol.* **196**:183-221.

Rashevsky, N. (1938). *Mathematical Biophysics*, 1st ed. University of Chicago Press, Chicago.

Redfearn, P. A. (1970). Neuromuscular transmission in new-born rats. *J. Physiol.* **209**:701-709.

Renshaw, B. (1946). Central effects of centripetal impulses in axons of spinal ventral root. *J. Neurophysiol.* **9**:191-204.

Reuter, H. and Beeler, G. W. (1969a). Sodium current in ventricular myocardiac fibers. *Science* **163**:397-399.

Reuter, H. and Beeler, G. W. (1969b). Calcium current and activation of contraction in ventricular myocardial fibers. *Science* **163**:349-401.

Ritchie, J. M. (1979). A pharmacological approach to the structure of sodium channels in myelinated axons. *Ann. Rev. Neurosci.* **2**:341-362.

Roberts, T. D. M. (1978). *Neurophysiology of Postural Mechanisms*, 2nd ed. Butterworths, London.

Robertson, J. D. (1956). The ultrastructure of a reptilian myoneural junction. *J. Biophys. Biochem. Cytol.* **2**:381-394.

Robertson, J. D. (1960). The molecular structure and contact relationships of cell membranes. *Prog. Biophys. Biophys. Chem.* **10**:343-418.

Robinson, G. A., Butcher, F. R. W. and Sutherland, E. W. (1971). *Cyclic AMP*. Academic Press, New York.

Rosenberg, Th. and Wilbrandt, W. (1955). The kinetics of membrane transport involving chemical reactions. *Exp. Cell Res.* **9**:49-67.

Rosenfalck, P. (1969). Intra- and extracellular potential fields of active nerve and muscle fibres. *Acta Physiol. Scand.* Sup. 321.

Rosenthal, J. (1969). Post-tetanic potentiation at the neuromuscular junction of the frog. *J. Physiol.* **203**:121-133.

Ross, T. R. Jr. and Sobel, B. E. (1972). Regulation of cardiac contraction. *Ann. Rev. Physiol.* **34**:47-90.

Rothman, J. E. and Lenard, J. (1977). Membrane asymmetry. *Science* **195**:743-753.

Rougier, O., Vassort, D., Garnier, D., Gargouil, Y. M. and Coraboeuf, E. (1969). Existence and role of a slow inward current during the frog atrial action potential. *Pflügers Arch.* **308**:91-110.

Rougier, O., Vassort, D. and Stämpfli, R. (1968). Voltage clamp experiments on frog atrial heart muscle fibres with the sucrose gap technique. *Pflügers Arch.* **301**:91-108.

Ruch, T. C. and Patton, H. D. (1965). *Physiology and Biophysics*. W. B. Saunders, Philadelphia.

Rudjord, T. (1972). Model study of muscle spindle primary endings subjected to dynamic fusimotor activation. *Kybernetik* **11**:148-153.

Rushton, W. A. H. (1951). A theory of the effects of fibre size in medullated nerve. *J. Physiol.* **115**:101-122.

Rushton, W. A. H. (1961). Peripheral coding in the nervous system. In *Sensory Communication* (Rosenblith, W. A., ed.) M. I. T. Press, Cambridge, pp. 169–181.

Rushton, W. A. H. (1972). Review lecture: pigments and signals in colour vision. *J. Physiol.* **220**:1–31P.

Ryall, R. W., Piercey, M. F., and Polosa, C. (1971). Intersegmental and intrasegmental distribution of mutual inhibition of Renshaw cells. *J. Neurophysiol.* **34**:700–707.

Rymer, W. Z., Houk, J. C., and Crago, P. (1979). Mechanisms of the clasp-knife reflex studied in an animal model. *Exp. Brain Res.* **37**:93–113.

Salmons, S. and Vrbová, G. (1969). The influence of activity on some contractile characteristics of mammalian fast and slow muscles. *J. Physiol.* **201**:535–549.

Sandow, A. (1944). Studies on the latent period of muscular contraction method. General properties of latency relaxation. *J. Cell. Comp. Physiol.* **24**:221–256.

Sato, M. (1961). Response of Pacinian corpuscles to sinusoidal vibration. *J. Physiol.* **159**:391–409.

Schäfer, S. S. (1967). The acceleration response of a primary muscle spindle ending to ramp stretch of the extrafusal muscle. *Experientia* **23**:1026–1027.

Schatzmann, H. J. and Bürgin, H. (1978). Calcium in human red blood cells. In *Calcium Transport and Cell Function* (Scarpa, A. and Carafoli, E., eds.). *Ann. N.Y. Acad. Sci.* **307**:125–146.

Schiaffino, S., Hanzilikóvá, V. and Peirobon, S. (1970). Relations between structure and function in rat skeletal muscle fibers. *J. Cell. Biol.* **47**:108–119.

Schnapp, B. and Mugnaini, E. (1978). Membrane architecture of myelinated fibers as seen by freeze-fracture. In *Physiology and Pathobiology of Axons* (Waxman, S. G., ed.). Raven Press, New York, pp. 83–123.

Scott, J. G. and Mendell, L. M. (1976). Individual EPSP's produced by single triceps surae Ia afferent fibers in homonymous and heteronymous motoneurons. *J. Neurophysiol.* **39**:679–692.

Selby, S. M. (1975). *C.R.C. Standard Mathematical Tables*, 23rd ed. Chemical Rubber Company Press, Cleveland.

Shannon, C. E. (1948). A mathematical theory of communication. *Bell System Tech. J.* **27**:379–423, 623–656.

Shepherd, G. M. (1979). *The Synaptic Organization of the Brain. An Introduction.* Oxford University Press, New York and London.

Sherrington, C. S. (1906). *The Integrative Action of the Nervous System.* Yale, New Haven (repr. 1961).

Shik, M. K. and Orlovsky, G. N. (1976). Neurophysiology of locomotor automatism. *Physiol. Rev.* **56**:465–501.

Singer, S. J. and Nicholson, G. L. (1972). The fluid mosaic model of the structure of cell membranes. *Science* **175**:720–723.

Sjölin, J. and Grampp, W. (1975). Membrane noise in slowly adapting stretch receptor neurone of lobster. *Nature* **257**:696–697.

Skilling, H. H. (1959). *Electrical Engineering Circuits.* Wiley, New York.

Skou, J. C. (1965). Enzymatic basis for active transport of $Na^+$ and $K^+$ across cell membrane. *Physiol. Rev.* **45**:596–617.

Skou, J. C. (1974). The $(Na^+ + K^+)$ activated enzyme and its relationship to transport of sodium and potassium. *Quart. Rev. Biophys.* **7**:401–434.

Smith, D. R. and Smith, G. K. (1965). A statistical analysis of the continual activity of single cortical neurones in the cat unanaesthetized isolated forebrain. *Biophys. J.* **5**:47.

Smith, J. L., Betts, B., Edgerton, V. R., and Zernicke, R. F. (1980). Rapid ankle extension during paw shakes: selective recruitment of fast ankle extensors. *J. Neurophysiol.* **43**:612–629.

Smith, R. S. (1966). Properties of intrafusal muscle fibres. In *1st Nobel Symposium, Muscular*

*Afferents and Motor Control* (Granit, R., ed.). Almquist and Wiksell, Stockholm, pp. 69–80.

Smythies, J. R., Bennington, F., Bradley, R. J., Bridgers, W. F. and Morin, R. D. (1974). The molecular structure of the sodium channel. *J. Theor. Biol.* **43**:29–42.

Snyder, S. H., Banerjee, S. P., Yamamura, H. I., and Greenberg, D. (1974) Drugs, neurotransmitters, and schizophrenia. *Science* **184**:1243–1254.

Sokolnikoff, I. S. and Redheffer, R. M. (1958). *Mathematics of Physics and Modern Engineering*. McGraw-Hill, New York.

Sokolove, P. (1972). Computer simulation of after-inhibition in crayfish slowly adapting stretch receptor. *Biophys. J.* **12**:1429.

Sotelo, C., Llinás, R. and Baker, R. (1974). Electrotonic coupling between neurons in cat inferior olive. *J. Neurophysiol.* **37**:560–571.

Spencer, A. W. (1977). In *Handbook of Physiology, Section 1, Vol. 1* (Brookhart, J. M. and Mountcastle, V. B., eds.). *Am. Physiol. Soc.*, Bethesda, MD., pp. 969–1021.

Staehelin, L. A. and Hull, B. E. (1978). Junctions between living cells. *Sci. Amer.* **302**:140–152.

Stauffer, E. K., Watt, D. G. D., Taylor, A., Reinking, R. M., and Stuart, D. G. (1976). Analysis of muscle receptor connections by spike-triggered averaging. 2. Spindle group II afferents. *J. Neurophysiol.* **39**:1393–1402.

Steim, J. M. Tourtellote, M. E., Reinert, J. C., McElhaney, R. N., and Radar, R. L. (1969). Calorimetric evidence for the liquid-crystalline state of lipids in a biomembrane. *Proc. Nat. Acad. Sci. U.S.A.* **63**:104–109.

Stein, R. B. (1965). A theoretical analysis of neuronal variability. *Biophys. J.* **5**:173.

Stein, R. B. (1967). The frequency of nerve action potentials generated by applied currents. *Proc. Roy. Soc. London B* **167**:64–86.

Stein, R. B. (1970). The role of spike trains in transmitting and distorting sensory signals. In *The Neurosciences: Second Study Program* (Schmitt, F. O., ed.). Rockefeller University Press, New York, pp. 597–604.

Stein, R. B. (1974). The peripheral control of movement. *Physiol. Rev.* **54**:215–243.

Stein, R. B. and Bertoldi, R. (1980). The size principle: a synthesis of neurophysiological data. In *Recruitment Patterns of Motor Units and the Gradation of Muscle Force* (Desmedt, J. E., ed.). *Prog. Clin. Neurophysiol.*, Vol. 9, Karger, Basel, pp. 85–96.

Stein, R. B. and French, A. S. (1970). Models for the transmission of information by nerve cells. In *Excitatory Synaptic Mechanisms* (Andersen, P. and Jansen, J., eds.). Universitetsforlaget, Oslo, pp. 247–257.

Stein, R. B. and Lee, R. G. (1980). Tremor and clonus. In *Motor Control* (Brooks, V. B. ed.) American Physiological Soc., Bethesda, M., *Hand. Physiol.* **2**: 325–343.

Stein, R. B., Lee, R. G., and Nichols, T. R. (1978). Modifications of ongoing tremors and locomotion by sensory feedback. *Electroen. Clin. Neurophysiol.* Suppl. **34**:511–519.

Stein, R. B., Leung, K. V., Mangeron, D., and Oğuztöreli, M. N. (1974). Improved neuronal models for studying neural networks. *Kybernetik* **15**:1–9.

Stein, R. B., Nichols, T. R., Jhamandas, J., Davis, L., and Charles, D. (1977). Stable long-term recordings from cat peripheral nerves. *Brain Res.* **128**:21–38.

Stein, R. B. and Oğuztöreli, M. N. (1976). Tremor and other oscillations in neuromuscular systems. *Biol. Cybern.* **22**:147–157.

Stein, R. B. and Parmiggiani, F. (1979). Optimal motor patterns for activating mammalian muscle. *Brain Res.* **175**:372–276.

Stein, R. B. and Pearson, K. G. (1971). Predicted amplitude and form of extracellularly recorded action potentials from unmyelinated nerve fibres. *J. Theor. Biol.* **32**:539–558.

Stein, R. B. and Wong, E. Y. M. (1974). Analysis of models for the activation and contraction of muscle. *J. Theor. Biol.* **46**:307–327.

Stein, W. D. (1967). *The Movement of Molecules Across Cell Membranes*. Academic Press, New York.

Stein, W. D., Eilam, Y., and Lieb, W. R. (1974). Active transport of cations across biological membranes. *Ann. N.Y. Acad. Sci.* **227**:328–336.

Stephens, J. A. and Taylor, A. (1972). Fatigue of maintained voluntary muscle contraction in man. *J. Physiol.* **220**:1–18.

Stephens, R. E. and Edds, K. T. (1976). Microtubules: structure, chemistry and function. *Physiol. Rev.* **56**:709–777.

Stevens, C. F. (1979). The neuron. *Sci. Amer.* **318**:54–65.

Stevens, S. S. (1961). The psychophysics of sensory function. In *Sensory Communication* (Rosenblith, W. A., ed.). M.I.T. Press, Cambridge, pp. 1–33.

Sugi, H. (1974). Inward spread of activation in frog muscle fibres investigated by means of high-speed microcinematography. *J. Physiol.* **242**:219–235.

Sunderland, S. (1972). *Nerve and Nerve Injuries.* Churchill Livingstone, London.

Tada, M., Yamamoto, T., and Tonomura, Y. (1978). Molecular mechanism of active calcium transport by sarcoplasmic reticulum. *Physiol. Rev.* **58**:1–79.

Tamar, H. (1977). The organization and some major processes of sensory systems. *Biosci. Comm.* **3**:287–330.

Tasaki, I. (1964). A new measurememt of action currents developed by single nodes of Ranvier. *J. Neurophysiol.* **27**:1199–1206.

Terzuolo, C. A. and Washizu, Y. (1962). Relation between stimulus strength, generator potential and impulse frequency in stretch receptor of crustacea. *J. Neurophysiol.* **25**:56–66.

Thomas, R. C. (1969). Membrane current and intracellular sodium changes in snail neurone during extrusion of injected sodium. *J. Physiol.* **201**:495–514.

Thomas, R. C. (1972). Electrogenic sodium pump in nerve and muscle cells. *Physiol. Rev.* **52**:563–594.

Tonomura, Y. and Yamada, S. (1973). Molecular mechanism of $Ca^{++}$ transport through the membrane of sarcoplasmic reticulum. In *Organization of Energy-Transducing Membranes* (Nakao, M. and Parker, L., eds.). University Park Press, Baltimore, pp. 107–115.

Trautwein, W. (1973). Membrane currents in cardiac muscle fibres. *Physiol. Rev.* **53**:793–835.

Tsien, R. W., Giles, W., and Greengard, P. (1972). Cyclic AMP mediates the effects of adrenaline on cardiac Purkinje fibres. *Nature New Biol.* **240**:181–183.

Tsien, R. W. and Nobel, D. (1969). A transition state theory approach to the kinetics of conductance changes in excitable membranes. *J. Memb. Biol.* **1**:248–273.

Uchizono, K. (1965). Characteristics of excitatory and inhibitory synapses in the central nervous system of the cat. *Nature* **207**:642–643.

Ulbricht, W. (1969). The effect of veratridine on excitable membranes of nerve and muscle. *Ergeb. Physiol.* **61**:18–71.

Urry, D. W. (1978). Basic aspects of calcium chemistry and membrane interaction: on the messenger role of calcium. In *Calcium transport and Cell Function* (Scarpa, A. and Carafoli, E., eds.). *Ann. N.Y. Acad. Sci.* **307**:3–26.

Urry, D. W., Goodall, M. C., Glickson, J. D., and Mayers, D. F. (1971). The gramicidin A transmembrane channel: characteristics of head-to-head dimerized $\pi_{(L, D)}$ helices. *Proc. Nat. Acad. Sci. U.S.A.* **68**:1907–1911.

Ussing, H. H. (1949). The distinction by means of tracers between active transport and diffusion. The transfer of iodide across the isolated frog skin. *Acta Physiol. Scand.* **19**:43–56.

Vallbo, Å. B. (1971). Muscle spindle response at the onset of isometric voluntary contractions in man. Time difference between fusimotor and skeletomotor effects. *J. Physiol.* **219**:405–431.

Vandlen, R. L., Wu, W. C.-S., Eisenach, J. C., and Raftery, M. A. (1979). Studies of the composition of purified *Torpedo californica* acetylcholine receptor and of its subunits. *Biochemistry* **18**:1845–1854.

Villegas, R., Caputo, C., and Villegas, L. (1962). Diffusion barriers in squid nerve. *J. Gen. Physiol.* **46**:245–255.

Vogel, F., Meyer, H. W. Grosse, R., and Repke, K. R. H. (1977). Electron microscopic visualization of the arrangement of the two protein components of $(Na^+ + K^+)$-ATPase. *Biochim. Biophys. Acta* **470**:497–502.

Vrbová, G., Gordon, T., and Jones, R. (1978). *Nerve–Muscle Interaction*. Blackwell and Hall, London.

Watt, D. G. D., Stauffer, E. K., Taylor, A., Reinking, R., and Stuart, D. G. (1976). Analysis of muscle receptor connections by spike-triggered averaging. I. Spindle primary and tendon organ afferents. *J. Neurophysiol.* **39**:1375–1392.

Waxman, S. G. and Swadlow, H. A. (1977). The conduction properties of axons in central white matter. *Prog. Neurobiol.* **8**:297–324.

Weber, A. and Murray, J. M. (1973). Molecular control mechanisms in muscle contraction. *Physiol. Rev.* **53**:612–673.

Weeds, A. G., Trentham, D. R., Kean, C. J. C., and Buller, A. J. (1974). Myosin from cross-reinnervated cat muscle. *Nature* **247**:135–139.

Weingrad, S. (1968). Intracellular calcium movements of frog skeletal muscle during recovery from tetanus. *J. Gen. Physiol.* **51**:65–83.

Wendler, L. (1963). Ueber die Wirkungskette zwischen Reiz und Erregung (Versuche an den abdominalen Streckreceptoren dekapoder Krebse). *Z. Vergl. Physiol.* **47**:279–315.

Werblin, F. S. and Dowling, J. E. (1969). Organization of the retina of the mudpuppy, *Necturus maculosus*. II. Intracellular recording. *J. Neurophysiol.* **32**:339–355.

Werman, R., Davidoff, R. A., and Aprison, M. H. (1968). Inhibitory action of glycine on spinal neurons in the cat. *J. Neurophysiol.* **31**:81–95.

Wernig, A. (1975). Estimates of statistical release parameters from the crayfish and frog neuromuscular junctions. *J. Physiol.* **244**:207–221.

White, D. C. S. and Thorson, J. (1974) The kinetics of muscle contraction. *Prog. Biophys. Mol. Biol.* **27**:175–258.

Whittaker, P. (1970). The vesicle hypothesis. In *Excitatory Synaptic Mechanisms* (Andersen, P. and Jansen, J. K. S., eds.). Universitetsforlaget, Oslo, pp. 67–76.

Widdas, W. F. (1952). Inability of diffusion to account for placental glucose transfer in the sheep and consideration of the kinetics of a possible carrier transfer. *J. Physiol.* **118**:23–39.

Wiener, N. (1958). *Nonlinear Problems in Random Theory*. Wiley, New York.

Wiersma, C. A. G. (1953). Neural transmission in invertebrates. *Physiol. Rev.* **33**:326–355.

Wilson, D. B. (1978). Cellular transport mechanism. *Ann. Rev. Biochem.* **47**:933–965.

Wilson D. M. and Waldron, I. (1968). Models for the generation of the motor output pattern in flying locusts. *Proc. I.E.E.E.* **56**:1058–1064.

Wong, R. K. S., Prince, D. A., and Busbaum, A. I. (1979). Intradendritic recordings from hippocampal neurons. *Proc. Nat. Acad. Sci. U.S.A.* **76**:986–990.

Woodbury, J. W. (1965). The cell membrane: ionic and potential gradients and active transport. In *Physiology and Biophysics* (Ruch, T. C. and Patton, H. D., eds.). Saunders, Philadelphia, pp. 1–25.

Woodbury, J. W., Gordon, A. M., and Conrad, J. T. (1965). Muscle. In *Physiology and Biophysics* (Ruch, T. C. and Patton, H. D., eds.). Saunders, Philadelphia, pp. 113–152.

Wuerker, R. B., McPhedron, A. M., and Henneman, E. (1965). Properties of motor units in a heterogeneous pale muscle (m. gastrocnemius) of the cat. *J. Neurophysiol.* **28**:85–99.

Wyman, R. J. (1977). Neural generation of the breathing rhythm. *Ann. Rev. Physiol.* **39**:417–448.

Wyman, R. J., Waldron, I., and Wachtel, G. M. (1974). Lack of fixed order of recruitment in motoneuron pools. *Exp. Brain Res.* **20**:101–114.

Yu, L. C., Dowben, R. M., and Kornacker, K. (1970). The molecular mechanism of force generation in striated muscle. *Proc. Nat. Acad. Sci. U.S.A.* **66**:1199–1205.

# Index